X-Ray Spectroscopy
in Atomic and Solid
State Physics

NATO ASI Series

Advanced Science Institutes Series

A series presenting the results of activities sponsored by the NATO Science Committee, which aims at the dissemination of advanced scientific and technological knowledge, with a view to strengthening links between scientific communities.

The series is published by an international board of publishers in conjunction with the NATO Scientific Affairs Division

A	Life Sciences	Plenum Publishing Corporation
B	Physics	New York and London
C	Mathematical and Physical Sciences	Kluwer Academic Publishers Dordrecht, Boston, and London
D	Behavioral and Social Sciences	
E	Applied Sciences	
F	Computer and Systems Sciences	Springer-Verlag
G	Ecological Sciences	Berlin, Heidelberg, New York, London,
H	Cell Biology	Paris, and Tokyo

Recent Volumes in this Series

Volume 180—The Physics of Submicron Semiconductor Devices
edited by Harold L. Grubin, David K. Ferry, and
C. Jacoboni

Volume 181—Fundamental Processes of Atomic Dynamics
edited by J. S. Briggs, H. Kleinpoppen, and H. O. Lutz

Volume 182—Physics, Fabrication, and Applications of Multilayered Structures
edited by P. Dhez and C. Weisbuch

Volume 183—Properties of Impurity States in Superlattice Semiconductors
edited by C. Y. Fong, Inder P. Batra, and S. Ciraci

Volume 184—Narrow-Band Phenomena—Influence of Electrons with Both
Band and Localized Character
edited by J. C. Fuggle, G. A. Sawatzky, and J. W. Allen

Volume 185—Nonperturbative Quantum Field Theory
edited by G. 't Hooft, A. Jaffe, G. Mack,
P. K. Mitter, and R. Stora

Volume 186—Simple Molecular Systems at Very High Density
edited by A. Polian, P. Loubeyre, and N. Boccara

Volume 187—X-Ray Spectroscopy in Atomic and Solid State Physics
edited by J. Gomes Ferreira and M. Teresa Ramos

Series B: Physics

X-Ray Spectroscopy in Atomic and Solid State Physics

Edited by
J. Gomes Ferreira and
M. Teresa Ramos

Atomic Physics Center
University of Lisbon
Lisbon, Portugal

Plenum Press
New York and London
Published in cooperation with NATO Scientific Affairs Division

Proceedings of a NATO Advanced Study Institute
on X-Ray Spectroscopy in Atomic and Solid State Physics,
held August 30–September 12, 1987,
in Vimeiro, Portugal

Library of Congress Cataloging in Publication Data

NATO Advanced Study Institute on X-ray Spectroscopy in Atomic and Solid State
  Physics (1987: Vimeiro, Lisbon, Portugal)
  X-ray spectroscopy in atomic and solid state physics / edited by J. Gomes
Ferreira and M. Teresa Ramos.
  p.   cm.—(NATO ASI series. Series B, Physics; vol. 187)
  "Proceedings of a NATO Advanced Study Institute on X-ray Spectroscopy in
Atomic and Solid State Physics, held August 30–September 12, 1987, in Vimeiro,
Portugal"—T.p. verso.
  "Published in cooperation with NATO Scientific Affairs Division."
  Bibliography: p.
  Includes index.
  ISBN 0-306-43029-0
  1. X-ray spectroscopy—Congresses. 2. Molecular spectroscopy—Congresses.
3. Solid State physics—Congresses. I. Ferreira, J. Gomes. II. Ramos, M. Teresa.
III. North Atlantic Treaty Organization. Scientific Affairs Division. IV. Title. V.
Series: NATO ASI series. Series B, Physics; v. 187.
QC482.S6N37 1987   88-23929
539.7′222—dc19   CIP

© 1988 Plenum Press, New York
A Division of Plenum Publishing Corporation
233 Spring Street, New York, N.Y. 10013

All rights reserved

No part of this book may be reproduced, stored in a retrieval system, or transmitted
in any form or by any means, electronic, mechanical, photocopying, microfilming,
recording, or otherwise, without written permission from the Publisher

Printed in the United States of America

PREFACE

The fields of X-Ray Spectroscopy in Atomic and Solid State Physics have undergone spectacular growth, sometimes rather anarchic, during the past decade. The old mold of X-ray spectroscopy has been burst, and this ASI provided an in-depth exploration of theory and recently developed techniques; however, some work still needs to be done to create a new frame and reduce anarchy in the field.

The purpose of this Institute was to gather atomic and solid state physicists working in theoretical and new experimental techniques recently developed. The lectures were concerned with, among others, the following fields: theory of X-ray near-edge structure, XPS and AES with conventional and synchrotron radiation sources, PIXE, EXAFS, SEXAFS, XRF, SXS, and molecular spectroscopy.

The Institute considered in detail some of these experimental techniques and the pertinent theoretical interpretations by selecting an important list of lectures which summarize the scientific contents of the ASI.

The truly international character of this NATO ASI, its size, and the high quality of the lecturers contributed to make this school a very fruitful scientific meeting.

Two to four general lectures were given each working day and three afternoons were reserved for presentation of current work in the form of posters. We think that these poster presentations reflect the current research work of the participants.

The lectures given were of high quality, and some of them were followed by tutorials in order to clarify some points. We think that the critical discussion of current developments and ideas between scientists of some seniority and younger researchers working in the same field is very important.

This 12-day Institute was attended by 73 persons from 16 countries, and it was very gratifying to receive such an enthusiastic response from scientists belonging to a broad spectrum of countries.

The organizing committee noted with pleasure that some of the participants in the Advanced Study Institute wrote to offer their congratulations on the content and organization of the course.

The main lectures at the Institute were given by nineteen scientists, including, in addition to those listed in the table of contents, Prof. G.A. Sawatzky and Prof. J. Fuggle (Netherlands), Prof. J.P. Briand (France), and Dr. M. Krause (USA). Interesting introductory and concluding remarks were offered, respectively, by Prof. Fuggle and Prof. L. Watson (Glasgow).

In the organization of this Institute I had the cooperation of a local committee including M. Teresa Ramos, Fernando C. Parente, and Carlos P. Cardoso of the University of Lisbon.

D. Berenyi (Debrecen), J.P. Briand (Paris), B. Craseman (Oregon), J. Fuggle (Nijmegen), M. Krause (Oak Ridge), A. Policarpo (Coimbra), D. Urch (London), and L. Watson (Glasgow) made valuable suggestions for the scientific contents of the meeting. To them I wish to express my deep gratitude.

J. Gomes Ferreira

CONTENTS

AUGER ELECTRON SPECTROSCOPY

Auger Spectroscopy of Free Atoms: Experimental 1
   Seppo Aksela

Auger Electron Spectroscopy of Free Atoms:
   Theoretical Analysis 15
   Helena Aksela

Ion-Induced Auger Electron Processes in Gases. 25
   D. Berényi

X-RAY ABSORPTION SPECTROSCOPY

Inner Shell X-Ray Photoabsorption as a Structural
   and Electronic Probe of Matter 67
   C. R. Natoli

Trends of EXAFS and SEXAFS in Solid State Physics. 107
   A. Fontaine

X-RAY PHOTOELECTRON SPECTROSCOPY

PAX (Photoelectron and X-Ray Emission) Spectroscopy:
   Basic Principles and Chemical Effects 155

Electronic Structure of Molecules, Complexes and Solids,
   using PAX Spectroscopy 177
   David S. Urch

MOLECULAR SPECTROSCOPY

Importance of Intra- and Inter-Atomic Contributions to
   Molecular X-Ray Emission Processes 201
   Frank P. Larkins

The Electronic Decay of Core Hole Excited States in
   Free and Chemisorbed Molecules 215
   W. Eberhardt

Fragmentation of Small Molecules Following Soft
   X-Ray Excitation 227
  W. Eberhardt

X-RAY FLUORESCENCE

Synchrotron Radiation Applied to X-Ray Fluorescence
   Analysis 237
  Pierre Chevallier

SOFT X-RAY SPECTROSCOPY AND APPLICATIONS TO SOLIDS

Soft X-Ray and X-Ray Photoemission Studies of
   Light Metals and Alloys 255
  L. M. Watson

X-Ray Inelastic Scattering Spectroscopy and
   Its Applications in Solid State Physics 279
  Nikos. G. Alexandropoulos and Irini Theodoridou

PROTON INDUCED X-RAY EMISSION

Particle-Induced X-Ray Emission: Basic Principles,
   Instrumentation and Interdisciplinary
   Applications 301
  Ede Koltay

ELECTRON ENERGY LOSS

Electron Energy Loss Spectroscopy in Reflection
   Geometry 335
  Falko P. Netzer

SYNCHROTRON RADIATION

Properties of Synchrotron Radiation 367
  George S. Brown

INSTRUMENTATION

Gaseous X-Ray Detectors 375
  A. J. P. L. Policarpo

Posters 393

Participants 419

Index 421

AUGER SPECTROSCOPY OF FREE ATOMS:
EXPERIMENTAL

Seppo Aksela

Department of Physics
University of Oulu
SF-90570 Oulu, Finland

INTRODUCTION

In the initial state of the normal Auger process the atom is ionized in a core level. The core hole is filled by an outer electron and a second electron is emitted as the Auger electron (Fig. 1a). There are commonly used symbols for the emitted Auger electrons which are based on x-ray energy level notations and indicate the energy levels of the initial and the two final state core holes. The two final state holes can combine to form several spectroscopic terms with slightly different energies. Within this simple picture wherein we neglect the effects of electrons associated with the primary ionization to the Auger process, the energy of the outgoing Auger electron is given by the difference of the total energy of the system (atom, molecule) in the initial single and the final double hole states,

$$E_{Aug}(ijk;X) = |E(i, N-1) - E(jk; X, N-2)|.$$

Due to its double hole final state property Auger spectra are always characterized, also in the case of closed shell atoms, by several line components in comparison the usual single core level photolines. This is demonstrated in Fig. 2, where the Auger spectra of Zn and Ag are shown. Especially for the outer level Auger transitions the splitting between numerous line components is so small that a very high electron energy resolution is needed to obtain reliable positions and intensities for more or less overlapping lines. Because molecular and solid state effects always cause considerable broadening to the Auger lines the molecular or solid state spectra are often of rather limited use for studies of fine structures of the Auger spectra. Therefore, for detailed studies of Auger spectra free atoms are most suitable.

The experimentally observed spectrum is always a convolution of inherent spectrum emitted by the sample and the instrument function representing the broadening caused by the spectrometer used. The resolution of energy analyzers should be such that the line broadening caused by the analyzer is typically smaller than the inherent life time width of the Auger lines. In practice this usually means that the energy resolution should be better than 0.05 % of the initial kinetic energy of Auger electrons.

Within this simplified stepwise model of the Auger process the energies of the Auger electrons are independent of the primary ionization process and thus it can equally be done e.g. by photons or electrons. Photon excitation produces considerably less background from inelastically scattered electrons but the intensity of the usual laboratory soft x-ray sources is much lower than what can be easily obtained by simple electron guns. Especially in the case of gaseous atom targets the atom density in the interaction volume is so low that high intensity electron beam excitation is usually a very practical solution.

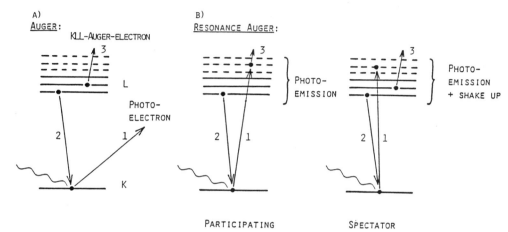

Fig. 1. Schematic representation of normal Auger (A) and resonance Auger (B) processes.

Synchrotron radiation provides a very elegant and unique excitation method also for normal Auger spectroscopy. The photon flux from modern storage rings after effective monochoromators (e.g. toroidal grating monochoromators) is so high that gaseous samples and also their Auger spectra can be successfully studied. The unique advantage of the synchrotron radiation is that the energy of the the exciting radiation can be selected to fullfill optimal conditions. Thus e.q. excitation of a deeper core level than the one considered can be avoided. A high energy (1 – 5 keV) electron beam often causes ionization with some probability also in deeper levels. These cause Auger cascades and corresponding satellite lines. As an example of this the LMM Auger spectra of Ar excited by two photon energies of synchrotron radiation are displayed in Fig. 3. The 350 eV photons can ionize also 2s electrons causing additional structure to the spectrum and higher continuous background.

The most powerful use of synchrotron radiation is when it is used not to ionize a given core level but to excite its electrons selectively to different unfilled orbitals (Fig. 1b). These resonantly excited states decay via Auger type processes producing so called resonance Auger spectra. These spectra can provide very useful complementary information to the normal Auger spectra.

EXPERIMENTAL SET-UPS

Electronspectrometers

The most commonly used types of electron energy analyzers are the electrostatic analyzers applying either spherically or cylindrically symmetric electrodes. In principle any kind of analyzer with some energy dispersion can achieve a high energy resolution if the electron optical slits, defining the path length and angular variations of the detected electrons, are made small enough. Therefore, such simple and electron optically less advantageous devices as the parallel plate and 127° cylindrical deflector analyzers have been used successfully in cases where the intensity has not been a problem. The properties of the analyzer become more critical when the intensity of the electrons to be analyzed is low. This is typically the case when electron spectra of free atoms are studied.

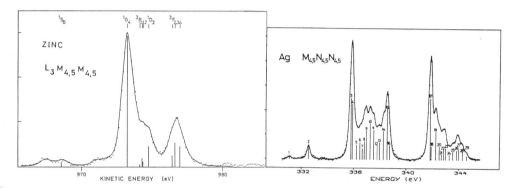

Fig. 2. Auger electron spectra of free Zn and Ag atoms.

The important quantity is therefore the transmission to resolution ratio. The transmission is taken simply to mean the fraction of the emitted electrons from the source which can be detected by the analyzer. It is closely related to the solid angle accepted by the limiting slits of the analyzer.

Let us consider the cylindrical mirror analyzer as an example, shown schematically in Fig. 4. In order to have some electrons detected we should let some range $\Delta\theta$ for the emission angles θ of the electrons with respect to the symmetry axis. The transmission is proportional to $\Delta\theta$ and the azimuthal angle ϕ. The path length L is /see e.g. Refs. 1–5/ a function of the kinetic energy of the electron, of the voltage between the cylinders and of the emission angle θ with a given geometry described by the parameter n, or $L = L(E,V,\Delta\theta, n)$. In order to minimize the dependence of L (and E) from $\Delta\theta$ it is highly advantageous to find a situation where as many partial derivates of L with respect to θ vanish as possible, or

$$\frac{\partial L}{\partial \theta} = \frac{\partial^2 L}{\partial \theta^2} = \ldots = \frac{\partial^n L}{\partial \theta^n} = 0.$$

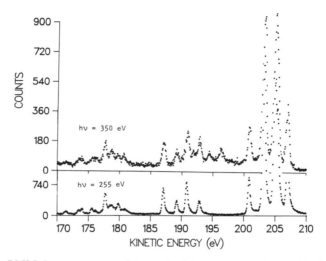

Fig. 3. LMM Auger spectra of Ar excited by 350 eV and 255 eV photons.

For the cylindrical mirror analyzer the first two derivatives vanish e.g. when $\theta = 42.3°$ and the source and image points are on the symmetry axis. It is said that the system then has the second order focusing property. It has been found /6/ that also a third order focus can be realized by this type of analyzer. For the spherical deflector analyzer, which is most commonly used in electron spectroscopy, only first order focus can be obtained. In practice this means with the usual optimal choice $\Delta E(\Delta L) = \Delta E(\Delta\theta)$ and the considered energy resolution of 0.05 % that the full angular spreads $\Delta\theta = 2$–$3°$ and $\Delta\theta = 8$–$10°$ can be used for the first and second order focusing cases, respectively. If the symmetry of the electrodes causes focusing in the direction ϕ perpendicular to θ angle the system is said to have the space focusing property. It is usually the most important factor to the transmission. For cylindrical geometry full 2π range of ϕ can be principally used whereas for spherical analyzers it is much more resticted.

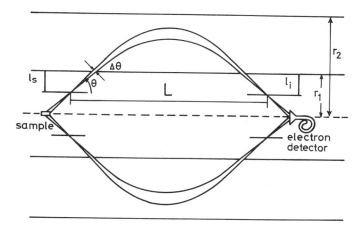

Fig. 4 Schematic diagram of a cylindrical mirror analyzer.

Redartation

The energy resolution of electron energy analyzers can be improved by retarding the electrons before they enter into the analyzing electric field, because the energy resolution of the analyzer is a given percentage of the kinetic energy of the electrons in the analyzing field. The retardation can be done either by means of special electron lenses /5/, which are also focusing the divergent incoming beam of electrons or by homogenous electric fields generated by grids in different voltages. The basic problem in the retardation with lenses is that the solid angle which can be effectively used is often very small. For example if the full effective angular opening of the lens is 4° it corresponds to the transmission of 0.03 %. For comparison the transmission of a cylindrical mirror analyzer with full angular spread of 10° and a mean θ-value of 42.3° is 4.3 %. There is a very large difference of more than two orders of magnitudes. Therefore, it is very advantageous for intensity reasons, if the retardation can be done in cylindrical analyzers with cylindrical or spherical elements still making use of the full 2π geometry possible.

Position sensitive detectors

An interesting new method to increase the intensity of registered electrons is provided by applications of position sensitive detectors /5,7-14/. Then the slit in the image plane is replaced by rather large ($\phi \sim 25$ mm) microchannel plate which is followed by a resistive anode or diode array from which the hitting points of the individual electrons can be decoded and arranged into proper energy channels of the spectrum based on the known relations between the position and energy differences. Gain factors between 10 and 100 in the data collection speed have been typically obtained. For this application spherical analyzers are most commonly used because they have a well defined focal plane. The use of position sensitive detectors in cylindrical mirror analyzers is more complicated because there is no simple focal plane but a cone-shaped focal surface. However, successful applications also in cylindrical mirror analyzers have been reported /12,15/.

Atom sources

At room temperature only rare gases appear as free atoms. A very useful method to produce free atoms is the vapourization of solid samples. Most solid elements vapourize mainly as free atoms and only a few form molecular clusters (Sb_4, Te_2). The typical vapour pressure used in source volume is about 10^{-3} torr. The temperatures needed to generate these vapour pressures vary from some hundred centigrades up to 3000 °C. Resistive heating applying bifilar heating wires can be done conveniently until around 1200 °C. For higher temperatures mainly two kinds of heating are used, namely inductive heating /7,15–17/ or electron bombardment heating /18/. The inductive heating is technically a rather simple method to realize. Its main problem is the radio frequency noise which is often induced into electronic circuits, especially in data collection systems. Therefore, gating must usually be applied so that during the short heating periods (10 – 100 ms) no data are collected and during the data collection periods (10 – 100 ms) heating is off. This of course causes reduction in the data collection efficiency, which increases with the needed temperatures because the heating to data collection time ratio must be increased. Electron bombardment heating /18/ has been applied successfully by Prof. Sonntag's group at DESY. It is much more noise-free and allows continuous data collection which is important especially in measurements with synchrotron radiation because the beam time is very valuable.

The main problem is very high temperature evaporation is extremely high reactivity of many metals as liquids and vapours. For several elements graphite or ceramic oven materials are best suitable but for some Ta, Mo or W crucibles have to be used.

As an axample of high temperature Auger electron spectrometers the measuring system /17,19/ used by our research group at the University of Oulu is shown in Fig. 5. It comprises an inductively heated sample oven, a four element retarding lens /20/ between the vapour sample and a double-pass simulated /21/ spherical-field electron energy analyzer. The system has been used to measure electron excited Auger spectra e.g. from vapours of Au /17/, Pd /19/ and 3d transition metals /22,23/.

COMPARISON BETWEEN ATOMIC AND SOLID STATE SPECTRA

Solid state Auger spectra appear systematically at 10 – 20 eV higher kinetic energies than corresponding free atom spectra. This arises mainly from large extra–atomic relaxation energies in solid state. Core holes in solid metal are effectively screened by the easily moveable conduction electrons in attemp to lower the total energy of core hole state. This relaxation or polarization energy is in the linear polarization approximation proportional to the square of the positive charge to be screened by the conduction electrons. The initial state of the Auger process is a single core hole state and the final state a double hole state. Thus, the relaxation energy is in the final state four times the single core hole initial state relaxation energy. The Auger electron energies experience their difference or the extra–atomic relaxation energy which is thus three times the relaxation energy pro single core hole state of photoemission.

We have studied solid state shifts also applying /24,25/ as special experimental set-up shown in Fig. 6. A partially cooled needle is set inside the vapour oven on which some vapour is condensing and forming a fresh and clean metal surface. With proper adjustment both the solid state spectrum from the needle and the vapour phase spectrum from the atoms surrounding the needle can be obtained simultaneously under identical experimental conditions. As an example the spectra of Zn are shown in Fig. 7. The observed solid state shifts 3.2 eV and 13.1 eV for binding and Auger energy shifts, respectively clearly demonstrate the greater effects of extra–atomic relaxation energies on Auger transitions.

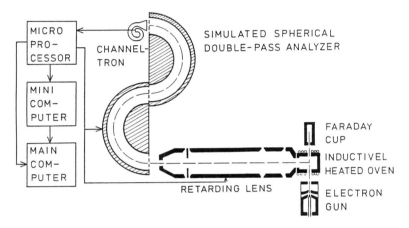

Fig. 5. Schematic of the experimental set-up used to measure Auger electron spectra at the University of Oulu.

As a second example $L_{2,3}M_{4,5}M_{4,5}$ Auger spectra of free Ni /23/ atoms and solid Ni, now taken with different spectrometers, are displayed in Fig. 8. The spectra show a large kinetic energy shift and very different spectral shapes. The vapour phase spectrum is characterized by a double structure, which can explained /23/ by two different initial state electronic configuration $3d^8 4s^2$ and $3d^9 4s$ which are both almost equally populated in high vapour temperatures. Both configurations give Auger spectra which are separated in kinetic energies by about 5 eV. The solid state is characterized by a $3d^{10}$ electron structure giving rather sharp L_3VV Auger spectrum. A drastic change can be seen in the intensity of the L_2VV group. The explanation of which are strong L_2L_3V Coster-Kronig transitions which are energetically possible, due to extra-atomic relaxation energy in solid state but not in free atoms.

Auger and recombination spectra of lanthanides

The photon absorption and electron energy loss spectra of the lanthanides are characterized by very strong and broad peaks above the 4d ionization threshold. They are called 4d → (4f,εf) giant resonances. The important decay channel of these resonance excitations is a so called direct recombination (autoionization) process where the created 4d core hole is filled by the excited electron and a second electron e.g. a 4f electron is emitted. The intensity of these 4d → 4f recombination lines, also in the case of electron beam excitation is comparable to the intensity of $N_{4,5}O_{2,3}N_{6,7}$ Coster-Kronig or $N_{4,5}N_{6,7}N_{6,7}$ super-Coster-Kronig transitions. In kinetic energies these recombination lines are on the high energy side of $N_{4,5}N_{6,7}N_{6,7}$ Auger lines partly overlapping with them in the solid state spectra. A point of interest is that the direct recombination processes create single core hole final states in comparison to double hole final states of Auger type transitions. Therefore, their solid state shifts should be different.

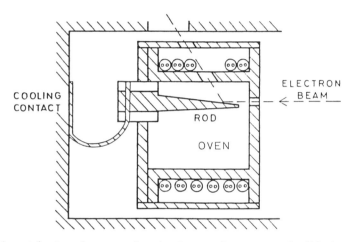

Fig. 6. Experimental set-up for measuring simultaneously vapour and solid phase Auger spectra.

This is actually the case, as can be seen from the spectra /26/ of Sm and Eu displayed in Fig. 9. The figure shows solid and vapour phase spectra and by white synchrotron radiation excited vapour phase spectra both for Sm and Eu. The structures indicated by a, b and c are from Auger transitions whereas peaks d originate 4d → 4f recombination processes. The energy scales are adjusted so that peaks b coincide. For vapour spectra the energy separation between b and d is clearly larger. In going to solid state Auger peaks a, b and c are shifting more than peak d causing peaks c and d to overlap much more in the solid state spectra. Thus, comparison between vapour and solid state spectra can be used for the verification of the recombination structures. Another pronounced feature of the spectra is the very high enchancement of recombination peaks when white synchrotron radiation is used. This gives further reinforcement to the identification of different types of processes.

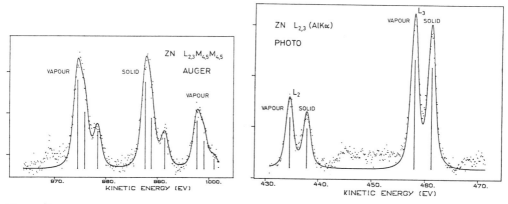

Fig. 7. Vapour-metal $L_{2,3}$ (AlK$_\alpha$) photoelectron and $L_{2,3}M_{4,5}M_{4,5}$ Auger electron spectra of Zn.

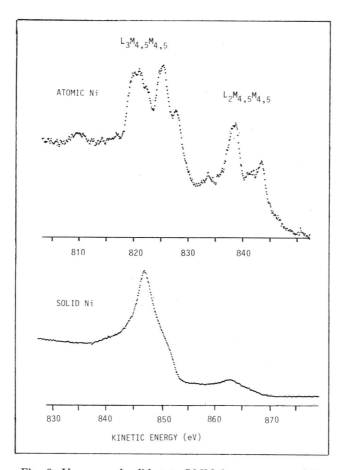

Fig. 8. Vapour and solid state LMM Auger spectra of Ni

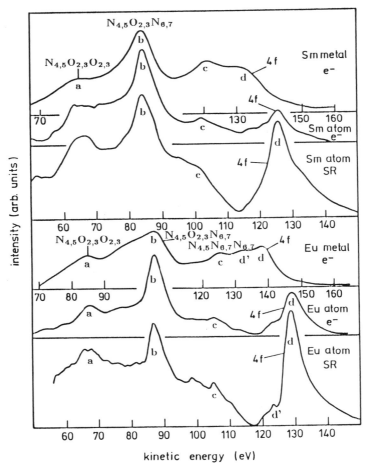

Fig. 9. 4d based Auger and recombination spectra of Sm and Eu metals and free atoms excited by electron beam and white synchrotron radiation.

RESONANCE AUGER SPECTRA

In the resonance Auger process a given core electron is excited selective to some unfilled atomic or molecular orbital. This selective excitation is done by the monochromatized synchrotron radiation tuned to correspond to just the excitation probably by means of a Auger type process, where one electron fills the created core hole and another electron is emitted. In the primary excitation event to the outer orbital transferred electron either participates in the core hole filling process or alternatively remains as the spectator. The former is the usual autoionization phenomena also commonly called the participating resonance Auger process. The latter is known as the spectator resonance phenomena, or shortly as the resonance Auger transition if there is no danger of confusion.

The resonance Auger spectroscopy is still a very new field of research and therefore the number of carefully studied systems is rather limited. Resonance Auger spectra of rare gases /27–34/ are best known experimentally and also their theoretical interpretation is on a rather sound basis.

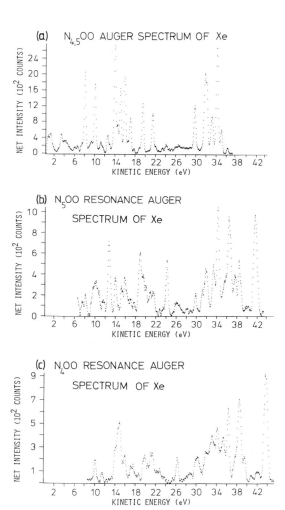

Fig. 10. Normal and resonance Auger spectra of Xe.

As an example the normal and resonance Auger spectra of Xe /30/ are shown in Fig. 10. These resonance Auger spectra correspond to cases where $4d_{5/2}$ or $4d_{3/2}$ electrons are excited to a 6p unfilled Rydberg orbit. As a first thought one should expect in these resonance Auger spectra only half the number of lines of the normal Auger spectrum, because only electrons from one 4d spin-orbit components are excited whereas in the normal Auger process both of the components are ionized. A brief inspection of the spectra in Fig. 10 shows that this is not the case. The number of peaks in resonance Auger spectra is at least as high as in the normal Auger spectrum. The main reason for that is that many electron phenomena now play a much stronger role in resonance Auger spectra. Especially the shake up of another electron during the resonance Auger emission process has been found to take place with rather high probability. Hence, resonance Auger spectroscopy clearly provides a new method to investigate electronic structure of atoms.

The second remarkable feature in the spectra of Fig. 10 is that lines have shifted to higher kinetic energies from corresponding reference lines in the normal Auger spectrum. This arises from the spectator electron on the outer orbital. A closer inspection of spectra reveals that due to the spectator electron also the fine structure is different. The parent lines of normal Auger spectrum are split into daughter lines.

REFERENCES

1. S. Aksela, M. Karras, M. Pessa and E. Suoninen, Rev. Sci. Instrum. **41**, 351 (1970).

2. H.Z. Sar-El, Rev. Sci. Instrum. **38**, 1210 (1967).

3. S. Aksela, Rev. Sci. Instrum. **42**, 810 (1971).

4. J.S. Risley, Rev. Sci. Instrum. **43**, 95 (1972).

5. E.H.A. Granneman and M.J. Van der Wiel, pp. 367–462 in Handbook on Synchrotron Radiation, Vol. 1, edited by E.E. Koch, North Holland, 1983.

6. W. Franzen and J. Taaffe, J. Phys. E **13**, 719 (1980).

7. A. Morris, N. Jonathan, J.M. Dyke, P.D. Francis, N. Keddar and J.D. Mills, Rev. Sci. Instrum. **55**, 172 (1984).

8. A. Bosch, H. Feil and G.A. Sawatzky, J. Phys. E **17**, 1187 (1984).

9. L.J. Richter and W. Ho, Rev. Sci. Instrum. **57**, 1469 (1988).

10. H.A. Van Hoof and M.J. Van der Wiel, J. Phys. E **13**, 409 (1980).

11. G.J.A. Hellings, H. Ottevanger, S.W. Boelens, C.L.C.M. Knibbeler and H.H. Bongersma, Surf. Sci. **162**, 913 (1985).

12. O. Benka, Nucl. Instr. Methods **203**, 547 (1982).

13. V. Schmidt, private communications.

14. A. Yagishita, Japan. J. Appl. Phys. **25**, 657 (1986).

15. V. Schmidt, Comment. At. Mol. Phys. **17**, 1 (1985).

16. R. Malutzki and V. Schmidt, J. Phys. B **19**, 1035 (1986).

17. S. Aksela, M. Harkoma, M. Pohjola and H. Aksela, J. Phys. B **17**, 2227 (1984).

18. T. Prescher, M. Richter, B. Sonntag and H.E. Wetzel, Nucl. Instr. Phys. Res. A **254**, 627 (1987).

19. S. Aksela, M. Harkoma and H. Aksela, Phys. Rev. A **29**, 2915 (1984).

20. B. Wannberg ans A. Sköllermo, J. electron Spectrosc. **10**, 45 (1977).

21. K. Jost, J. Phys. E **12**, 1006 (1979).

22. S. Aksela, T. Pekkala, H. Aksela, M. Wallenius and M. Harkoma, Phys. Rev. A **35**, 1426 (1987).

23. H. Aksela, S. Aksela, T. Pekkala and M. Wallenius, Phys. Rev. A **35**, 1522 (1987).

24. R. Kumpula, J. Väyrynen, T. Rantala and S. Aksela, J. Phys. C **12**, L 809 (1979).

25. S. Aksela, R. Kumpula, H. Aksela, J. Väyrynen, R.M. Nieminen and M. Puska, Phys. Rev. B **23**, 4362 (1981).

26. S. Aksela, M. Richter, B. Sonntag, to be published.

27. W. Eberhardt, G. Kalkoffen and C. Kung, Phys. Rev. Lett. **41**, 156 (1978).

28. V. Schimidt, S. Krummacher, F. Wuilleumier and P. Dhez, Phys. Rev. A **24**, 1803 (1981).

29. S. Southworth, U. Becker, C.M. Truasdale, P.H. Kobrin, D.W. Lindle, S. Owaki and D.A. Shirley, Phys. Rev. A **28**, 261 (1983).

30. H. Aksela, S. Aksela, G.M. Bancroft and K.H. Tan, Phys. Rev. A **33**, 3867 (1986).

31. U. Becker, T. Prescher, E. Schmidt, B. Sonntag and K.H. Tan, Phys. Rev. A **33**, 3891 (1986).

32. D.W. Lindle, P.A. Heimann, T.A. Ferret, M.N. Piancastelli and D.A. Shirley, Phys. Rev. A **35**, 4605 (1987).

33. H. Aksela, S. Aksela, H. Pulkkinen, G.M. Bancroft and K.H. Tan, Phys. Rev. A **33**, 3876 (1986).

34. H. Aksela, S. Aksela, H. Pulkkinen, G.M. Bancroft and K.H. Tan, to be published.

AUGER ELECTRON SPECTROSCOPY OF FREE ATOMS:

THEORETICAL ANALYSIS

Helena Aksela

Department of Physics
University of Oulu
SF-90570 Oulu, Finland

INTRODUCTION

When an atom is ionized in an inner shell by electron, ion or photon impact, the resulting hole state can be filled by means of X-ray emission or a nonradiative Auger process. In the latter case an outer-shell electron fills the hole and another outer-shell electron is ejected leaving the atom in a doubly ionized state. Auger electron spectra show a complicated fine structure because the spectra display the electronic structures of single vacancy and double vacancy states of an atom. The theoretical analysis of the spectra can result in a better understanding of the electronic structure of the periodic table elements. Theoretical profiles of Auger decays can be produced for comparison with experiments since the evaluation of energies and transition probabilities has been carried out.

EVALUATION OF ENERGIES

The energies of Auger transitions can be obtained by the energy difference between separately optimized total energies of the singly ionized initial and doubly ionized final-state levels of the emitting system (the ΔSCF approach). The total energies in electronic states can be calculated with the multiconfiguration Dirac-Fock (MCDF) code of Grant et al.[1,2] The computer program is described in detail by Grant et al. [1,2] and therefore only a brief summary will be given here.

Atomic-state wave function (ASF) is presented as a linear combination of configuration-state functions (CSF). The ASF of the νth electronic state of an atom with total angular momentum JM is thus given by

$$\Psi_\nu(JM) = \sum_{\beta=1}^{n} c_{\nu\beta}^J \Phi_\beta(JM) \qquad (1)$$

where n is the number of CSF included in the expansion and $c_{\nu\beta}^J$ are the mixing coefficients for the state ν. The CSF are constructed from antisymmetric products of Dirac central-field spinors

$$u_a(r) = \frac{1}{r}\begin{pmatrix} P_a(r)\,\chi_a \\ iQ_a(r)\,\bar\chi_a \end{pmatrix} \qquad (2)$$

where $P_a(r)$ and $Q_a(r)$ are radial functions and χ_a, $\bar{\chi}_a$ are two-component spin orbitals.

For the MCDF method, the mixing coefficients and the orbitals $u_a(r)$ are determined by applying the variational method to the expectation value of the Hamiltonian with respect to AFS $\Phi_\beta(JM)$ subject to orbital orthogonality constraints. The Hamiltonian used in the variational principle is taken to consist of a sum of single-particle Dirac Hamiltonians plus the purely Coulomb interelectronic repulsion

$$H = \sum_{i=1}^{N} H_i + \sum_{\substack{i,j \\ i<j}}^{N} \frac{1}{r_{ij}} \qquad (3)$$

Using the standard notation, H_i is given by

$$H_i = c\boldsymbol{\alpha} \cdot \mathbf{p}_i + (\beta - 1)c^2 + V_{nuc}(r_i) \qquad (4)$$

where $V_{nuc}(r_i)$ is the potential due to the nucleus. The Breit interaction, along with other quantum-electrodynamic (QED) contributions, are added to the Hamiltonian matrix once the orbitals have been determined, and the complete matrix of the Hamiltonian is diagonalized to determine the corrected energy levels and mixing coefficients.

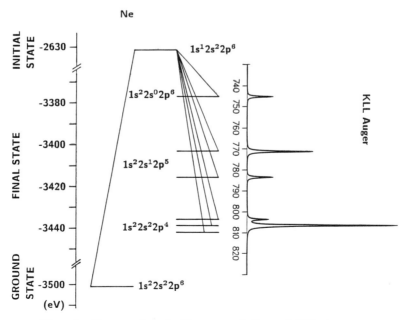

Fig. 1. Energy level diagram of Ne and KLL Auger transitions.

Figure 1 displays the energy levels of neutral Ne and of the initial and final states of the KLL Auger transitions predicted by the single-manifold DF calculations. (A manifold is defined as the set of jj-coupled states which reduce to a single nonrelativistic configuration in the nonrelativistic limit.) Vertical lines connecting the energy levels of the initial and final states represent the Auger transitions, and the corresponding spectrum is depicted on the right hand side of the figure.

In a closed shell atom the energy levels are due to the hole states caused by the Auger decay. The spectrum of an open shell atom is more complicated and reveals more fine structure due to the coupling of the participating hole states to the outermost open shell where the vacancies stay passive during the decay. This causes the splitting of the parent levels to daughters, as can be seen from figure 2, where the final state energy level structures of Pd, Ag and Cd, predicted by the single-manifold Dirac-Fock calculations, are depicted. Since the open-shell elements make up a larger part of the periodic table their understanding will be important for fruitful progress in atomic physics.

The correlation effects may cause large energy shifts and the redistribution of intensity, as is nicely demonstrated in the case of ns hole states of real gases and their neighbour elements.[3] Figure 3 displays the energy level structure of the doubly ionized krypton obtained from single-manifold and multiconfiguration Dirac-Fock calculations, Auger spectroscopy and optical spectroscopy.[4] Figure 4, furthermore, shows a comparison between calculated single-manifold and multiconfiguration profiles and

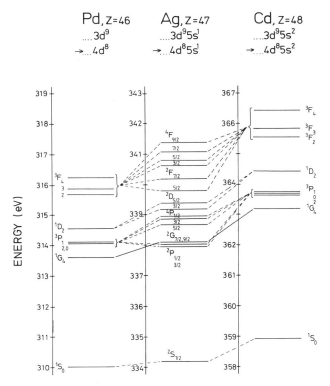

Fig. 2. Energy levels of Pd, Ag and Cd with two 3d vacancies.

FINAL-STATE ENERGY LEVELS

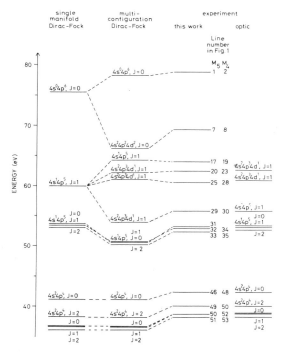

Fig. 3. Energy levels of doubly ionized Kr.[4]

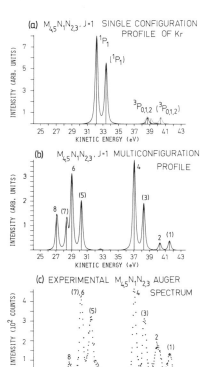

Fig. 4. Comparison between calculated and experimental Auger spectra of Kr.[5]

the experiment.[5] The single-particle picture is not able to provide a good description of the $M_{4,5}N_1N_{2,3}$ transitions. The calculations carried out by taking into account the correlation between the $4s4p^5$ and $4s^24p^34d$ final-state configurations reproduce the experiment fairly well, although they slightly overestimate the shift between the main (1–4) and the satellite (5–8) structures.

In addition to the electron correlation also the thermal population of the energy levels lying in close proximity to each other may result in an anomalous spectral structure. The free iron group atoms at the end of the transition elements are good examples.[6] Single-manifold predictions of the energies of the ground state and of the initial and final state configurations of the $2p^{-1} \to 3d^{-2}$ Auger transitions of Ni are shown in figure 5.[7] Due to the near degeneracy of the configurations, a mixing between them is expected. The multiconfiguration calculations do not predict any strong mixing, however. The near degeneracy makes the thermal population of the two lowest-lying configurations possible in the ground

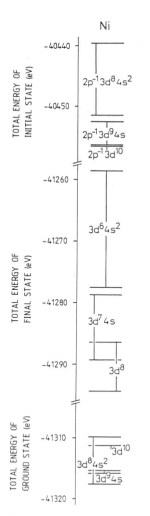

Fig. 5. Energy level diagram of Ni.[7]

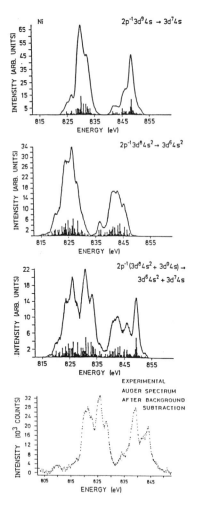

Fig. 6. Comparison between calculated and experimental Auger spectra of Ni.[7]

state of the atom, especially at the experiment temperatures, which results in the two-fold spectral structure. The calculated profile assuming 60:40 populations for the $3d^84s^2$ and $3d^94s$ ground state configurations agrees fairly well with the experiment as shown in figure 6.

The calculated KLL Auger spectrum of Ne depicted in figure 1 was obtained by taking the configuration mixing between the $2s^{-2}$ and $2p^{-2}$ hole states into account. The theory was modified to better agree with the experiment by introducing the interchannel mixing in the final state (see eg. ref. 8 for details). Experimental spectrum, however, contains rich fine structure that cannot be reproduced by the normal KLL Auger transitions alone. The observed satellite structure is mainly due to shake-up processes that take place before the Auger decay. The Auger process thus takes place in an atom with an outermost electron excited to a Rydberg State (shake-up) or into the continuum (shake-off). Figure 7 shows the energy level diagram of the initial and final states of the normal KLL and corresponding satellite transitions. Experimental and calculated KLL

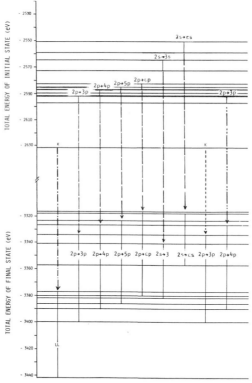

Fig. 7.   KLL Auger and satellite transitions of Ne:
— · — · —   Normal KLL transitions.
— — —   Ordinary shake-up and shake-off satellite transitions where the shake process accompanies the primary ionization and the shaken electron remains unchanged during the decay.
— — — — —   Shake-up transitions with singly ionized initial states and excited final states.
— · · — · · —   Shake-up transitions from excited initial states. The shaken electron is further excited to another Rydberg state during the Auger decay.

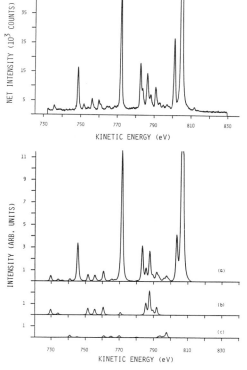

Fig. 8. Upper part: Experimental KLL spectrum of Ne.
Lower part: (a) Calculated KLL spectrum with satellites. The ordinary shake-off 2p → εp (b) and shake-up 2p → 3p (c) transitions gave the biggest contributions.

spectra with satellites are shown in figure 8. It is also possible that shake processes take place during the Auger transition. Their contribution is rather small in the case of normal KLL spectrum of Ne.

Synchrotron radiation can be used to excite the initial electron to a Rydberg state, which then may stay as a spectator during the Auger-type decay. There is, however, a possibility for the excited electron to be shaken up or off during the decay. The energy levels of the Kr $3d^9 \to 4p^4$ decay with 5p or 6p spectator electron are shown in figure 9.[5] Due to the occurrence of a process where the spectator electron is shaken up during the decay, extra structure appears in the spectrum (figure 10). The effect is most pronounced in the case of Ar.[9] When the electron is excited to a 3d Rydberg state, the collapse of the 3d orbital during the decay (figure 11) makes it likely that the excited electron will jump to the next Rydberg state.

EVALUATION OF TRANSITION PROBABILITIES

The Auger transition probability is calculated from the perturbation theory. The transition rate in frozen orbital approximation is given by

$$T = \frac{2\pi}{\hbar} \left| \langle \Psi_f | \sum_{\alpha < \beta} V_{\alpha\beta} | \Psi_i \rangle \right|^2 \tag{5}$$

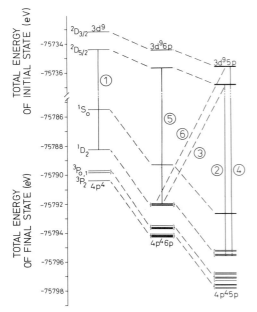

Fig. 9. Energy levels of Kr.[5] Vertical lines display the following transitions. 1: $3d^9(^2D_{5/2}) \to 4p^6(^1D_2)$. 2: $3d^9(^2D_{5/2})5p \to 4p^4(^1D)5p$ resulting in peak ^1D of Fig. 10. 3: $3d^9(^2D_{5/2})5p \to 4p^4(^1D)6p$ resulting in peak (^1D) of dashed line in Fig. 10. 4: $3d^9(^2D_{3/2})5p \to 4p^4(^1D)5p$. 5: $3d^9(^2D_{5/2})6p \to 4p^4(^1D)6p$. 6: $3d^9(^2D_{3/2})5p \to 4p^4(^1D)6p$.

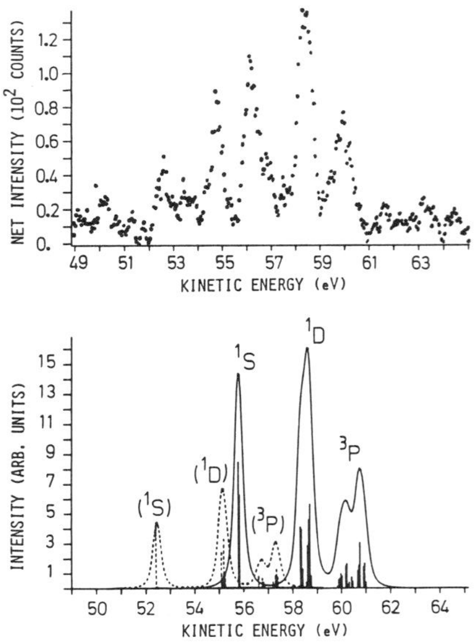

Fig. 10. Upper part: Experimental $M_5N_{2,3}N_{2,3}$ resonance Auger spectrum of Kr.[5] Lower part: Calculated $3d^5_{5/2}5p \to 4p^45p$ (solid line) and $3d^5_{5/2}5p \to 4p^46p$ (dashed line) profiles.

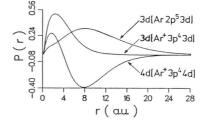

Fig. 11. Radial wave functions of Ar Rydberg orbitals in the initial and final states of Auger-like decay.[9]

where Ψ_i and Ψ_f are the wave functions of the initial and final states of the Auger decay in any nonclosed-shell atom. Final state also contains the continuum electron wave function which is assumed to be normalized per unit energy range. $V_{\alpha\beta}$ is the two-electron operator, that is taken to be the sum of the Coulomb and generalized Breit operators[10]

$$V_{12} = \frac{1}{r_{12}} - \boldsymbol{\alpha}_1 \cdot \boldsymbol{\alpha}_2 \frac{\cos(\omega r_{12})}{r_{12}} + (\boldsymbol{\alpha}_1 \cdot \nabla_1)(\boldsymbol{\alpha}_2 \cdot \nabla_2) \frac{\cos(\omega r_{12}) - 1}{\omega^2 r_{12}} \qquad (6)$$

where ω is the wave number of the exchanged virtual photon and standard notations have been used elsewhere.

The wave function of the νth singly ionized initial state is given by equation (1) and the wave function of the ηth doubly ionized final state, having total angular momentum J'M', by

$$\Psi_\eta(J'M') = \sum_{\alpha=1}^{m} c_{\eta\alpha}^{J'} \Phi_\alpha(J'M') . \qquad (7)$$

By using jj-coupling between the ηth final state and the emitted Auger electron (angular momentum j,m) we obtain for the Auger component transition probability that a singly ionized initial state $(n_1 l_1 j_1)^{-1}$; J decays into any of the doubly ionized final states $(n_2 l_2 j_2)^{-1}(n_3 l_3 j_3)^{-1}$; J' with the emission of an electron with energy ε

$$T_{\nu\eta} = \frac{2\pi}{\hbar} \sum_j \left| \sum_\alpha \sum_\beta c_{\eta\alpha}^{J'} c_{\nu\beta}^{J} <\Phi_\alpha(J'M')\varepsilon j ; JM | \sum_{\alpha,\beta} V_{\alpha\beta} | \Phi_\beta(JM)> \right|^2 . \qquad (8)$$

The matrix elements of the two-electron operator between two CSF's can be separated by tensor algebra into angular parts multiplied by radial integrals. The angular factors can be evaluated by using computer code MCP and MCBP routines from Grant's MCDF program.

The radial integrals in frozen core approximation can be calculated with the use of bound state Dirac-Fock wave functions that correspond to the initial one hole-state configuration. The continuum wave functions can be obtained by solving the Dirac equations in the initial state potential without[10] or with[11] the exchange contribution. The continuum wave functions are then orthogonalized to the bound state wave functions.

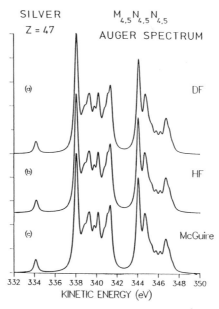

Fig. 12. Intensity distribution of the $3d^9 5s^1 \rightarrow 4d^8 5s^1$ transitions of Ag calculated with the Coulomb operator and DF or HF wave functions. Lower part shows the profile obtained using McGuires radial integrals.[12]

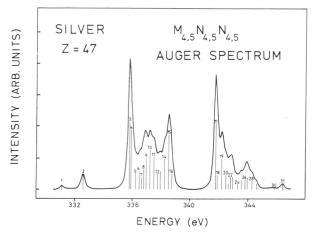

Fig. 13. Experimental spectrum of Ag.

Theoretical profile for the $M_{4,5}N_{4,5}N_{4,5}$ Auger decay of Ag predicted by relativistic DF calculations is shown in figure 12. The importance of the Breit interaction was tested by producing profiles with and without it. The introduction of the Breit operator or even the Möller operator did not improve visibly the results obtained by Coulomb operator. Pure Coulomb operator yielded the profile (a) when obtained with relativistic wave functions, whereas nonrelativistic wave functions gave the profile (b). An experimental spectrum is depicted in figure 13. Relativistic single-manifold ΔSCF calculations used throughout to predict Auger energies overestimate the splitting between final state energy levels, which causes the main disagreement between experiment and theory. The intensity distribution obtained with McGuires radial integrals[12] is also shown for comparison.

REFERENCES

1. I. P. Grant, B. J. McKenzie, P. H. Norrington, D. F. Mayers, and N. C. Pyper, Comput. Phys. Commun. 21, 207 (1980).
2. B. J. McKenzie, I. P. Grant, and P. H. Norrington, Comput. Phys. Commun. 21, 233 (1980).
3. W. Mehlhorn, in: Atomic Inner-Shell Physics, G. Crasemann (ed), New York: Plenum Press (1985).
4. H. Aksela, S. Aksela, and H. Pulkkinen, Phys. Rev. A30, 2456 (1984).
5. H. Aksela, S. Aksela, H. Pulkkinen, G. M. Bancroft, and K. H. Taw, Phys. Rev. A33, 3876 (1986).
6. S. Aksela, T. Pekkala, H. Aksela, M. Wallenius, and M. Harkoma, Phys. Rev. A35, 1426 (1987).
7. H. Aksela, S. Aksela, T. Pekkala, and M. Wallenius, Phys. Rev. A35, 1522 (1987).
8. G. Howat, T. Åberg, O. Goscinski, S. C. Soong, C. P. Bhalla, and M. Ahmed, Phys. Lett. 60A, 404 (1977).
9. H. Aksela, S. Aksela, H. Pulkkinen, G. M. Bancroft, and K. H. Taw, to be published.
10. M. H. Chen, and B. Crasemann, Phys. Rev. A35, 4579 (1987).
11. J. Tulkki, private communication.
12. E. J. McGuire, Sandia Laboratories Research Report No. SC-RR-71-0835 (1972) unpublished.

ION-INDUCED AUGER ELECTRON PROCESSES IN GASES

Dénes Berényi

Institute of Nuclear Research (ATOMKI) of the Hungarian Academy of Sciences, Debrecen, Hungary

1. INTRODUCTION

1.1. Complementarity of X-ray and Auger transitions

The Auger transition is a spontaneous de-excitation process of an atom having a vacancy in one of its shells. It was discovered by P. Auger in 1923 (Auger, 1923, 1925) as an alternative of the radiative process.

An atom with vacancies in its shells can be de-excited by different channels which are

- radiative processes
- Auger processes
- Coster-Kronig transitions
- autoionization processes.

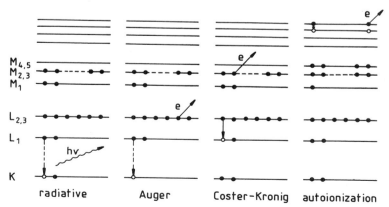

Fig. 1. Schematic illustration of the different nonradiative de-excitation processes in comparison with the radiative one.

The above processes are illustrated in Fig. 1 and in the present survey we will deal with the last three ones which have their own special names each but they are all in fact Auger processes and there is no real difference between them in principle. The autoionization however is a process for the outer shells. In this an emission of a low energy photon is also possible as in the case of Coster-Kronig transitions, too.

In the case of the Auger-electron emission when an inner-shell vacancy is filled by an electron from a higher shell, the excess energy released is given to another electron of a higher shell to transfer it to a continuum state. The <u>Coster-Kronig transition</u> is that special case of the Auger-transition when the level having the vacancy and the level from which the electron fills the vacancy are within the same shell. As the energy differences of the subshells in a shell are relatively small, the electron will be emitted from the highest level of the atom concerned and so the energy of the Coster-Kronig electrons is rather small. Before <u>autoionization</u> two electrons in the highest levels will be excited simultaneously to even higher bound but unfilled states. Then one of the electrons will return to its original state and the other electron will be ejected. In such a process even more than two electrons can be included. The name of the autoionization is autodetachment if the process takes place in a negative ion.

Having two possible decay modes of an atomic state with a vacancy, namely the radiative and nonradiative transitions, the branching ratios of these competing processes can be given.

For K-shell vacancies

$$\omega_K = \frac{N_{KX}}{N_K} = \frac{P_{KX}}{P_{KX}+P_{KA}} \qquad a_K = \frac{N_{KA}}{N_K} = \frac{P_{KA}}{P_{KX}+P_{KA}} \qquad (1)$$

$$\omega_K + a_K = 1 \qquad (2)$$

where N_K is the number of K-vacancies, N_{KX} is the number of the X-rays in the

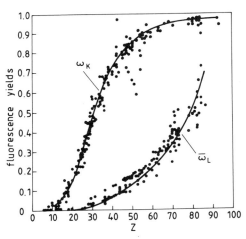

Fig. 2. K-shell (Bambynek et al., 1972) and average L-shell (Venugopala Rao, 1975) fluorescence yields as a function of the atomic number.

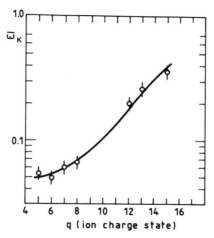

Fig. 3. Mean Ne K-shell fluorescence yield versus Cl projectile-ion charge state in Cl^{q+} (50 MeV)-Ne collision (Burch et al.,1974)

K-series, N_{KA} is the number of Auger electrons in the K-series, P_{KX} and P_{KA}, however the corresponding probabilities, ω_K is the <u>fluorescence yield</u> and a_K is the <u>Auger yield</u> for the K-shell.

For L-shell vacancies not only Auger but Coster-Kronig transitions are also possible, the so called <u>Coster-Kronig yields</u> are as follows

$$f_{L_1L_2} = \frac{N_{L_1L_2}}{N_{L_1}}, \quad f_{L_1L_3} = \frac{N_{L_1L_3}}{N_{L_1}}, \quad f_{L_2L_3} = \frac{N_{L_2L_3}}{N_{L_2}} \tag{3}$$

$$\omega_{L_1} + a_{L_1} + f_{L_1L_2} + f_{L_1L_3} = 1 \tag{4}$$

$$\omega_{L_2} + a_{L_2} + f_{L_2L_3} = 1 \tag{5}$$

$$\omega_{L_3} + a_{L_3} = 1 . \tag{6}$$

The number of L X-rays emitted per primary vacancy in the L subshells will be

$$\nu_{L_1} = \omega_{L_1} + f_{L_1L_2}\omega_{L_2} + (f_{L_1L_3} + f_{L_1L_2}f_{L_2L_3})\omega_{L_3} \tag{7}$$

$$\nu_{L_2} = \omega_{L_2} + f_{L_2L_3}\omega_{L_3} \tag{8}$$

$$\nu_{L_3} = \omega_{L_3}$$

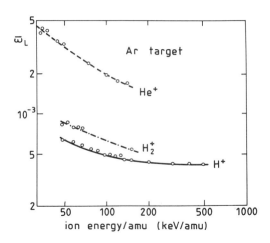

Fig. 4. Average Ar L-shell fluorescence yield versus the projectile velocity (total energy divided by the mass of the projectile) for different light ion projectiles (Stolterfoht et al., 1973)

and

$$\omega_L = n_1 \nu_{L_1} + n_2 \nu_{L_2} + n_3 \nu_{L_3} \tag{9}$$

where n_i is the relative probability of primary vacancy creation of the corresponding subshells. The notations are obvious.

It is customary to define <u>average (mean) fluorescence yields</u> for the individual shells, without specifying subshells:

$$\bar{\omega}_L = \frac{N_{LX}}{N_L} \tag{10}$$

Here N_{LX} is the total number of L X-rays, N_L is the total number of primary L vacancies.

Fig. 2 shows the ω_K and $\bar{\omega}_L$ as a function of the atomic number. As can be seen in the figure the probability of the Auger process in case of K vacancy is especially high for the lightest atoms (below Z number $\simeq$ 10-12), and that of L vacancy for atoms below Z number $\simeq$ 35-38.

Similar yields can be defined also for higher (M,N,...) shells correspondingly, for M shell e.g. the Auger electron emission is prevailing below 80 (Z number).

It should be noted here that the fluorescence yields and the corresponding other yields are not unchanged in the collision processes. Fig. 3 shows the mean fluorescence yield $\bar{\omega}_K$ of Ne as a function of the projectile charge state in the Cl^{q+} (50 MeV)-Ne collision. The mean K-shell fluorescence yield $\bar{\omega}_K$ should be used here because there is an X-ray branching over the states of multiple ionization and excitation produced in the collision for which one should make an average. The fluorescence yield is changing not only in heavy ion collisions but in those of light ions. Fig. 4 shows the Ar L-shell average fluorescence yield as a function of the impact energy for different light-ion projectiles.

Shortly should be dealt here with the notation and classification of the Auger transitions. The emitted Auger electrons (Auger lines) and the corresponding transitions are denoted by three letters and three numbers characterising the three levels included in the actual process itself (see Fig. 1). Thus e.g. KL_2L_3 denotes the Auger transition when the K-shell vacancy (the lower index is missing here because the K-shell has no subshells) is filled by an electron from the L_2 shell and one electron is expelled from the L_3 shell.

According to the primary vacancy we can speak about K,L... Auger electtrons (these are the so called K,L etc series) and also about KLL, KLM etc.

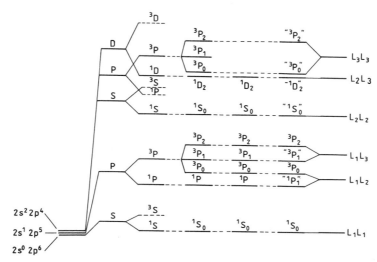

Fig. 5. Configurations and levels for the final state in the KLL Auger groups if different electron interactions and coupling types are taken into consideration. The different perpendicular set of levels corresponds to a different interaction and/or coupling (Bergström et al., 1968).

Auger groups. To give a full description of an Auger transition, the total number of the final states resulting from two electron vacancies should be taken into consideration. Fig. 5 shows all the configurations ($2s^2 2p^4$, $2s^1 2p^5$, $2s^0 2p^6$) and levels for the KLL group with various suppositions for the electron interaction and the type of coupling.

It is very important to differentiate between diagram (or normal) and satellite lines (transitions). An Auger transition with only one single vacancy in the initial state and just two vacancies in the final state is called diagram or normal transition or line. All the other transitions (lines) are satellites having so either more than one single vacancy in the initial and/or more than two vacancies in the final states. The satellites are very important in the ion induced Auger spectroscopy.

There is not enough time and place here to treat such phenomena as super-Coster-Kronig transitions, double Auger effect etc. For further informations on the Auger effect I could suggest e.g. the following surveys: Bergström et al. (1968), Burhop and Asaad (1972), Rao (1975), Chattarji (1976), Mehlhorn (1978), Stolterfoht (1987).

1.2. Short history of the studies on ion-induced Auger electron processes

Vacancies as initial states to Auger transitions can be produced by electromagnetic radiations, electrons and by ions. The latter is the subject of the present paper and studies of this kind were started relatively late some twenty years ago.

In the middle of the sixties Rudd and his coworkers measured the first autoionization and Auger spectra produced in ion-atom collision, namely at light ion impact on light target: H^+, H_2^+(75 keV)-He (Rudd,1964; Rudd,1965), as well as at light and heavy ion impact on heavy target: H^+, Ar^+(100keV)-Ar, (Rudd et al., 1966a and 1966b) and H^+, He^+, Ne^+(150-250 keV)-Ne (Edwards and Rudd, 1968). In the second half on the sixties similar other measurements were also published: e.g. Kessel et al., 1966; Kessel et al., 1967; McCaughey et al., 1968.

In the seventies the research work in this field was broadened very much. The experimental techniques were improved in several respects. High resolution electron spectrometers were started to use ($\simeq$1 eV or less FWHM line width was attained), coincidence relations were more and more intensively studied (e.g. impact parameter dependence of Auger production cross section) and high energy multiple charged ions were used as projectiles including ions of the highest Z number and with energy up to 10-20 MeV/amu.

In the recent years the study of the multiple electron aspect in Auger processes come into prominence both in theory and in experiment, and at the same time the correlation phenomena between the electrons concerned are also investigated. Techniques of zero degree electron spectroscopy as well as special equipments for Auger electron angular distribution studies were built. More and more complex coincidence arrangements are also put into operation.

To have more information in the whole field of ion-induced Auger processes there are many older and recent surveys. Some of them are the following: Rudd and Macek (1974), Rudd (1975), Richard (1975), Moore (1976), Stolterfoht (1978), Matthews (1980), Berényi (1981), Rudd (1982), Berényi (1983), Berényi (1984), Stolterfoht (1986), Stolterfoht (1987).

It should be noted here that in the case of ion-induced Auger processes, the situation is more complex relative to the vacancy production by electrons or photons, because Auger electron emission is possible both from the projectile ("scattered ion") and the target ("recoil ion") in general. Thus the Auger electrons are usually emitted 10^{-13}- to 10^{-16} sec after the vacancy creation (after the collision event). However much longer lifetimes are also observed and there are cases when the Auger electrons are emitted from the so-called quasi-atom or quasi-molecule if the velocity of the projectile is much slower than that of the shell-electrons concerned. One of the recent examples for this phenomenon was given by Shergin (1987) for the He^+-He collision system. In that case a broad "quasi-continuous" hump is observable in the electron spectrum (or correspondingly in the X-ray spectrum) which is in fact not characteristic of either the target for the projectile. Such X-rays were found by Saris et al.(1972) for the first time.

1.3. Some remarks on instrumentation

In this section only some short remarks will be given on the nowadays more and more complicated experimental techniques of this field. More detailed surveys on this issue can be found in some of the review-papers referred in Section 1.2, as well as e.g. in Berényi (1976), Stolterfoht (1983), Berényi (1987).

I would not intend dealing here with the otherwise so important sources of ion-beams (ion-sources, accelerators etc.) in this field, with the target systems, even less with the scattered and recoil ion selectors which are important from the point of view of the Auger electrons in the related coincidence measurements.

My short comment here concerns the spectrometers, namely the X-ray spectrometers and the electron spectrometers used in Auger electron spectrum measurements. There is no dramatic difference in the most important parameters of the X-ray and electron spectrometers (e.g. target size, size of the target seen by the spectrometer, the solid angle and energy resolution of the spectrometer etc.) except in the over-all luminosity of them. It is orders of magnitudes higher for electron spectrometers which is a very important advantage. It is mainly due to the relatively small reflectivity ratio of dispersion on crystals or gratings in the case of X-ray spectrometers.

Regarding the resolution of the X-ray and electron spectrometers, there is a permanent competition between them. In both cases the high resolution attained is nearly the same in orders of magnitude, namely several tenth of eV (line-width). In general, however, the actual resolution depends on many factors namely ion energy region, type of spectrometer etc. In addition, at crystal diffraction spectrometers it also depends on the actual crystal, in electron spectroscopy, however, on the acceleration or deceleration of the electrons in the spectrometer. Fig. 6 shows a very rough comparison for a more or less "standard" X-ray and electron spectrometer. Stolterfoht (1983), however, gives a tabulated comparison for a number of parameters in case of two spectrometers of better resolution. Fig. 7 gives an actual example showing the Ne K X-ray and a nearly corresponding Auger electron spectrum. It would be easy to find examples where the X-ray spectrum has a better resolution.

The attained best instrumental resolution in both cases is somewhere at 0.1-0.01 eV in orders of magnitude now. In X-ray and Auger studies, however, the natural width Stolterfoht (1976) is the most important factor giving the actual limit for the resolution. The natural width increases with Z number and decreases towards the higher shells. E.g. it is 0.24 eV for KL_2L_3 Auger line, and the same value for K_{α_1} X line in the case of Ne but for Sn 14.1 eV for KL_2L_3 and 11.2 eV for K_{α_1} (see Krause and Oliver, 1979).

In case of electron spectroscopy there is a possibility, however, to improve the resolution in the low energy region of autoionization electrons. Here the autoionization lines - superimposed on the so called „cusp" at 0° in the laboratory frame (see Fig. 8) - can be studied by a much better resolution if the spectrum measured in the laboratory rest frame is transformed into the projectile rest frame (Stolterfoht et al., 1984). It is the utilization of the advantage of the "moving laboratory". Recently e.g. Atan et al.(1987) investigated the Ar autoionization spectra with a resolution in the meV region.

1.4. Kinematic considerations

Without entering into particulars the importance of the kinematic effects in Auger spectra should be emphasized because the emitter (i.e. the scattered projectile or the recoiled target ion) is a moving system in every case. We are only mentioning here the influence of the kinematic effects for the line-width as stretching, enhancing and broadening and for the position of the line as shifting and doubling and refer the papers concerned, e.g. Stolterfoht et al., 1975; Stolterfoht, 1978; Stolterfoht, 1984; Stolterfoht, 1987, Rudd and Macek, 1973.

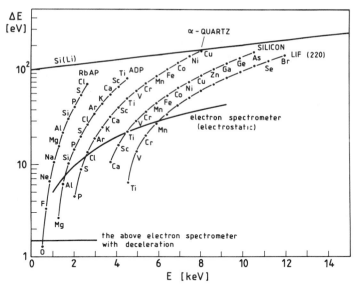

Fig. 6. A rough comparison of the resolution (ΔE=full line width at the half maximum) of a more or less "standard" X-ray spectrometer (with different diffraction crystals) and of an electron spectrometer (with and without deceleration) (Török and Varga, 1987) for about 0.5 keV. The X-ray spectrometer data are related for first order reflection and a 0.3° angular divergence of the Soller-slits.

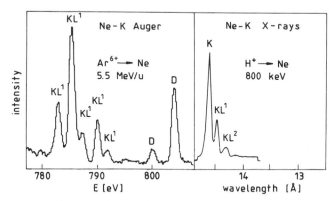

Fig. 7. Corresponding regions of Ne K X-ray (Schneider, 1975) and Ne K Auger spectra (Kádár et al., 1986).

The Doppler broadening of the Auger electron line is the cause why high-resolution studies are practically impossible in the impact energy region of several MeV except the zero-degree measurements where the broadening effect practically disappears.

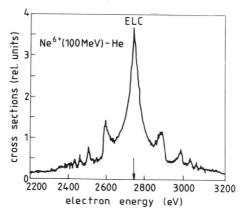

Fig. 8. Ne autoionization lines superimposed on the so called ELC (electron loss to the continuum) peak (Schneider et al., 1986)

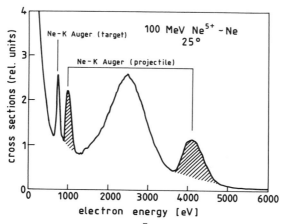

Fig. 9. Electron spectrum from the Ne^{5+}(100 MeV)-Ne collision (Prost, 1980)

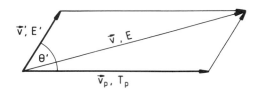

Fig. 10. Velocity vectordiagram for the Auger electron emission from moving emitter (Stolterfoht et al.,1984)

Another important consequence of the kinematic effect is that the projectile Auger peak can appear twice in the electron spectrum — if the projectile is faster than the Auger electron — taken in the laboratory frame (see Fig. 9). It is based on the following relation obtained from the velocity diagram (Fig.10) at small angles

$$E = E' + \frac{T_p m_e}{M_p} + 2(E'\frac{T_p m_e}{M_p})^{\frac{1}{2}}\cos\theta', \qquad (11)$$

which has two solutions with respect to E. Here E and E' are the energy of the Auger electron in the laboratory and the projectile frame, respectively, $T_p = v_p^2/2$ is the projectile kinetic energy, m_e and M_p the electron and the projectile rest mass, θ' is the ejection angle of the electron in the projectile rest frame.

2. AUGER ELECTRON EMISSION FROM THE TARGET ATOM

The investigation of ion-induced electron spectroscopy was started by the study of the Auger electrons emitted from the target atom in the midsixties as it was already mentioned in Section 1.2. These can be found among others in Table II and III of the survey of Berényi (1981).

In the treatment of the target Auger electron spectroscopy, different subdivisions of the materials are possible. In the present survey, the results for light and heavy ion impact will be treated separately and in these subsections the characteristics of the spectra, the satellite structure, the production cross sections, the recoil Auger spectroscopy, statistical analysis of the spectra will be given.

2.1. Light ion projectile

In Fig.11 a comparison of the X-ray, electron and proton induced Ne KLL Auger spectrum is represented. A part of the spectra in Fig. 11 is shown for e⁻ impact in Fig. 12 at an even higher resolution. Practically no difference can be observed between the different spectra. In the spectra the diagram lines (D) can be found first of all while the others are satellite lines. The well visible lines are so called shake-of (or ionization) satellites when in addition to the vacancy on the K-shell another vacancy was also produced simultaneously either in the 2s or in the 2p shell.

The very close similarity of the spectra induced by protons to that by electrons and X-rays is standing only if the energy of the protons in relatively high as in the case of Fig. 11. The two spectra in Fig. 13 shows the difference between the Auger electron spectra induced by 6 MeV and by 150 keV

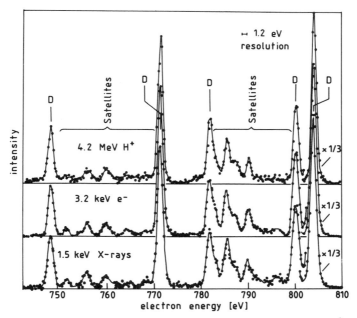

Fig. 11. A comparison of the Ne K Auger spectrum induced by H^+, e^- and X-rays (Stolterfoht, 1987)

protons. If we define the <u>satellite-to-total ratio</u> as

$$\frac{N_{sat}}{N_{tot}} = R_{s/t} \qquad (12)$$

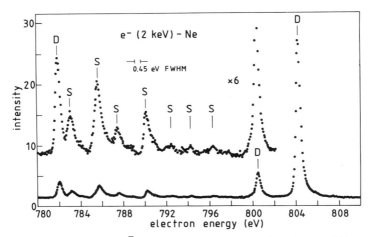

Fig. 12. A part of the e^- induced Ne K Auger Spectrum with very high resolution (Ricz et al., 1987).

where N_{sat} denotes the sum of the intensity of the satellite lines in the region of the Auger spectrum concerned, and N_{tot} the corresponding total spectrum intensity, then Fig. 14 shows that this ratio increases towards lower energies at proton impact, i.e. at lower impact energies the intensity of the satellite lines is higher. The value $R_{S/T}$ for electron is indicated in the figure.

In Fig. 14 in addition to the $R_{S/T}$ for H^+ and e^- impact, a value for He^+ impact is also indicated. This latter value and the data in Table I show that the intensity of the satellites is increasing with the increase of the charge. It is an effect which will be very important at heavy ion impact.

Table 1

Satellite-to-total ratio

in Ar L spectrum (Stolterfoht, 1987)

Impact energy keV/u	Projectile		
	H^+	He^+	He^{2+}
100	0.58	0.82	
250	0.49	0.74	0.83
500	0.41		0.67

Not only the intensity of the satellites changes with the energy of the projectile but the width (FWHM) <u>of the spectrum lines,</u> too (cf. Fig. 13). The diagram in Fig. 15 (made on the basis of Matthews et al., 1974) shows the decrease of the line-width as a function of the impact energy. It is due to the recoil of the target atom because the recoil energy decreases as the

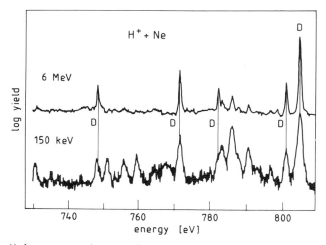

Fig. 13. Ne K Auger spectra produced by 6 MeV and 150 keV proton impact (Matthews et al., 1974)

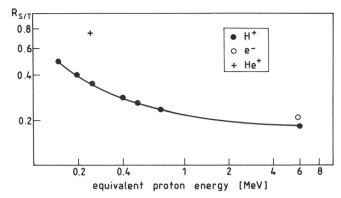

Fig. 14. Satellite-to-total ratio as a function of the impact energy (velocity) for protons. A single value for e⁻ and another for He⁺ is also indicated (Matthews et al., 1974).

projectile energy increases (kinematic collision paradoxon). May I mention here, however, that the line-width is practically independent on the charge state of the projectile.

Having an instrumental technique of very high resolution the natural width of the lines can be studied (cf. Section I.3). It gives a possibility

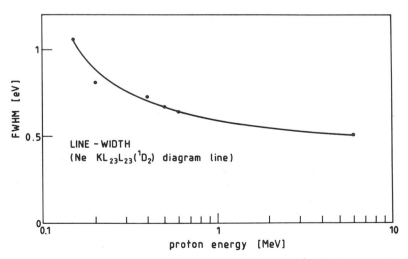

Fig. 15. The decrease of the width of the $KL_{23}L_{23}(^1D_2)$ line in the Auger electron spectrum produced in H⁺-Ne collision (on the basis of the data of Matthews et al., 1974). The solid line serves only for guiding the eye.

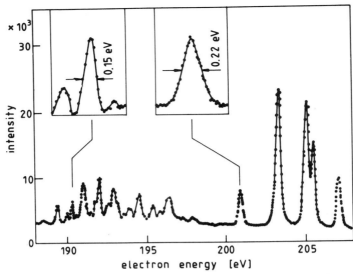

Fig. 16. Ar L Auger electron spectrum produced by 200 keV H$^+$ impact (Ridder et al., 1975 in: Stolterfoht, 1976)

to determine the lifetimes of the vacancy states in the range cca. 10^{-14} - 10^{-15} sec. Figure 16 gives one of the best examples for natural line-width measurements. The instrumental width in this case was 0.090±0.015 eV.

By the measurement of the Auger electron spectra the Auger production cross section can be determined. Using the corresponding fluorescence yield (see in Section 1.1) values one can obtain the vacancy production (ionization) cross section for the shell concerned. The total $L_{2,3}$ Auger electron production cross section is shown in Fig. 17. Here two different theoretical curves are also indicated. The agreement with the calculation of Basbas et al. (1973) is rather good.

In Fig. 18 the K-shell ionization cross section is given for C, N, O, F and Ne in a scaled diagram at proton impact. The scaling parameters are the K binding energy I_K and the mass ratio λ for protons and electrons, E_p denotes the impact energy of the protons. As it can be seen the scaling is perfect but at lower energies neither the binary encounter theory nor PWBA can describe the experimental data.

The cross section measurements are very suitable to check the theoretical models (for the collision mechanism) of the vacancy production and the cross sections in the case of the L-subshells are more sensitive for the calculations than the K-shell or the total L-shell cross sections. The three most important models for the collision mechanism are the direct Coulomb ionization, the quasi-molecular model and the capture or charge transfer (exchange) mechanism. We are not going into the details here, we will stay a little only at the direct Coulomb ionization.

To describe the direct Coulomb ionization different concrete approximations are possible as e.g. the binary encounter approximation (BEA) or plane-wave Born approximation (PWBA). In both cases the cross section concerned is proportional to the square of the projectile charge Z_p. It is the so called Z^2 rule which has only an approximate validity. Both the

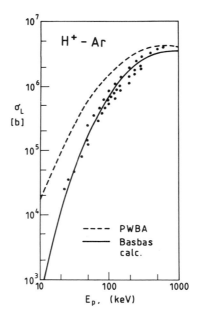

Fig. 17. L_{23} Auger electron production cross section by proton impact. Details on the plotted experimental values and on the theoretical calculations see in Basbas et al., 1973. The figure is taken from this paper and somewhat modified.

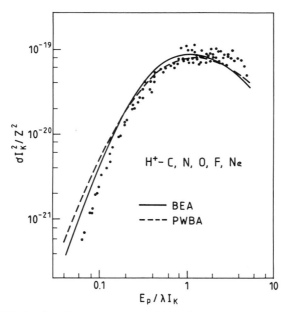

Fig. 18. K-shell ionization cross section for proton impact from Auger spectrum measurements in a scaled plot. See for further details of the plot in Kobayashi et al., 1976 from where the figure is taken with some modification.

experimental data and the more detailed theoretical calculations deviate from this rule as it is shown e.g. in Fig. 19.

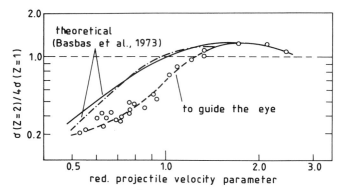

Fig. 19. Deviation from the Z^2 rule for the direct Coulomb ionization cross section. In the plot the cross section ratios are given for H^+, D^+ ($Z_p=1$) and He^+, He^{2+} ($Z_p=2$) projectiles at C, N, Ne target atoms. The figure is based on Stolterfoht, 1978 (see Berényi, 1981).

2.2. Heavy ion projectiles

There is no exact border between light and heavy ions as projectiles. Primarily the hydrogen and helium isotope ions are regarded as light projectiles. The others up to the uranium isotope ions are already heavy ions customarily. The main characteristic of the heavy ion collision processes is the involvement of a number of electrons into the collision including the electrons accompanying the projectile ion. Another important feature is the higher charge of the projectile.

The study of heavy ion induced Auger electron spectra was started at the beginning of the ion-induced Auger spectrum investigations in the second half of the sixties (see Table III in the survey of Berényi, 1981) In these cases, however, the energy of the Ar^+ and Ne^+ projectiles is relatively low (100-300keV) and the ions are only single ionized.

The Auger spectra produced in high-energy heavy ion collision processes (at about 2 MeV/u) were studied in the seventies. In the first investigations the projectiles were oxigen ions (in 5+ and 7+ charge states) and the target was Ne in these and several other experiments at the beginning (Burch et al., 1972; Matthews et al., 1973; see Table IV in Berényi, 1981 and the survey: Berényi, 1984).

Fig. 20 shows the main characteristics of the heavy ion induced Auger spectra. First of all a number of satellite lines are present with much overlapping due to the large number of vacancies produced in the different shells. In addition, a kinematic broadening effect (the so called azimuthal broadening) is also appearing at the lines. The latter is even more expressed at lower impact energies as e.g. in Rudd et al. (1966b) measurements for the Ar^+ - Ar collision where the projectiles energy is in the range from 100 to 300 keV or in the case of Ne⁺ - Ne collision at 500 keV impact energy (Stolterfoht et al., 1975).

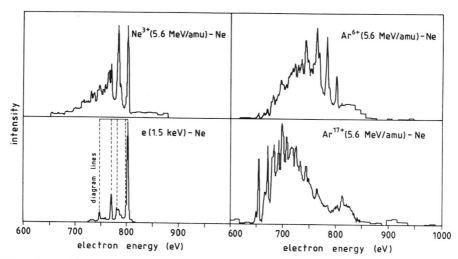

Fig. 20. Ne K Auger-spectra at different projectiles. For comparison, the corresponding spectrum at electron impact is also shown (made on the basis of Kádár et al., 1985).

In the case of more energetic projectiles (as in Fig. 20), the above broadening is smaller due to the collision paradoxon (the recoil energy is smaller if the impact energy is higher). However, even at the energies of the projectiles of Fig. 20, the Auger lines are so overlapped and blended that the study of the individual lines is rather difficult. So it is quite justified to determine certain <u>average quantities</u> from the spectra and to interpret them on the basis of some statistical considerations. It should be mentioned here, however, that if the projectile is in an even higher charge state, then the target atom may be stripped to a few electron system and thus e.g. Li-, Be- like etc. ions will be produced. In these cases the individual Auger lines originating from these very highly ionized atoms can be studied without any difficulty. Such an Auger spectrum is shown in Fig. 21.

Taking the heavy ion induced Auger spectra (e.g. Fig. 20) it is not difficult to state qualitatively that the centroid value of the spectrum is smaller at higher charge of the projectile at the same impact energy. In fact, there is a relation between the centroid energy and the degree of the multiple ionization. The linear relationship between the centroid energy E_C and the mean number of vacancies in the L shell (together with a K shell vacancy) $\bar{n}_L$ for Ne K Auger spectra is given in Fig. 22.

If P_n is the probability for n vacancy production in the L-shell together with a K shell vacancy production in a collision process, then from statistical considerations (Burch, 1973; Hansteen and Mosebeck, 1972) follows

$$P_n = \binom{8}{n} p_L^n (1-p_L)^{8-n} \tag{13}$$

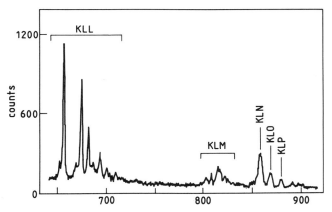

Fig. 21. Ne K Auger spectrum of a Li-like Ne atom produced in Kr^{18+} (1.4 MeV/u)-Ne collision (from Mann et al., 1981).

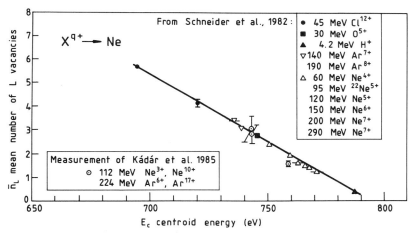

Fig. 22. Relationship between the mean number of L vacancies $\bar{n}_L$ and the centroid energy of the Ne K Auger spectrum at different heavy ion impact X^{q+} with various charge state and energy. (For corresponding theoretical calculations see Matthews, 1980; Végh, 1984)

Here p_L is the probability for a single-electron ionization in the L shell, P_o denotes the probability for the production of the diagram lines. An important relationship is here

$$\bar{n}_L = \sum_{n=0}^{8} n_L P_n = 8 p_L \qquad (14)$$

and also

$$\sum_{n=0}^{n=8} P_n = 1. \qquad (15)$$

If we have an experimental value of P_n, e.g P_o by determining it from the ratio of the sum of all the diagram lines to the total Auger K spectrum or similarly P_6 by means of the lithium-like lines (see in Fig. 20 on the low energy region of the spectrum), then one can obtain the distribution of P_n as a function of n_L (number of L vacancies) by using equ.(13). This binomial distribution in general has a maximum at the higher n_L value the higher the projectile charge state at the same velocity.

At the detailed analysis of an Auger spectrum of high resolution, p_L can be determined from the ratio of the intensity of the different satellite groups (e.g. with one or two holes in the L shell in addition to the K shell vacancy) or that of the diagram group to the KL^1 group. Kádár et al. (1986) experienced, however, that the values determined by the diagram--to-total ratio and by the ratio of the Auger-lines from KL^0 and KL^1 groups are not equals outside the limits of the errors. The reason was that at such a good resolution one can differentiate between satellites where the vacancies in the initial state were produced by ionization or by excitation to a higher level. They also could determine the p_{2s} and p_{2p} for the "ionization satellites", namely the subshell ionization probabilities.

Sulik et al. (1984 and 1987) (see also Kádár et al. 1985 and Sulik and Hock, 1985) worked out a simple binary encounter model to interpret the experimental p_L values and to give a scaling for them. Fig. 23 shows that the agreement is very good between these calculations and the experimental values of rather high precision. It can be seen that a different scaling is valid for p_{2s}, p_{2p} and p_L. The model is able to interpret the different p_L values in a broad range of experimental parameters if the impact velocity is high enough (higher than $\frac{1}{3}$ of v_e of the atomic electrons concerned). Respective calculations were made using coupled channel method by Becker et al., (1980) (further references see in Sulik et al., 1987). They are comprehensive and physically realistic but the calculations concerned rather complicate and do not render any simple scaling.

It should be mentioned here a very simple but very characteristic quantity in heavy ion collisions, namely the satellite-to-total ratio $R_{S/T}$. Fig. 24 shows $R_{S/T}$ for Ne K Auger electron spectra for different impact ions. The Z_p^{eff}/v_p scaling of $R_{S/T}$ seems to be very good.

The cross section which some way can be regarded as an "average" or "integral" parameter can also be determined by means of Auger electron spectra in heavy ion-atom collisions. A typical example is given in Fig. 25 in comparison with the corresponding H^+ impact values on Ne target. The capture and ionization contributions to the total cross section are also indicated. The number of absolute cross section measurements for heavy ion impact especially in the high-energy region is rather low.

The so called hypersatellites should be still mentioned. They are appearing in the K Auger-spectra when not only a single but a double

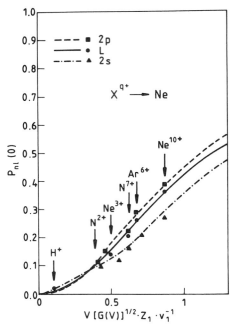

Fig. 23. Comparison of the simple binary encounter model for multiple ionization with the corresponding experimental values for Ne L subshells (Sulik et al., 1987). Here $p_{n\ell}(0)$ is the simple ionization probability for the $n\ell$ subshell at zero impact parameter, $G(V)$ is the BEA scaling function (McGuire, J. H. and Richard, 1973) and $V = v_1/v_2$ (v_1 - the projectile velocity, v_2 - the velocity of the electron in the atomic shell concerned).

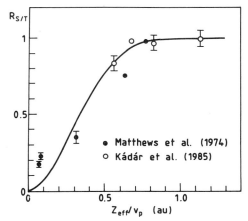

Fig. 24. Satellite-to-total ratio $R_{S/T}$ for Ne K Auger spectra as a function of Z_{eff}/v_p (Kádár et al., 1985). The curve is calculated according to a simple binary encounter model (Sulik et al., 1984).

vacancy is produced in the collision. In Fig. 20 in the spectrum at Ar^{17+} impact there is some increase around 915 eV but the hypersatellite group is well visible in Fig. 26 and it was studied by Woods et al. (1975).

If the instrumental resolution is enough and the line broadening allows (see earlier) the <u>study of the individual Auger electron lines</u> becomes possible. In Fig. 20 it can be seen that the intensity of the diagram lines (look at the most intensive one at 804.3 eV at e^- impact) is much changing as a function of the atomic number and charge state of the impact ion at the same velocity. In the case of Ar^{17+} e.g. the diagram lines are not seen any more. On the other hand at the latter impact ion another very expressed line (more correctly: group of lines) appears in the Auger electron spectrum (see the lowest energy region in the spectrum in Fig. 20). These are the so called Li-like Auger lines. They originate from an ionic state in which the Ne atom has already only three electrons.

The Li-like states between 650 and 680 eV in Ne have been studied relatively thoroughly both experimentally and theoretically. The results are well summarized by Stolterfoht (1987). We should like to mention here that an even more detailed analysis is going on now in our Institute (Kádár et al, 1987; Kádár et al., 1987b). Fig. 27 shows the section concerned of the Ne K Auger electron spectrum from Ar^{16+} (5.5 MeV/u) -Ne collision with the identification of the individual lines.

The study of angular distribution of Auger electrons is important to obtain information about the alignment of the inner-shell vacancy states in the collision process. As it is well known an alignment is expectable in a particle impact if the ionized atom from which the Auger electron is emitted, has an angular momentum $J \geq 1$. I refer here the review article of Mehlhorn (1982a, 1982b), and Kabachnik (1987).

An indication of non-isotropic angular distribution of Auger satellites in heavy ion collisions was made by Stolterfoht et al. (1977). Very expressed non-isotropic angular distributions at several satellites in the Auger electron spectra from Ne^{q+}, Ar^{q+} (5.5 MeV/u)-Ne collisions were observed and the anisotropy parameters determined and analysed by Ricz et al. (1984, 1985, 1986, 1987a, 1987b). Recently, there should be supposed an A_4 term in fitting the angular distribution for $(1s^1 2s^2 2p^4)$ ^2D-$(1s^2 2s^0 2p^4)$ ^1D transition (720.8 eV satellite line in the Ne K Auger spectrum produced in Ne^{3+}(5.5 MeV/u)-Ne collision; see the corresponding angular distribution in Fig. 28). The effect of the charge state of the projectile on the alignment parameter has also been studied and a definite dependence on that was found (Ricz et al. 1987; see also in Fig. 28) in contradiction with the theoretical prediction.

In the Ne KLL Auger spectrum from Xe^{31+}(200 MeV/u)-Ne collision in addition the Li-like lines, some other lines corresponding to the configurations 1s2s4p and 1s2p4p were observed (Stolterfoht et al. 1977). Mann et al. (1981) (further references see in Stolterfoht 1982) clarified that these configurations are produced in electron capture processes by the recoil target ion. Namely, He-like target atoms are produced by fairly high cross section in the primary collision with relatively long life time states and this low energy moving recoil ions (eV in orders of magnitude) can capture electrons in collisions with the neutral target atoms. This discovery open a new research field, namely the study of the collision of very low multiply charged ions with neutral atoms. Such processes can be investigated only by this way.

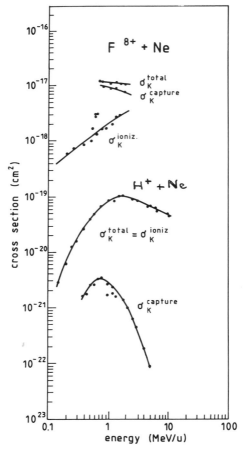

Fig. 25. The total and capture as well as ionization cross sections for F^{8+} - Ne and H^+-Ne collisions as a function of the impact energy measured by the study of the respective Auger spectra. (Richard, 1980, the data of Woods et al., 1976).

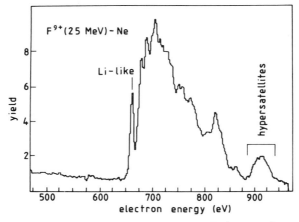

Fig. 26. Ne KLL Auger spectrum produced in F^{9+}(25 MeV) impact (Woods et al., 1975).

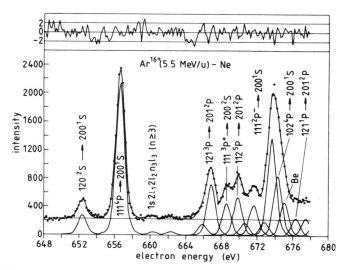

Fig. 27. Auger lines from the Li-like states of Ne induced in Ar^{16+}(5.5 MeV/u)-Ne collision (Kádár et al., 1987b).

Finally, the observation of molecular explosion in heavy-ion-atom collisions is mentioned here. Fig. 29 shows the carbon K Auger spectrum for three different molecular targets excited in Ar^{12+}-CX collision. The "truncated" line in the case of CO has a line-width about five times bigger than the corresponding one from the CH$_4$ target. The phenomenon can be understood from the steriometric structure of the molecules concerned.

3. AUGER TRANSITIONS IN PROJECTILE IONS

Auger transitions, Auger electron emission is possible not only from the atom excited in the collision but also from the projectile ion if it has accompanying electrons. This is, however a quite new promising branch of Auger electron spectroscopy. For good reviews see e.g. Stolterfoht et al. (1984), Itoh and Stolterfoht (1985), Stolterfoht (1987), Stoterfoht et al. (1987).

3.1. Kinematics and main characteristics

In contrast to the target Auger spectroscopy, in the projectile Auger spectroscopy the Auger electrons are emitted from a fast moving ion. It means, however, the presence of different kinematic effects.

Supposing that the scattering angle of the projectile is very small, then the relation between the energy of the Auger electron in the laboratory frame E and that in the projectile rest frame E' is as follows

$$E=[t_p^{\frac{1}{2}} \cos\theta + (E' - t_p \sin^2\theta)^{\frac{1}{2}}]^2 . \qquad (16)$$

Here θ is the electron observation angle relative to the direction of the

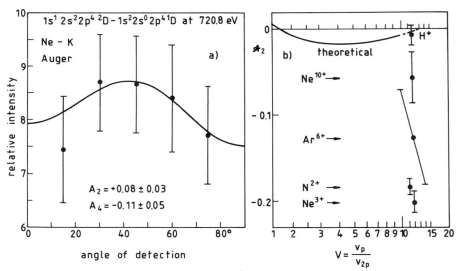

Fig. 28. a) Angular distribution of the $(1s^12s^22p^4)^2D - (1s^22s^02p^4)^1D$ satellite line in the KLL Auger spectrum from the Ne^{3+} (5.5 MeV/u)--Ne collision (Ricz et al., 1987b).
b) Alignment parameter for the $125\,^1P$ and $125\,^3P$ states of Ne produced in X^{q+}(5.5 MeV/u)-Ne collisions (Ricz et al., 1987a).

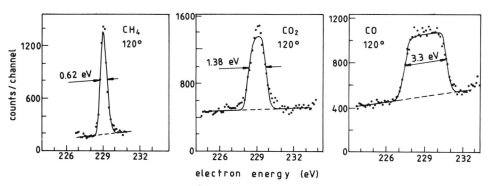

Fig. 29. The $(1s2s2p)^4P - (1s^2)^1S$ carbon Auger line produced by Ar^{12+}(56 MeV) impact for different molecular targets (Mann et al., 1978).

48

impact ion and $t_p = T_p \frac{m_e}{M_p}$ (the meaning of the letters here see in p.10). Equ.(11) is the version of equ.(16) for small angles. When the projectile is faster than the emitted electron, then a negative expression will be in equ.(16) under square root and so no solution exists for the equation. It means that corresponding Auger peak is not observable at angles larger than $\Theta = \arcsin(E'/t_p)^{\frac{1}{2}}$, at smaller angles, however, two peaks appear in the spectrum because of the two possible signs of the square root (see Fig. 9).

Kinematic broadening effects have already been mentioned (Section 1.4). Here, however, the line broadening $\Delta E(\Theta)$ is varied rather strongly with Θ, (see e.g. Fig. 30). The first-order term for that is the following

$$\Delta E(\Theta) = \Delta\Theta \, 2 \, (Et_p)^{\frac{1}{2}} \sin\Theta [1 - (t_p/E)^{\frac{1}{2}} \cos\Theta]^{-1} \quad (17)$$

where $\Delta\Theta$ is the acceptance angle of the spectrometer. It can be seen that the line-width also depends on the incident energy.

It is clear that the first-order term is equal 0 at $0°$ and here

$$E = (t_p^{\frac{1}{2}} + E'^{\frac{1}{2}})^2 . \quad (18)$$

In this way, there is a possibility for high resolution spectroscopy of the Auger electrons emitted from the projectile ion if we measure the Auger spectrum at $0°$ relative to the direction of the projectile with a reduced as possible acceptance angle of the spectrometer $\Delta\Theta$. This is the so called zero-degree projectile Auger electron spectroscopy. The intensive utilization of it was initiated recently by Stolterfoht and his collaborators (Itoh et al., 1983).

The projectile Auger spectroscopy has several advantageous features. Some of them are as follows:

- At heavy ion impact spectroscopy one of the main problem is the line blending, in the spectra there is a non-resolved mixture of the Auger lines originated from the various charge states in general (see e.g. Fig. 20). If light targets as He, H₂ are used in projectile Auger spectroscopy then it becomes possible the study of the Auger lines from the selected charge states, mainly the lines from one charge state can be found (details see later) in the spectrum.

- The study of a large variety of atomic species are possible because molecular compounds break apart in the ion sources of the accelerators.

- Due to kinematics, the very small energy lines (e.g. autoionization type) at the projectile rest frame are shifted to higher energies in the laboratory frame (see equ. 18; this is the so-called shifting effect) where then the detection is much easier. In this way, the study of very low energy lines is possible till the meV region (see also later).

- The above effect makes also possible a very high accuracy in the determination of the line position, too. Namely, the relative importance of the contact potential (which does not allow to obtain a more accurate absolute determination of the position of the Auger line than about 1-2 eV) will be much diminished when the lines shifted to higher energies will be transformed back to the projectile rest frame.

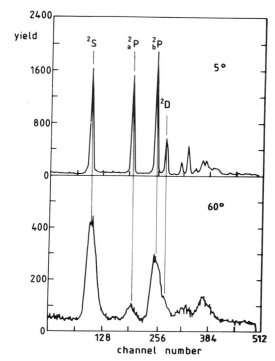

Fig. 30. Li autoionization spectra from Li$^+$ (250 keV)-He at 5° and 60° (Bisgard et al., 1981).

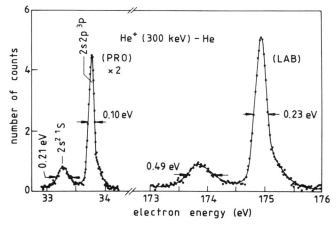

Fig. 31. Autoionization line taken in the laboratory frame and the same after transformation into the projectile rest frame (Stolterfoht, 1984).

- Because of kinematic effects (stretching), the line width is much reduced in the rest frame of the projectile. So the $0°$ Auger spectra measured in the laboratory frame after transformation into that of the projectile will have lines with smaller width. (It is well demonstrated in Fig. 31). In this way <u>measurements of very good resolution</u> (till the natural line width) are possible.

In the evaluation of the $0°$ projectile Auger spectroscopy, the following procedure is used in general. Because of the doubling effect, the same group of Auger lines will appear at two different places in the spectrum (see equ. 16 and Fig. 9). The low energy "appearance" is a "reflected image" of the high energy one (see Fig. 32). The former is denoted by L or $180°$, the latter by H or $0°$ spectrum.

If we use equ.(18), the energy of the Auger lines can be determined precisely by adjusting the value of the projectile energy (which is not known accurately enough in general) until the energy values E' of the corresponding lines coincide.

Although numerous measurements in projectile Auger spectroscopy were carried out by using foil target (see a good review e.g. Sellin, 1978; see also in Stolterfoht, 1987), we will deal only with the projectile Auger spectroscopy at gaseous target furthermore in this survey. It should be mentioned here, however, that the good localization at foil target is especially advantageous in time delay measurements (for life time determination in nanosecond region).

3.2 Auger spectroscopy of selected charge states of the impact ion - "ion surgery"

As we have seen the main problem is in the production of Auger spectra in heavy ion collisions the presence of the many satellite lines (which are also broadened in certain cases). It makes very difficult the analysis of these spectra. It is true both in the target and the projectile Auger spectroscopy in general. Although the problem is alleviated to a certain extent by using electron spectrometers with very high resolution or electron-ion coincidence arrangement, their effect is rather limited.

It seems to be a very suitable solution to diminish the overlapping of the Auger lines initiated by Stolterfoht and his co-workers (Itoh et al., 1983 and other references at the beginning of Section 3). We only mention here (first of all because of the rather limited application possibility) another possibility, when the number of charge states (and so the complexity of the Auger spectrum) from which Auger emission takes place is decreased by single or double capture of target electrons by the projectile ion which has no or at most one or two electrons (see e.g. Dillingham et al., 1984).

As regards the method when the Auger spectrum emitted by a heavy projectile ion which was excited by the collision with a light target ion (He, H_2), then if an inner shell vacancy in the projectile ion is produced, the outer shell structure is practically undisturbed. This is the phenomenon of the "needle ionization" which is a realization of "ion surgery". If we have e.g. a Ne ion let us say in the charge state 5, Ne^{5+}, after the needle ionization we will obtain mainly Ne^{6+} from which the Auger electrons will be emitted. In contrast to this at the excitation of Ne target by a heavy ion (e.g. Ar^{6+}), a number of charged states will be produced and the result in the Auger spectrum will be a mixture of the Auger lines from the different charge states with a lot of overlapping and line blending (Fig. 20). Fig. 33 shows how the Auger spectra looks like at the impact of the

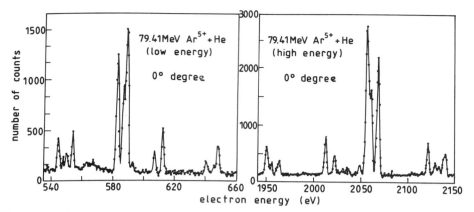

Fig. 32. Low and high energy appearance of a group of Auger lines in 0° projectile Auger spectroscopy (Stolterfoht, 1987).

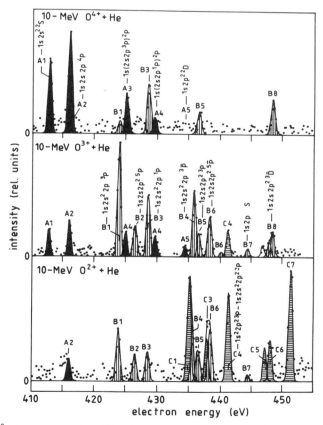

Fig. 33. 0° degree projectile Auger spectrum at O^{2+}, O^{3+}, O^{4+} (10 MeV) impact on He (Stolterfoht et al., 1987).

same ion (O^{q+}) in different charge states. In the case of O^{3+} mainly Be-like lines can be found in the Auger spectrum while at O^{4+} projectile in addition to Li-like lines only very few Be-like lines are present. In the work under discussion not only excitation and loss by one-electron processes but also transfer excitation and transfer loss involving two electrons were also studied (Stolterfoht et al., 1987). Energies of Li-like, Be-like and B-like Ne K Auger lines were determined by this method and the identification of the individual lines were performed (Itoh et al., 1987b).

3.3 Measurements of the very low energy electrons - "moving laboratory"

We have seen (see Section 3.1) that the energy E' in the projectile rest frame is kinematically shifted to higher energies in the laboratory frame where the measurements are carried and (see equ.18). It makes possible the study of Auger electrons at very low energies (from about 10 eV downwards to the meV region in the projectile frame). The study of the electron spectra at such low energies would be practically impossible without the effect of shifting because of instrumental difficulties at low energies.

The main characteristics in general of the measurements of the very low energy electrons in the above energy region by zero degree projectile electron spectroscopy is that the low energy electron lines are observable on the wings of the so called ELC peak in both sides simmetrically (Fig.8).The appearance of the very low energy lines on the wings of the cusp is understandable because the position of the top of the cusp measured in the laboratory frame corresponds to the zero energy point in the projectile frame (see equ. 19). It was first observed by Lucas and Harrison, (1970), later by Suter et al., (1979). The electron loss to the continuum (ELC) peak, which has a so-called "cusp shape", is produced in the electron spectrum in the laboratory frame at an electron energy which corresponds to the velocity which is equal to the projectile velocity (see some more details on ELC peak e.g. in Breinig et al., 1982; Berényi, 1985 or Kövér et al., 1983).

The evaluation procedure here is the same in principle as in other 0^0 spectra (Section 3.2). First, however, the smoothly varying "cusp background" should be subtracted, it follows the calculation of the values E' on the basis of equ. (18) using the lower energy appearance (low energy wing of the ELC cusp) and the high energy appearance of the lines (high energy wing of the cusp) by adjusting the projectile energy (which here corresponds to the position of the top of the cusp). After heaving received the energy values in the projectile rest frame E', also the spectrum can be obtained in the latter frame by transforming the spectrum taken in the laboratory frame according to equ. (19):

$$\frac{d\sigma'}{dE'd\Omega'} = \left(\frac{E'}{E}\right)^{\frac{1}{2}} \frac{d\sigma}{dEd\Omega} . \qquad (19)$$

The lines to be studied by the above method are originating from autoionization states. These are produced by double excitation in heavy ion--atom collisions. E.g. in Ne^{6+} (100 MeV)-He collision one of the 2s electrons will be excited to the 2p subshell and the other to an $n\ell$ Rydberg level producing the $1s^2 2pn\ell$ autoionizing state. So then, if it is energetically allowed, there will take place a 2p-2s transition and the emission of the electron from the $n\ell$ shell (Coster-Kronig transition!). It is also possible that one of the electrons is in the $n\ell$ shell and a vacancy is existing in the 2s subshell before the collision and only the other 2s electron will be excited to the 2p in the collision process. Another example is the

Ne⁴⁺ (70 MeV)-He. Here the 1s²2s2p nℓ or 1s²2s² 3ℓnℓ autoionization states will be produced. Autoionization states involving loosely bound Rydberg electrons can also be produced in processes in which electron(s) is (are) captured by the projectile ion from the target.

The above studies provides information about Rydberg states of the atoms (ions) and during the last two or three years very intensive research has been conducted in this field.

First of all, the positions (energies) of the autoionization peaks attributed to a series of Rydberg states and the intensities (production cross section) were studied (Stolterfoht et al., 1986; Schneider et al., 1987; Atan et al., 1987).

Fig. 34 shows a series of autoionization peaks to which Rydberg states can be assigned. The energies of the line of the spectra can be given by

$$\varepsilon_{n\ell} = \Delta E_{2s2p} - \frac{Q^2}{2(n-\mu_\ell)^2} \qquad (20)$$

not regarded the term and spin-orbit splitting. Here Q is the effective charge seen by a Rydberg electron, μ_ℓ is its quantum defect associated with the angular momentum ℓ and ΔE_{2s2p} is the energy difference between the indicated orbitals (calculated values in Stolterfoht, 1987). The second term in (20) is the binding energy of the respective electron B_n. The spacing of the lines follows an n^{-3} rule where n is the quantum number of the Rydberg electron which is associated to the autoionization lines. For the intensity variation of the lines also the n^{-3} rule was proved (Fig. 35). In the same work (Stolterfoht et al., 1986) the structure of the Rydberg lines (assigned to a given n value) was also analysed. It was shown that a term splitting is produced by the coupling of the Rydberg electron with the 2p electron.

In these investigations as low energy of the autoionization line (electron) was measured as 8 meV (Atan et al., 1987) and so high Rydberg states were studied for which $n \geq 100$ (Schneider, 1987).

3.4 Transfer and excitation, uncorrelated and correlated double-electron capture

Recently a number of works were published on the phenomenon of non-resonant and resonant transfer and excitation, on double-electron capture with and without correlation between the two electrons, using the techniques of X-ray spectroscopy (e.g. Tanis et al., 1982; Andriamonje et al., 1984; Pepmiller et al., 1985; Tanis, 1987).

These phenomena can be studied by the above methods and instrumentation of Auger spectroscopy and only the actuality and importance of the issue justifies to treat them here in a separated subsection.

In the nonresonant mode of transfer and excitation (NTE) one (or more) projectile electron is excited by the screened target nucleus and one (or more) target electron is captured by the projectile. The two processes are here independent and they are initiated by electron-nucleus interactions. In the resonant transfer and excitation (RTE), however, the interaction of two electrons, namely a bound projectile electron and a loosely bound target electron, is the prevailing. In such an interaction the projectile electron will be excited and the target electron will be transferred to (captured by) the projectile. The result will be a doubly excited state of the projectile. It is to be mentioned here that RTE process is regarded as a correlated while NTE as an uncorrelated process. If the captured electron is initially free, than the name of the process is dielectronic recombination. In the latter case, the process should be of sharply resonant nature because it is in fact an inverse Auger process (the energy released at the electron capture to a bound state of the impact ion excites an electron of the projectile from one bound state to another). According to this, a resonance behaviour for the incident ion energy is expectable. The resonance is existing in the case of RTE, too, even if it is not so sharp. The probability of NTE and RTE and their dependence on the impact energy is, of course, different (see e.g. in Fig. 37).

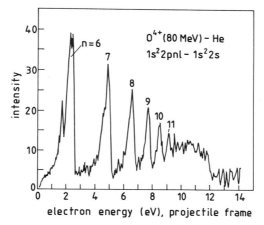

Fig. 34. A series of autoionization lines from the O^{4+} (80 MeV)-He collision attributed to a series of Rydberg states (Schneider et al. 1987).

The above processes have been studied first by the methods of X-ray spectroscopy in a coincidence arrangement between the X-ray from the projectile and the scattered projectile which capture one electron, in general. By utilization of the techniques of high-resolution Auger electron spectroscopy, one can obtain information about the NTE and RTE contributions in the case of the individual states even without any coincidence measurement while in X-ray studies the contributions concerned at most for the individual configurations can be determined.

First, the above processes were studied in $He^+ \rightarrow He$ collision taken the autoionization spectrum in the 50-500 keV impact energy region by an electron spectrometer of very good resolution (0.2 eV) (Itoh et al., 1985c).

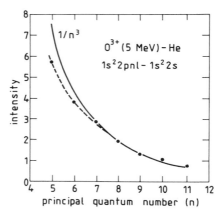

Fig. 35. Intensity of the autoionization lines assigned to a series of Rydberg state with principal quantum number from n=5 to n=11. The solid calculated curve is normalized to the experimental data for n=8. (Stolterfoht et al., 1986).

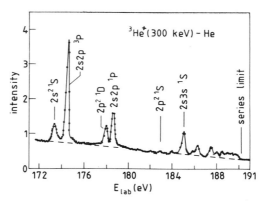

Fig. 36. Spectrum of autoionization electrons at $0°$ as taken in the laboratory frame (Itoh et al., 1985).

As can be seen in Fig. 36 the lines corresponding to the different states are resolved. Itoh et al., found no resonance behaviour of the transfer--excitation cross section for the states ^1S, ^3P and ^1P as well as for the total cross section as a function of the projectile ion impact energy. For ^1D state, however, a rather broad peak can be seen centered about 350 keV which can be attributed to RTE process. It is in qualitative agreement with the theoretical prediction.

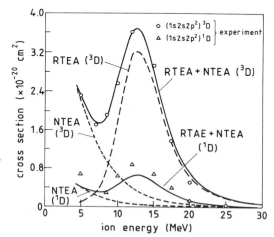

Fig. 37. Auger electron production cross section for the transition of the $1s2s2p^2$ ^3D and the $1s2s2p^2$ ^1D states to the $1s^22s$ ^2S state as a function of the impact energy of O^{5+} on He. (Swenson et al., 1986) The theoretical curves were calculated by Brandt, 1983.

Another experiment in which the high-resolution Auger electron spectroscopy at $0°$ was used to identify the states involved in electron transfer and excitation processes, is the O^{5+} - He collision in the energy range from 5 to 25 MeV. In this case Be-like excited intermediate states are produced and the cross sections for the production of Auger electrons from the decay of these states, namely the $1s2s2p^2$ ^3D and $1s2s2p^2$ ^1D states to the $1s^22s$ ^2S ionic ground state were determined. The two broad peaks in Fig. 37, where the cross section is plotted as a function of the incident energy of the projectile, show the presence of RTE process. The agreement with the corresponding theoretical calculations is also satisfactory even if some multiplication factors were applied for the calculated values. Corresponding calculations for RTE followed by Auger electron emission are abbreviated as RTEA and similarly for NTE (correspondingly NTEA).

The presence of Coster-Kronig electrons (series of lines attributed to Rydberg states) in the spectrum of electrons from slow, highly ionized collision systems is a proof for the involvement of the correlated double-electron capture mechanism. Namely, if the <u>capture of the two electrons</u> take place sequentially then they will be captured in the same or nearly the same shell. Even in the case of correlated double-electron capture there is a probability for formation of states with equivalent or near equivalent configurations but the observation of non-equivalent electron configurations with one of the electrons in a high Rydberg state is regarded as being real reference of electron correlation. Such an evidence was given by Stolterfoht et al. (1986) in O^{6+} (60 keV)-He and C^{4+} (40 keV)-He collisions. The evidence is especially strong in the case of O^{6+} (see Fig. 38). The total cross section for Coster-Kronig electron production was determined here, too, but it is rather difficult to reconcile with earlier results of Crandall et al., (1976).

A number of other studies were carried out on the issue of double-electron capture by means of the techniques of electron spectroscopy: Bordenave-Montesquieu et al., 1982; 1984; 1984b; 1985; Mann and Schulte, 1987; Meyer et al., 1987. From one of the recent studies we give here a tabulation (Table 2) which proves very clearly the role of correlation effect in two-electron transfer processes.

Table 2. Auger electron production cross sections (Meyer et al., 1987)

Projectile	Energy (keV)	Target	n	Cross section (10^{-17} cm^2)	
				(core)$2pn\ell$ [1]	(core)$3\ell n\ell^2$ [2]
C^{4+}	40	He	4	< 2.0	–
C^{4+}	40	H_2	4	15	15
N^{5+}	50	He	5	8.8	8.8
N^{5+}	50	H_2	5	32	100
O^{6+}	60	He	6	13	22
O^{9+}	90	He	9	13	68

[1] $L_1L_{23}X$ type Coster-Kronig transitions
[2] L M X type Auger transitions

4. EXPERIMENTS WITH AUGER ELECTRONS IN COINCIDENCE ARRANGEMENTS

It is quite disputable why to devote a separated section to the study of Auger electrons in coincidence with ions from the collision processes which induced the Auger electron emission. It is quite true that this is a mere instrumental approach and some processes which may be mentioned here because coincidence measurements were carried out on them, have already been treated in another section of this survey on the basis of works without using the coincidence techniques.

It is fact, however, that by using coincidence set-up in Auger studies, one can obtain more or more accurate information on the processes concerned than without utilization of it. It should be also mentioned here that the number of these experiments is not very high relative not only to the Auger electron studies without involving coincidence techniques but also to coincidence measurements at the study of X-rays originated in ion-atom collisions. In the present section, however, only some examples will be given, there is no real possibility here for a true survey.

DuBois et al., (1984) studied the <u>ratio between 2p capture and total 2p vacancy production</u> (in addition to the capture also the ionization mechanism for vacancy production is present here) cross sections for 30-80 keV protons on Mg by using the coincidence techniques. Detecting the neutralized projectiles H^0 (which were neutralized during the collision) in coincidence with the Auger electrons, only those vacancy production processes will be counted which were produced by electron capture from the target. A rather high capture fraction was found, varying between 50 and 35 % over the impact energy range concerned.

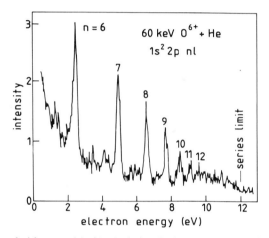

Fig. 38. Series of lines attributed to Rydberg states from Coster-Kronig transition after correlated double electron capture in O^{6+} (60 keV)-He collision (Stolterfoht et al., 1986).

The measurement of very low energy autoionization electrons utilized the kinematic shifting effect was already treated in Subsection 3.3 and also its application to the study of the double-electron capture was dealt with (Subsection 3.4). We have seen that one can have information on the electron transfer and ionization phenomena as well as on the double--electron capture without any coincidence measurement but by having very high resolution electron spectrometer.

In the case Au^{q+} projectiles in the q region from 5 to 19, very little is known about the ion-core structure and for the identification of the autoionization lines a coincidence technique was used. (Andersen et al., 1984; Andersen et al., 1984b). The question is how we can <u>differentiate among the different excitation mechanisms</u> by means of coincidence measurements.

The simplest processes which lead to the formation of doubly excited states are in the case of Au projectile ions as follows

- double-core excitation:

$$Au^{q+} + T \rightarrow Au^{q+**} + T \rightarrow Au^{(q+1)} + T + e^- ; \qquad (21a)$$

- single capture to an excited state plus core excitation (TE):

$$Au^{q+} + T \rightarrow Au^{(q-1)+**} + T^+ \rightarrow Au^{q+} + T^+ + e^- ; \qquad (21b)$$

- double-electron capture to excited states:

$$Au^{q+} + T \rightarrow Au^{(q-2)**} + T^{2+} \rightarrow Au^{(q-1)+} + T^{2+} + e^- \qquad (21c)$$

where T denotes a target atom (in this actual case they were H_2, He, Ar).

Looking at Fig. 39, on the basis of (21b) one can see that the TE mechanism is responsible for the autoionization electron lines because they are appearing in the spectra when electrons are measured in coincidence with Au ions in the same charge state which they have before the collision (cf. 21b). The energies and production cross sections of the observed autoionization lines were determined and the total cross sections for TE process (21b), too. This latter was found to be much smaller than that for single electron capture (Damgaard et al., 1983).

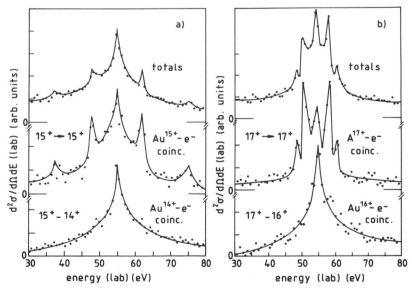

Fig. 39. Double differential cross section as a function of the energy of the emitted electrons:
a) Au^{15+} (20 MeV)-He (Andersen et al., 1984);
b) Au^{17+} (20 MeV)-He (Andersen et al., 1984b);

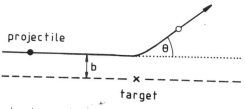

Fig. 40. A sketch to demonstrate the impact parameter b and the deflection angle Θ in collisions.

In other coincidence measurements (Shanker et al., 1982; Schiwietz et al., 1987) the impact parameter dependence of different yields in heavy ion collisions were determined.

As is well known, the impact parameter b is the distance at which the projectile would pass if there were no interaction between the projectile and the target (see Fig. 40). At the same time there is a unique relation between the impact parameter and the laboratory deflection angle of the projectile Θ as a consequence of the collision event. Thus, if we measure the Auger electron (and similarly the X-ray) spectra in coincidence with the deflected impact ions as a function of the deflection angle, then one can have information about how far the process concerned will proceed from the collision center (target) in a classical way of thinking. In the work of Shanker et al., both the X-ray and the Auger yield was measured in Xe^{3+} (1.05 MeV)-Xe collision as the function of the impact parameter. Thus the impact parameter dependence of the average Xe M shell fluorescence yield $\bar{\omega}$ in given in Fig. 41. One can see that going nearer to the collision center, the probability for Auger electron emission is smaller which reflects the increase of outer shell ionization at closer collisions.

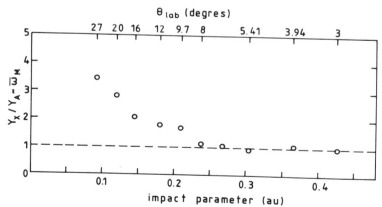

Fig. 41. The average Xe M shell fluorescence yield as a function of the impact parameter measured by taking the M Auger and X ray spectra in coincidence with the deflected projectile at different deflection angles (Shanker et al., 1982)

The impact parameter dependence of the electron yields emitted in H^+(30,100,350 keV), $^3He^{2+}$(100 keV)-Ar collisions was measured by Schiwietz et al., (1987) in a principally similar coincidence arrangement as above. Table 3 gives the vacancy production ratio for Ar L shell ionization by $^3He^{2+}$ and H^+ as a function of the impact parameter at an impact energy of 33 keV/u.

Table 3. Vacancy production ratio for Ar L shell by $^3He^{2+}$ and H^+ as a function of impact parameter (33 keV/u), Schiwietz et al., 1987

Impact parameter (a.u.)	Vacancy production ratio
0.499	1.67±1.0
0.341	2.09±0.7
0.266	1.81±0.5
0.221	1.96±0.4

It can be seen that - as expected - the vacancy production yield is higher at $^3He^{2+}$ at every impact parameter value studied.

It should be mentioned finally that the research is going on now very intensively in the field of heavy ion induced Auger spectroscopy and so the picture of the actual stage of our knowledge is changing very quickly.

REFERENCES

Andersen, L. H., Frost, M., Hvelplund, P., Knudsen, H., and Datz, S., 1984, Phys. Rev. Lett., 52:518.
Andersen, L. H., Frost, M., Hvelplund, P., and Knudsen, H., 1984b, J. Phys. B, 17:4701.
Andriamonje, S., Chemin, J. F., Roturier, J., Saboya, B., Scheurer, J. N., Gayet, R., Salin, A., Laurent, H., Auger, P., and Thibaud, J. P., 1984, Z. Phys. A, 317:251
Atan, H., Steckelmacher, W., and Lucas, M. W., 1987, in:"Proc. 3rd. Workshop on High-Energy Ion-Atom Collisions", D. Berényi and G. Hock, eds., Springer Vlg., Berlin.
Auger, P., 1923, Compt. Rend., 177:169.
Auger, P., 1925, J. Phys. Radium, 6:205.
Bambynek, W., Crasemann, B., Fink, R. W., Freund, H. V., Mark, H., Swift, C. D., Price, R. E., and Rao, P. V., 1972, Rev. Mod. Phys., 44:416.
Bordenave-Montesquieu, A., Gleizer, A., and Benoit-Cattin, P., 1982, Phys. Rev. A, 25:245.
Bordenave-Montesquieu, A., Benoit-Cattin, P., Gleizes, A., Marrakchi, A. I., Dousson, S., and Hitz, D., 1984, J. Phys. B, 17:L127.
Bordenave-Montesquieu, A., Benoit-Cattin, P., Gleizes, A., Marrakchi, A. I., Dousson, S., and Hitz, D., 1984b, J. Phys. B, 17:L223.
Bordenave-Montesquieu, A., Benoit-Cattin, P., Gleizes, A., Dousson, S., and Hitz, D., 1985, J. Phys. B, 18:L195.
Basbas, G., Brandt, W., Laubert, R., 1973, Phys. Rev. A, 7:983
Becker, R. L., Ford, A. L., and Reading, J. F., 1980, J. Phys. B, 13:4095.

Berényi, D., 1976, Adv. Electronics and El. Phys., 42:55.
Berényi, D., 1981, Adv. Electronics and El. Phys., 56:411.
Berényi, D., 1982, Investigations in Atomic Physics by Heavy Ion Projectiles, p.489, in: "Proc. International School-Seminars on Heavy Ion Physics, Alushta, 14-21 April, 1983", JINR, Dubna.
Berényi, D., 1984, p.207, in: "X84-Int. Conf. X-Ray and Inner-Shell Processes in Atoms, Molecules and Solids, Aug. 20-24, 1984", A Meisel, J. Finster, eds., Karl-Marx-Univ., Leipzig.
Berényi, D., 1986, p.161, in: "Atomic and Nuclear Interactions, Part. 1", G. Semenescu, L. A. Dorobantu, N. V. Zamfir, eds., Central Inst. of Phys., Bucharest.
Berényi, D., 987, Vacuum, 37:53
Bergström, I., Nordling, C., Snell, A. H., Wilson, R., and Pettersson, B. G., 1968, Some "Internal" Effect in Nuclear Decay, Chap. 25, p.1523, in: "Alpha-Beta- and Gamma-Ray Spectroscopy, Vol.2'", K. Siegbahn, ed., North Holland, Amsterdam.
Bisgård, P., Dahl, P., Fastrup, B., and Mehlhorn, W., 1981, J. Phys. B, 14:2023.
Brandt, D., 1983, Phys. Rev. A, 27:1314.
Breinig, M., Elston, S. B., Huldt, S., Liljeby, L., Vane, C. R., Berry, S. D., Glass, G. A., Schauer, M., Sellin, I. A., Alton, G. D., Datz, S., Overbury, S., Laubert, R., and Suter, M., 1982, Phys. Rev. A, 25:3015.
Burch, D., 1973, p.121, in: "Proc. Int. Conf. Inner-Shell Ionization Phenm.", R. W. Fink, S. T. Manson, J. M. Palms, P. V. Rao, eds., US Atomic Energy Commission, Oak Ridge, Tenn.
Burch, D., Stolterfoht, N., Schneider, D., Wieman, H., and Risley, J. S., 1974, Phys. Rev. Lett., 32:1151.
Burhop, E. H. S., and Asaad, W. N., 1972, Adv. Atom. Molec. Phys., 8:163
Chattarji, D., 1976, "The Theory of Auger Transitions", Academic Press, London.
Crandall, D. H., Olson, R. E., Shipsey, E. J., and Browne, J. C., 1976 Phys. Rev. Lett., 36:858.
Damgaard, H., Haugen, H. K., Hvelplund, P., and Knundsen, H., 1983, Phys. Rev. A, 27:112.
Dillingham, T. R., Newcomb, J., Hall, J., Pepmiller, P. L., and Richard, 1984, Phys. Rev. A, 29:3029.
Edwards, A. K., and Rudd, M. E., 1968, Phys. Rev. 170:140.
Hansteen, J. M., and Mosebeck, O. P., 1972, Phys. Rev. Lett., 29:1361.
Itoh, A., Schneider, T., Schiwietz, G., Doller, Z., Platten, H., Nolte, G., Schneider, D., and Stolterfoht, N., 1983, J. Phys. B, 16:3965.
Itoh, A., and Stolterfoht, N., 1985, Nucl. Instr. Meth. B, 10/11:97.
Itoh, A., Schneider, D., Schneider, T., Zouros, T.J.M., Nolte, G., Schiwietz, G., Zeitz, W., and Stolterfoht, N., 1985b, Phys. Rev. A, 131:684.
Itoh, A., Zouros, T. J. M., Schneider, D., Stettner, Zeitz, W., and Stolterfoht, N., 1985c, J. Phys. B, 18:4581.
Kabachnik, N. M., 1987, in: "Proc. XV ICPEAC, Brighton, July 21-29, 1987", in press.
Kádár, I., Ricz, S., Shchegolev, V. A., Sulik, B., Varga, D., Végh, J., Berényi, D., and Hock, G., 1985, J. Phys. B, 18:275.
Kádár, I., Ricz, S., Shchegolev, A. A., Varga, D., Végh, J., Berényi, D., Hock, G., and Sulik, B., 1986, Phys. Lett., 115:439.
Kádár, I., Ricz, S., Varga, D., Sulik, B., Végh, J., and Berényi, D., 1987, in course of publication (material of the conf. X-87).
Kádár, I., Ricz, S., Sulik, B., Varga, D., Végh, J., and Berényi, D., 1987b, "Proc. 3rd Int. Workshop on High Energy Ion-Atom Collisions", D. Berényi and G. Hock, eds., Springer Vlg., Berlin.
Kádár, I., Ricz, S., Sulik, B., Varga, D., Végh, J., and Berényi, D., 1987c, private communication.

Kessel, Q. C., McCaughey, M. P., and Everhardt, E., 1966, Phys. Rev. Lett., 16:1189.
Kessel, Q. C., McCaughey, M. P., and Everhardt, E., 1967, Phys. Rev. 153:57.
Kobayashi, N., Maeda, N., Nori, H., and Sakisaka, M., 1976, J. Phys. Soc. Jap., 40:1421.
Kövér, Á., Varga, D., Szabó, Gy., Berényi, D., Kádár, I., Ricz, S., Végh, J., and Hock, G., 1983, J. Phys. B, 16:1017.
Krause, M. O., and Oliver, J. H., 1979, J. Phys. Chem. Ref. Data, 8:329.
Lucas, M. W., and Harrison, 1972, J. Phys. B, 5:L20.
Mann, R., Folkmann, F., Peterson, R. S., Szabó, Gy., and Groeneveld, K.-O., 1978, J. Phys. B, 11:3045.
Mann, R., Beyer, H. F., and Folkmann, F., 1981a, Phys. Rev. Lett., 46:646.
Mann, R., Folkmann, F., and Beyer, M. F., 1981b, J. Phys. B, 14:1161.
Mann, R., and Schulte, H., 1987, Z. Phys. D, 4:343.
McCaughey, M. P., Knystautas, E. J., Hayden, H. C., and Everhardt, E., 1968, Phys. Rev. Lett., 21:65.
McGuire, J. H., and Richard, 1973, Phys. Rev. A, 8:1374.
Matthews, D. L., Johnson, B. M., Mackey, J. J., Smith, L. E., Hodge, W., and Moore, C. F., 1974, Phys. Rev. A, 10:1177.
Matthews, D. L., 1980, Ion-induced Auger Electron Spectroscopy, Chap.9, p.443, in: "Atomic Physics - Accelerators", P. Richard, ed., Academic Press, New York.
Mehlhorn, W., 1978, "Electron Spectroscopy of Auger States: Experiment and Theory", The Institute of Physics Aarhus, Aarhus.
Mehlhorn, W., 1982a, p.55, in: "X-Ray and Atomic Shell Physics". B. Crasemann, ed., AIP Conf. Proc., New York.
Mehlhorn, W., 1982b, p.83, in: "High-Energy Ion-Atom Collisions". D. Berényi, and G. Hock, eds., Akadémiai Kiadó, Budapest.
Meyer, F. W., Havener, C. C. Phaneuf, R. A., Swenson, J. K., Shafroth, S. M., and Stolterfoht, N., 1987, Nucl. Instr. Meth. B, 24/25:136.
Moore, C. F., 1976, p.447, in: "Proc. 9th ICPEAC, Seattle, 24-30 July, 1975", J. S. Risley and R. Geballe, eds., Univ. Washington Press, Seattle.
Pepmiller, P. L., Richard, P., Newcomb, J., Hall, J., and Dillingham, T. R., 1985, Phys. Rev. A, 31:734.
Prost, M., 1980, Doctoral Dissertation, Free University, Berlin.
Rao, V. P., 1975, p.1, in: "Atomic Inner-Shell Processes, Vol.2", B. Crasemann, ed., Academic Press, New York.
Richard, P., 1975, p.74, in: "Atomic Inner-Shell Processes, Vol.1", B. Crasemann, ed., Academic Press, New York.
Richard, P., 1980, p.125, in: "Electronic and Atomic Collisions", N. Oda and K. Takayamagi, eds., North-Holland, Amsterdam.
Ricz, S., Kádár I., Varga, D., Végh, J., Borbély. T., Shchegolev, V. A., Berényi, D., Hock, G., and Sulik, B., 1984, p.376, in: "X.84 Abstracts". Karl-Marx-Universität, Leipzig.
Ricz, S., Kádár, I., Varga, D., Végh, J., Borbély, T., Shchegolev, V. A., Berényi, D., Hock, G., and Sulik, B., 1985, p.171, in: "High-Energy Ion-Atom Collisions". D. Berényi, and G. Hock, eds., Akadémiai Kiadó, Budapest.
Ricz, S., Kádár, I., Shchegolev, V. A., Varga, D., Végh, J., Berényi, D., Hock, G., and Sulik, B., 1986, J. Phys. B, 19:L411.
Ricz, S., Hock, G., Kádár, I., Sulik, B., Végh, J., and Berényi, D., 1987, p.494, "Abstracts, XV ICPEAC, Brighton, 2-28, 1987", ICPEAC, Brighton.
Ricz, S., Végh, J., Kádár, I., Sulik, B., Varga, D., and Berényi, D., 1987b, in: "Proc. 3rd Workshop on High-Energy Ion-Atom Collision, Debrecen, Aug. 3-5, 1987". D. Berényi, and G. Hock, eds., Springer Verlag, Berlin, in course of publication.
Ricz, S., Kádár, I., Sulik, B., Varga, D., Végh, J., and Berényi, D., 1987c, private communication.

Ridder, D., Dieringer, J., and Stolterfoht, N., 1975, p.419, in: "Proc. 9th ICPEAC, Abstracts of Papers, J. S. Risley and R. Geballe, eds., Univ. of Washington Press, Seattle.
Rudd, M. E., 1964, Phys. Rev. Lett., 13:503.
Rudd, M. E., 1965, Phys. Rev. Lett., 15:580.
Rudd, M. E., Jorgensen, T. Jr., and Volz, D. J., 1966, Phys. Rev. Lett., 16:929.
Rudd, M. E., Jorgensen, T. Jr., and Volz, D. J., 1966b, Phys. Rev., 151:28.
Rudd, M. E., 1966c, Phys. Rev., 151:28.
Rudd, M. E., and Macek, J. H., 1974, Case Studies in Atom. Phys., 3:46.
Rudd, M. E., 1975, Rad. Res., 64:153.
Rudd, M. E., 1982, p.13, in: "High Energy Ion-Atom Collisions", D. Berényi and G. Hock, eds., Akadémiai Kiadó, Budapest.
Saris, F. W., van der Weg, W. F., Tamara, H., and Laubert, R., 1972, Phys. Rev. Lett., 28:717.
Schiwietz, G., Stettner, U., Zouros, T. J. M., and Stolterfoht, N., 1987, Phys. Rev. A, 35:598.
Schneider, D., 1975, Dissertation, Free University, Berlin.
Schneider, D., Prost, M., DuBois, B., and Stolterfoht, N., 1982, Phys. Rev. A, 25:3102.
Schneider, D., Stolterfoht, N., Itoh, A., Schneider, T., Schiwietz, G., Zeitz, W., and Zouros, T., 1986, p.671, in: "Electronic and Atomic Collisions", D. C. Lorents, W. E. Meyerhof, J. R. Peterson, eds., Elsevier Sci. Publ., New York.
Schneider, D., Stolterfoht, N., Schiwietz, G., Schneider, T., Zeitz, W., Bruch, R., and Chung, K. T., 1987, Nucl. Instr. Meth., 24/25:173.
Sellin, I. A., 1978, p.273, in: "Structure and Collision of Ions and Atoms", I. A. Sellin, ed., Springer Vlg., Berlin.
Shanker, R., Hippler, R., Wille, U., and Lutz, H. O., 1982, J. Phys. B, 5:2041.
Shergin, A. P., 1987, in: "Program of 10th ISIAC, Bad Soden, July 30-31, 1987".
Stolterfoht, N., 1976, IEEE Trans. Nucl. Sci., CH1175-9:311.
Stolterfoht, N., 1978, p.155, in: "Structure and Collisions of Ions and Atoms". I. A. Sellin, ed., Springer Vlg., Berlin.
Stolterfoht, N., 1983, p.295, in: "Fundamental Processes in Energetic Atomic Collins", H. O. Lutz, J. S. Briggs, H. Kleinpoppen, eds., Plenum Press, New York.
Stolterfoht, N., 1986, p. 85, in: "Atomic and Nuclear Interactions, Part 1", G. Semenescu, L. A. Dorobantu, N. V. Zamfir, eds., Central Inst. of Phys., Bucharest.
Stolterfoht, N., 1987, Phys. Rep., 146:317.
Stolterfoht, N., Havener, C. C., Phaneuf, R. A., Swenson, J. K., Shafroth, S. M., and Meyer, F. W., 1986, Phys. Rev. Lett., 57:74.
Stolterfoht, N., de Heer, F. J., and van Eck, J., 1973, Phys. Rev. Lett., 30:1159.
Stolterfoht, N., Itoh, A., Schneider, D., Schiwietz, G., Platten, H., Nolte, G., Glodde, R., Stettner, U., Zeitz, W., and Zouros, T., 1984, in: "X84-Int. Conf. X-Ray and Inner-Shell Processes in Atoms, Molecules and Solids", A. Meisel, and J. Finster, eds., Karl-Marx-Universität, Leipzig. p.193.
Stolterfoht, N., Miller, P. D., Krause, H. F., Yamazaki, Y., Dittner, P. F., Pepmiller, P. L., Sellin, I. A., and Datz, S., 1986, in: "Joint US-Mexico Symp. on Two Electron Processes", Cocoyoc.
Stolterfoht, N., Miller, P. D., Krause, H. F., Yamazaki, Y., Swenson, J. K., Bruch, R., Dittner, P. F., Pepmiller, P. L., and Datz, S., 1987, Nucl. Instr. Meth. B, 24/25:168.
Stolterfoht, N., Schneider, D., Bruch, D., Aagaart, B., Bøving, E., and Fastrup, B., 1975, Phys. Rev. A, 12:1313.

Stolterfoht, N., Schneider, D., Mann, R., Folkmann, F., 1977, J. Phys. B, 10:L281.
Sulik, B., Hock, G., and Berényi, D., 1984, J. Phys. B, 17:3239.
Sulik, B., and Hock, G., 1985, p.183, in: "Proc. 2nd Workshop on High-
    -Energy Ion-Atom Collisions, Aug. 27-28, 1984, ·Debrecen", D. Berényi and G. Hock, eds., Akadémiai Kiadó, Budapest.
Sulik, B., Kádár, I., Ricz, S., Varga, D., Végh, J., Hock, G., and Berényi, D., 1987, Nucl. Instr. Meth., in press.
Suter, M., Vane, C. R., Elston, S. B., Alton, G. D., Griffin, P. M., Thoe, R. S., Williams, L., Sellin, I. A., and Laubert, R., 1984, Z. Physik A, 289:433.
Swenson, J. K., Yamazaki, Y., Miller, P. D., Krause, H. F., Dittner, P. F., Pepmiller, P. L., Datz, S., and Stolterfoht, N., 1986, Phys. Rev. Lett., 57:3042.
Tanis, J. A., Bernstein, E. M., Graham, W. G., Clark, M., Shafroth, S. M., Johnson, B. M., Jones, K. W., and Meron, M., 1982, Phys. Rev. Lett., 49:1325.
Tanis, J. A., 1987, in: "Proc. 3rd Workshop on High-Energy Ion-Atom Collisions, 3-5 Aug, 1987, Debrecen; D. Berényi and G. Hock, eds., Springer Vlg., Berlin.
Török, I., and Varga, D., 1987, ATOMKI, Debrecen, private communication.
Végh, L., 1984, Phys. Rev. A, 30:2127.
Woods, C. W., Kauffman, R. L., Jamison, K. A., Stolterfoht, N., and Richard, P., 1975, Phys. Rev. A, 12:1393.
Woods, C. W., Kauffman, R. L., Jamison, K. A., Stolterfoht, N., and Richard, P., 1976, Phys. Rev. A, 13:1358.

INNER SHELL X-RAY PHOTOABSORPTION AS A STRUCTURAL AND ELECTRONIC PROBE OF MATTER

C.R. Natoli

INFN
Laboratori Nazionali di Frascati
P.O.Box, 13 00044 Frascati (Italy)

1. INTRODUCTION

Electromagnetic radiation has been historically the most widely used tool in the investigation of the properties of the physical state of matter. The reason lies in the smallness of the fine structure constant $\alpha=(e^2/\hbar c)=(1/137)$ that governs the coupling of the radiation with matter. The resulting weak interaction has a twofold advantage: on one hand the perturbation on the system under study is negligible so that one is able to investigate the properties of the unperturbed system; on the other hand from a theoretical point of view one can use the linear response theory as an interpretative scheme in which to frame the experimental observations.

The study of the electronic excitation dynamics in the various states of the matter benefits of this fortunate circumstance. There is however a price to pay for this simplification in the investigation of the structural properties of matter. Due to the smallness of the coupling constant scattering experiments can only probe the pair correlation function of observables that couple to the electromagnetic probe, like the local density $\rho(\mathbf{r})$ or the current density $j_i(\mathbf{r})$. Except for periodic systems, where this information is usually sufficient to reconstruct the spatial organization of the atoms, in any other instance one has no clue to the atomic geometrical arrangement in the system under study.

The advent of the extensive use of synchrotron radiation has given a tremendous impulse to both areas of research. The unique properties of this radiation source, like its intensity, brilliance, polarization, tunability and collimation, to cite a few, coupled with sophisticated data acquisition techniques have made possible the explosive development of all kinds of spectroscopic research.

On the side of electronic excitation dynamics a deeper understanding has been achieved in the way an excited system reacts to the excitation probe. Screening, polarization, relaxation, autoionization and decay mechanics have been elucidated in a variety of cases, both because of higher quality data and better theoretical treatment.

On the structural side the photoabsorption process has been progressively recognized and used as a technique capable of providing structural information beyond the pair correlation function relative to the absorbing atom even in non periodic systems. In fact it has been realized that, even though the primary probe, the radiation, couples weakly with matter, the secondary probe generated in the photoabsorption process, i.e. the photoelectron, can couple strongly with the atoms of the system and therefore can carry supplementary information through final state interactions.

As a consequence photoabsorption and photoemission measurements, especially from inner shell states, have been progressively used for structural purposes. The limitation to inner shells, with the inherent simplification brought about by the localized and dispersionless initial state, has made simpler the theoretical interpretation of the experimental results, which in turn have exploited the selective power of the incoming radiation both in terms of the type of atom to excite and the type of final state to reach.

Another reasons for using deep core states has been the reduction, in the final state, of the amount of electronic correlation effects which in general tend to obscure the informational content relating to the structural arrangement of the atoms in the system.

However relaxation processes and double excitations are, to some extent, always present in the final state of inner shell photoabsorption. Therefore a theoretical scheme for interpreting the interplay between structural properties and electronic correlation dynamics would be highly desirable. This scheme is provided by the multichannel multiple scattering (m.s.) theory[1,2] which forms the objects of these lecture notes.

2. - THE MULTICHANNEL MULTIPLE SCATTERING THEORY

We begin with the total absorption cross section, given by

$$\sigma(\omega) = 4\pi^2 \alpha \hbar \omega \sum_f | (\Psi_f^N | \varepsilon \cdot \sum_{i=1}^N \mathbf{r}_i | \Psi_i^N) |^2 \delta(\hbar\omega - E_f + E_i) \quad (2.1)$$

where $\Psi^N_{i,f}$ are the many-body initial and final state wave functions for N electrons in the system and the sum over the final states Σ is intended also over all directions of the photoemitted electrons. $\hbar\omega$ is the incoming photon energy and ε its polarization.

For transitions from a core state we assume that, to a good approximation,

$$\begin{aligned}\Psi_i^N &= \sqrt{N!}\, \mathcal{A} \phi_c(\mathbf{r}) \sum_n c_n \Phi_n^{N-1}(\mathbf{r}_1 \ldots \mathbf{r}_{N-1}) \\ &= \sqrt{N!}\, \mathcal{A} \phi_c(\mathbf{r}) \Psi_G^{N-1}(\mathbf{r}_1 \ldots \mathbf{r}_{N-1})\end{aligned} \quad (2.2)$$

where $\mathcal{A}$ is the usual antisymmetrizing operator $\mathcal{A} = (1/N!) \sum_P (-1)^P P$ ($\mathcal{A}^2 = \mathcal{A}$) and $\Phi_n^{N-1}(\mathbf{r}_1 \ldots \mathbf{r}_{N-1})$ are Slater determinants describing the configurations present in the initial state wave function Ψ_i^N. Normalization imposes $\sum_n |c_n|^2 = 1$, if $(\phi_c | \phi_c) = 1$.

Similarly we assume that, by expanding $\Psi_f^N(\mathbf{r}, \mathbf{r}_1 \ldots \mathbf{r}_{N-1})$ in terms of the complete set $\Psi_\alpha^{N-1}(\mathbf{r}_1 \ldots \mathbf{r}_{N-1})$

$$\Psi_f^N = \sqrt{N!}\, \mathcal{A} \sum_\alpha f_\alpha(\mathbf{r}) \Psi_\alpha^{N-1}(\mathbf{r}_1 \ldots \mathbf{r}_{N-1}) \quad (2.3)$$

We take the functions Ψ_α^{N-1} to be eigenstates of the N-1 electron Hamiltonian

$$H_{N-1} = -\sum_{i=1}^{N-1} \nabla_i^2 - \sum_{i=1}^{N-1}\sum_{k=1}^{P} \frac{2Z_k}{|\mathbf{r}_i - \mathbf{R}_k|} + \sum_{\substack{1 \le i,j \le N-1 \\ i<j}} \frac{2}{|\mathbf{r}_i - \mathbf{r}_j|} \quad (2.4)$$

with eigenvalues E_α^{N-1}:

$$H_{N-1} \Psi_\alpha^{N-1} = E_\alpha^{N-1} \Psi_\alpha^{N-1} \tag{2.5}$$

where $\sum_{k=1}^{P} Z_k = N$, $\mathbf{R}_k$ denotes the nuclear positions and Z_k are the associated charges.

We use throughout atomic units of length and Rydberg units of energy. The factor $\sqrt{N!}$ in Eq. (2.3) again assumes that we can approximate Ψ_α^{N-1} by a linear combination of Slater determinants, belonging to a continuum spectrum if Ψ_α^{N-1} does. In any case we assume for simplicity all continuum states normalized into a box enclosing the system: one may eventually take the limit of the box linear dimensions to infinity and transform the sum in Eq. (2.3) into an integral.

The final state wave function Ψ_f is an eigenstate, with energy $E = \hbar\omega + E_i^N$, of the N-electron Hamiltonian

$$H_N = -\nabla_r^2 + \sum_{i=1}^{N-1} \frac{2}{|\mathbf{r}-\mathbf{r}_i|} - \sum_{k=1}^{P} \frac{2Z_k}{|\mathbf{r}-\mathbf{R}_k|} + H_{N-1} \tag{2.6}$$

$$= -\nabla_r^2 + V(\mathbf{r},\mathbf{r}_i,\mathbf{R}_k) + H_{N-1}$$

Therefore

$$H_N \Psi_f^N = E \Psi_f^N \tag{2.7}$$

and we shall henceforth assume that $E_i^N = E_g^N$ is the ground state of the system.

The insertion of Eq. (2.3) into Eq. (2.7) gives

$$(-\nabla_r^2 + V(\mathbf{r},\mathbf{r}_i,\mathbf{R}_k) + H_{N-1}) \mathcal{A} \sum_\alpha f_\alpha(\mathbf{r}) \Psi_\alpha^{N-1}(\mathbf{r}_1\ldots\mathbf{r}_{N-1}) = $$
$$= E \mathcal{A} \sum_\alpha f_\alpha(\mathbf{r}) \Psi_\alpha^{N-1}(\mathbf{r}_1\ldots\mathbf{r}_{N-1}) \tag{2.8}$$

and by multiplying on the left by Ψ_α^{N-1} and integrating we obtain the set of equations

$$(\nabla^2 + E - E_\alpha^{N-1}) f_\alpha(\mathbf{r}) = $$
$$= \sum_{\alpha'} [V_{\alpha\alpha'}(\mathbf{r},\mathbf{R}_k) + W_{\alpha\alpha'}(\mathbf{r},\mathbf{R}_k)] f_{\alpha'}(\mathbf{r}) \tag{2.9}$$

where

$$V_{\alpha\alpha'}(\mathbf{r},\mathbf{R}_k) = \int \prod_{i=1}^{N-1} d^3r_i \Psi_\alpha^{N-1}(\mathbf{r}_1...\mathbf{r}_{N-1}) \\ V(\mathbf{r},\mathbf{r}_i,\mathbf{R}_k) \Psi_{\alpha'}^{N-1}(\mathbf{r}_1...\mathbf{r}_{N-1})$$ (2.10)

is a direct potential term and we have lumped all the exchange terms into the quantities $W_{\alpha\alpha'}(\mathbf{r},\mathbf{R}_k)$ which are thus complicated, non local, exchange potentials for which a suitable, local approximation has to be found. If we impose the condition, as we shall do, that the functions $f_\alpha(\mathbf{r})$ be orthogonal to all the one particle states present in the configurations making up the ground state wave function (so as to ensure the orthogonality condition $(\Psi_f^N|\Psi_g^N)=0$) as well as to those configurations that enter in all the Ψ_α^{N-1}, then the exchange term is given by

$$W_{\alpha\alpha'}(\mathbf{r},\mathbf{R}_k) = 1/f_\alpha(\mathbf{r}) \int \prod_{i=1}^{N-1} d^3r_i \Psi_\alpha^{N-1}(\mathbf{r}_1...\mathbf{r}_{N-1}) V(\mathbf{r},\mathbf{r}_i,\mathbf{R}_k) \\ \sum_{P(\neq E)} (-1)^P P f_{\alpha'}(\mathbf{r}_i) \Psi_{\alpha'}^{N-1}(\mathbf{r}_1...\mathbf{r}...\mathbf{r}_{N-1})$$ (2.11)

We refer to the appropriate literature for the transformation of this non local operator into a local one[3]. Henceforth we shall assume that this transformation has been performed and that our problem is to solve the coupled set of Schrödinger equations with local potentials.

Since $E = \hbar\omega + E_g^N$ we can write in Eq. (2.9)

$$E - E_\alpha^{N-1} = \hbar\omega + E_g^N - E_\alpha^{N-1} = \hbar\omega + E_g^N - E_g^{N-1} - (E_\alpha^{N-1} - E_g^{N-1})$$
$$= \hbar\omega - I_c - \Delta E_\alpha = k_\alpha^2$$ (2.12)

since $E_g^{N-1} - E_g^N = I_c$ is the ionization potential for the core state and $\Delta E_\alpha = E_\alpha^{N-1} - E_g^{N-1}$ is the excitation energy left behind to the (N-1)-particle system. Therefore k_α is the wave-vector of the final state photoelectron

Eqs. (2.9) can then be rewritten as

$$(\nabla^2 + k_\alpha^2) f_\alpha(\mathbf{r}) = \Sigma_{\alpha\alpha'} \mathbf{V}_{\alpha\alpha'}(\mathbf{r},\mathbf{R}_k) f_{\alpha'}(\mathbf{r})$$ (2.13)

where for sake of brevity we have put $\mathbf{V}_{\alpha\alpha'} = V_{\alpha\alpha'} + W_{\alpha\alpha'}$.

The functions $f_\alpha(\mathbf{r})$ have a simple physical meaning in the case of electron-molecule scattering. Through the asymptotic conditions

$$f_\alpha(\mathbf{r}) \underset{r\to\infty}{\sim} (e^{i\mathbf{k}_\alpha \cdot \mathbf{r}} \delta_{\alpha\underline{\alpha}} + f_\alpha(\hat{\mathbf{r}},\hat{\mathbf{k}}_\alpha) \frac{e^{ik_\alpha r}}{r}) N_\alpha \qquad (2.14)$$

where the factor $N_\alpha = (k_\alpha/\pi)^{1/2}/(4\pi)$ is necessary to ensure normalization to one state per Rydberg, they describe an electron in the incoming channel $\underline{\alpha}$ with wave vector $k_{\underline{\alpha}}$ which can be scattered in any outgoing channel α, with wave vector k_α, after loosing the energy $\Delta E_{\underline{\alpha}}$. In the photoemission process we have to take the time-reversed state of Eq. (2.3) (complex conjugate if spin is neglected) so that the outgoing channels become incoming channels which interfere constructively in the wave packet describing the photoelectron so as to give an asymptotic plane wave propagating out at infinity with wave number $k_{\underline{\alpha}}$.

Therefore Eqs. (2.13) are to be supplemented with the boundary conditions Eqs. (2.14) written by replacing $f_\alpha(\mathbf{r})$ with $f_\alpha^*(\mathbf{r})$.

It is fairly obvious then that in the expansion (2.3) the most important (N-1)-particle states are the excited states Ψ_α^{N-1} with a core hole corresponding to the photoejected electron, for which $E_g^{N-1} - E_g^N = I_c$, so that $k_\alpha^2 = \hbar\omega - I_c - \Delta E_\alpha$ is small compared to $V_{\alpha\alpha'}$. In this sense the Ψ_α^{N-1} are the relaxed excited states of H_{N-1}.

The argument runs as follows. If $k_\alpha^2 \gg \hbar\omega - I_c = k_0^2$ and $k_\alpha^2 \gg |V_{\alpha\alpha'}(r_c)|$, where r_c is the radius of the atomic core, then to a first approximation we can neglect the potentials in the r.h.s. of Eqs. (2.13), so that, together with the boundary conditions Eqs. (2.14), we obtain

$$f_\alpha(\mathbf{r}) \sim e^{i\mathbf{k}_\alpha \cdot \mathbf{r}} \delta_{\alpha\underline{\alpha}} . \qquad (2.15)$$

The procedure for solving Eqs. (2.13) with boundary conditions (2.14) (in the end we shall take the complex conjugate) closely follows Ref. 2. We first transform Eq. (2.10) into a Lippman-Schwinger equation

$$f_\alpha(\mathbf{r}) = N_\alpha e^{i\mathbf{k}_\alpha \cdot \mathbf{r}} \delta_{\alpha\underline{\alpha}} + \int G^\alpha_0(\mathbf{r}-\mathbf{r}') \Sigma_{\alpha'} V_{\alpha\alpha'}(\mathbf{r}') f_{\alpha'}(\mathbf{r}') d^3r'$$

$$= N_\alpha e^{i\mathbf{k}_\alpha \cdot \mathbf{r}} \delta_{\alpha\underline{\alpha}} + \sum_{k=1}^{P} \int_{\Omega_k} G^\alpha_0(\mathbf{r}-\mathbf{r}') \Sigma_{\alpha'} V^k_{\alpha\alpha'}(\mathbf{r}') f_{\alpha'}(\mathbf{r}') d^3r'$$

$$+ \int_{\Delta\Omega} G^\alpha_0(\mathbf{r}-\mathbf{r}') \Sigma_{\alpha'} V^I_{\alpha\alpha'}(\mathbf{r}') f_{\alpha'}(\mathbf{r}') d^3r' \qquad (2.16)$$

where we have partitioned the space in non overlapping spheres Ω_k around the atomic nuclei and an interstitial region $\Delta\Omega$. An outer sphere Ω_0 enclosing all atomic spheres can be added by replacing $\Sigma_{k=1}$ with $\Sigma_{k=0}$. Also $V^k_{\alpha\alpha'}(\mathbf{r}') \equiv V_{\alpha\alpha'}(\mathbf{r}')$ for $\mathbf{r}' \in \Omega_k$. Moreover

$$(\nabla^2 + k_\alpha) G^\alpha_0(\mathbf{r}-\mathbf{r}') = \delta(\mathbf{r}-\mathbf{r}') \qquad (2.17)$$

whose solution is[2]

$$G^\alpha_0(\mathbf{r}-\mathbf{r}') = -(1/4\pi) \frac{e^{ik_\alpha |\mathbf{r}-\mathbf{r}'|}}{|\mathbf{r}-\mathbf{r}'|} = -ik_\alpha \Sigma_L j_l(k_\alpha r_<) Y_L(\hat{\mathbf{r}}_<)$$
$$h_l^+(k_\alpha r_>) Y_L(\hat{\mathbf{r}}_>) = -ik_\alpha \Sigma_L J^\alpha_L(\mathbf{r}_<) H^{+\alpha}_L(\mathbf{r}_>) \qquad (2.18)$$

where L stands for (l,m), $r_>$ ($r_<$) refers to the greater (lesser) of $|\mathbf{r}|$ and $|\mathbf{r}'|$ and j_l, n_l, h_l^+ are spherical Bessel, Neumann and Hankel functions, respectively, with $h_l^+ = j_l + in_l$. We shall use real spherical harmonics and put for brevity $J^\alpha_L(\mathbf{r}) = j_l(k_\alpha r) Y(\hat{r})$, etc...$G^\alpha_0(\mathbf{r}-\mathbf{r}')$ is the free Green's function with momentum $\hbar k_\alpha$ and outgoing wave boundary conditions.

Use of Eq. (2.13) allows us to write

$$f_\alpha(\mathbf{r}) = N_\alpha e^{i\mathbf{k}_\alpha \cdot \mathbf{r}} \delta_{\alpha\underline{\alpha}}$$
$$+ \int G^\alpha_0(\mathbf{r}-\mathbf{r}')(\nabla_{r'}^2 + k_\alpha^2) f_\alpha(\mathbf{r}') d^3r' \qquad (2.19)$$

which, together with the Green's theorem

$$\int_V [G^\alpha_0(\mathbf{r}-\mathbf{r}')(\nabla_{r'}^2 + k_\alpha^2) f_\alpha(\mathbf{r}')$$
$$- f_\alpha(\mathbf{r}')(\nabla_{r'}^2 + k_\alpha^2) G^\alpha_0(\mathbf{r}-\mathbf{r}')] d^3r' \qquad (2.20)$$
$$= \int_{S_V} [G^\alpha_0(\mathbf{r}-\mathbf{r}') \nabla_{r'} f_\alpha(\mathbf{r}') - f_\alpha(\mathbf{r}') \nabla_{r'} G^\alpha_0(\mathbf{r}-\mathbf{r}')] \cdot \mathbf{n} \, d\sigma'$$

leads to the following equations

$$f_\alpha(\mathbf{r}) = N_\alpha e^{i\mathbf{k}_\alpha \cdot \mathbf{r}} \delta_{\alpha\underline{\alpha}} + \sum_{k=1}^{P} \int_{S_{\Omega_k}} [G_0^\alpha(\mathbf{r}-\mathbf{r}') \nabla_{\mathbf{r}'} f_\alpha(\mathbf{r}') -$$

$$- f_\alpha(\mathbf{r}') \nabla_{\mathbf{r}'} G_0^\alpha(\mathbf{r}-\mathbf{r}')] \cdot \mathbf{n} \, d\sigma' \qquad (2.21a)$$

$$+ \int_{\Delta\Omega} G_0^\alpha(\mathbf{r}-\mathbf{r}') \Sigma_{\alpha'} V_{\alpha\alpha'}^I(\mathbf{r}') f_{\alpha'}(\mathbf{r}') d^3 r' \qquad \text{if } \mathbf{r} \notin \Sigma_k \Omega_k$$

$$0 = N_\alpha e^{i\mathbf{k}_\alpha \cdot \mathbf{r}} \delta_{\alpha\underline{\alpha}} + \sum_{k=1}^{P} \int_{S_{\Omega_k}} [G_0^\alpha(\mathbf{r}-\mathbf{r}') \nabla_{\mathbf{r}'} f_\alpha(\mathbf{r}') -$$

$$- f_\alpha(\mathbf{r}') \nabla_{\mathbf{r}'} G_0^\alpha(\mathbf{r}-\mathbf{r}')] \cdot \mathbf{n} \, d\sigma' \qquad (2.21b)$$

$$+ \int_{\Delta\Omega} G_0^\alpha(\mathbf{r}-\mathbf{r}') \Sigma_{\alpha'} V_{\alpha\alpha'}^I(\mathbf{r}') f_{\alpha'}(\mathbf{r}') d^3 r' \qquad \text{if } \mathbf{r} \in \Sigma_k \Omega_k$$

In order to perform the surface integrals around the spheres Ω_k centered at $\mathbf{R}_k$ we make use of the usual expansion[2]

$$G_0^\alpha(\mathbf{r}-\mathbf{r}') = \Sigma_{LL'} j_l(k_\alpha r_i) Y_L(\hat{\mathbf{r}}_i) G_{iL,jL'}^\alpha j_{l'}(k_\alpha r_j) Y_{L'}(\hat{\mathbf{r}}_j)$$

$$= \Sigma_{LL'} J_L^\alpha(\mathbf{r}_i) G_{iL,jL'}^\alpha J_{L'}^\alpha(\mathbf{r}_j) \qquad (2.22)$$

where

$$G_{iL,jL'}^\alpha = 4\pi k_\alpha \Sigma_{L''} i^{l''+l-l'} C_L^{L'}{}_{L''} [-ih_{l''}^+(k_\alpha R_{ij})] Y_{L''}(\hat{\mathbf{R}}_{ij})$$

$$= N_{iL,jL'}^\alpha - i J_{iL,jL'}^\alpha \qquad (2.23)$$

with

$$C_L^{L'}{}_{L''} = \int Y_L(\Omega) Y_{L'}(\Omega) Y_{L''}(\Omega) d\Omega \qquad (2.24)$$

and putting $\mathbf{r}_j = \mathbf{r} - \mathbf{R}_j$, $\mathbf{R}_{ij} = \mathbf{R}_i - \mathbf{R}_j$. Unless explicitly stated, we shall henceforth assume this meaning for $\mathbf{r}_j$. The matrices N and J are defined by decomposing $-ih_l^+ = n_l - ij_l$.

Moreover we need also an expression for the solution of the system of Schrödinger equations (2.13) inside each sphere Ω_k. Writing

$$f_\alpha(\mathbf{r}) = \Sigma_L f_L^\alpha(r) Y_L(\hat{\mathbf{r}}) \qquad (2.25)$$

inserting into Eq. (2.13) and projecting onto Y_L we find

$$[1/r(d^2/dr^2)r + k_\alpha^2 - l(l+1)/r^2]f^\alpha_L(r) =$$

$$\Sigma_{\alpha'L'} \, V_{k;LL'}{}^{\alpha\alpha'}(r) \, f^{\alpha'}{}_{L'}(r) \qquad (2.26)$$

Here we have assumed that around each center k,

$$V^k_{\alpha\alpha'}(\mathbf{r}) = \Sigma_L \, V_{k;L}{}^{\alpha\alpha'}(r) \, Y_L(\hat{\mathbf{r}}) \qquad (2.27)$$

so that

$$V_{k;LL'}{}^{\alpha\alpha'}(r) = \Sigma_{L''} \, C_L{}^{L'}{}_{L''} \, V_{k;L''}{}^{\alpha\alpha'}(r) \qquad (2.28)$$

If α runs from 1 to n_α and l from 0 to l_{max}, this is a set of $n_\alpha(l_{max}+1)^2$ equations and consequently we can construct this number of linearly independent solutions $f_{LL'}{}^{\alpha\alpha'}(r)$ regular at the origin which, for given $\alpha'L'$ can be interpreted as vector solutions whose components are labelled by αL. To start the integration, we might take, for example, near the origin,

$$f_{LL'}{}^{\alpha\alpha'} \simeq r^l \, \delta_{LL'} \, \delta_{\alpha\alpha'} \qquad (2.29)$$

Consequently the general solution can be written as

$$f^\alpha_L(r) = \Sigma_{\alpha'L'} \, C_{L'}{}^{\alpha'} f_{LL'}{}^{\alpha\alpha'}(r) \qquad (2.30)$$

so that without loss of generality, inside the sphere Ω_i, we can write

$$f^i_\alpha(\mathbf{r}_i) = \Sigma_{\alpha'} \, \Sigma_{LL'} \, C_{iL'}{}^{\alpha'} \, f_{i;LL'}{}^{\alpha\alpha'}(r_i) \, Y_L(\hat{\mathbf{r}}_i) \qquad (2.31)$$

Inserting this expression into Eq. (2.21b), taken for $\mathbf{r}\in\Omega_i$, remembering Eq. (2.22), one obtains

$$0 = \Sigma_L \, J^\alpha_L(\mathbf{r}_i) \{ k_\alpha \rho_i^2 \, \Sigma_{\alpha'L'} \, W[-ih^+_{l'}, f_{i;LL'}{}^{\alpha\alpha'}] C_{iL'}{}^{\alpha'}$$

$$+ \Sigma_{k(\neq i)} \Sigma_{\alpha'} \Sigma_{L'L''} \, \rho_k^2 \, G^\alpha_{iL,kL'} \, W[j_{l'}, f_{k;L'L''}{}^{\alpha\alpha'}] C_{kL''}{}^{\alpha'} \} \qquad (2.32)$$

$$+ \int_{\Delta\Omega} G^\alpha_0(\mathbf{r}-\mathbf{r}') \Sigma_{\alpha'} \, V^I_{\alpha\alpha'}(\mathbf{r}') f_{\alpha'}(\mathbf{r}') d^3r' + N_\alpha \, e^{i\mathbf{k}_\alpha \cdot \mathbf{r}} \delta_{\alpha\underline{\alpha}}$$

Here we have introduced ρ_k, the radius of sphere Ω_k, and defined the wronskian

$$W[f,g] = f(r)(d/dr)g(r) - g(r)(d/dr)f(r)|_{r=\rho_k} \qquad (2.33)$$

calculated for $r = \rho_k$.

We now put

$$B^\alpha_{kL'} = \rho_k^2 \Sigma_{\alpha'L''} W[j_{l'}, f_{k;L'L''}{}^{\alpha\alpha'}] C_{kL''}{}^{\alpha'} =$$
$$= \rho_k^2 \Sigma_{\alpha'L''} W(j,f_k)_{L'L''}{}^{\alpha\alpha'} C_{kL''}{}^{\alpha'} \qquad (2.34)$$

and invert this relation to obtain

$$\rho_k^2 C_{kL'}{}^{\alpha'} = \Sigma_{\alpha''L''}[W(j,f_k)^{-1}]_{L'L''}{}^{\alpha'\alpha''} B^{\alpha''}{}_{kL''} \qquad (2.35)$$

with obvious notation.

Then Eq. (2.32) becomes

$$0 = \Sigma_L J_L^\alpha(\mathbf{r}_i) \{\Sigma_{\alpha'L'}\Sigma_{\alpha''L''} k_\alpha [W(-ih^+, f_i)]_{LL'}{}^{\alpha\alpha'}$$
$$[W(j,f_i)^{-1}]_{L'L''}{}^{\alpha'\alpha''} B^{\alpha''}{}_{iL''} + \sum_{k(\neq i)} \Sigma_{L'} G^\alpha_{iL,kL'} B^\alpha_{kL'} \} \qquad (2.36)$$
$$+ N_\alpha e^{i\mathbf{k}_\alpha \cdot \mathbf{r}} \delta_{\alpha\underline{\alpha}} + \int_{\Delta\Omega} G_0^\alpha(\mathbf{r}-\mathbf{r}') \Sigma_{\alpha'} V^I_{\alpha\alpha'}(\mathbf{r}') f_{\alpha'}(\mathbf{r}') d^3r'$$

We now introduce the generalized inverse atomic T_{ai}^{-1}-matrix whose meaning we shall discuss later

$$(T_{ai}^{-1})^{\alpha\alpha'}_{LL'} = k_\alpha \Sigma_{\alpha''L''} [W(-ih^+, f_i)]_{LL''}{}^{\alpha\alpha''} [W(j,f_i)^{-1}]_{L''L'}{}^{\alpha''\alpha'} \qquad (2.37)$$

and use the usual development (remember that $N_\alpha = (k_\alpha/\pi)^{1/2}/(4\pi)$)

$$N_\alpha e^{i\mathbf{k}_\alpha \cdot \mathbf{r}} = (k_\alpha/\pi)^{1/2} \Sigma_L i^l J^\alpha_L(\mathbf{r}) Y_L(\hat{\mathbf{k}}_\alpha) =$$
$$= (1/k_\alpha \pi)^{1/2} \Sigma_L i^l Y_L(\hat{\mathbf{k}}_\alpha) \Sigma_{L'} J^\alpha_{iL',oL} J^\alpha_{L'}(\mathbf{r}_i) \qquad (2.38)$$

where we have reexpanded the function $J^\alpha_L(\mathbf{r}) \equiv J^\alpha_L(\mathbf{r}_o)$, which is defined with respect to the origin of the coordinates o, around site i through the quantity $J^\alpha_{iL',oL}$ defined in Eq. (2.23)[2].

Since the solution of Eq. (2.36) is linear in the source term $N_\alpha e^{i\mathbf{k}_\alpha \cdot \mathbf{r}}$, we can put in Eq. (2.38)

$$(1/k_\alpha\pi)^{1/2} \, i^l \, Y_L(\hat{\mathbf{k}}_\alpha) = \delta_{L\underline{L}}(1/k_\alpha\pi)^{1/2} \qquad (2.39)$$

so that finally we can write

$$0 = \Sigma_L J^\alpha_L(\mathbf{r}) \{ \Sigma_{\alpha'L'} (T_{ai}^{-1})^{\alpha\alpha'}{}_{LL'} B^{\alpha'}{}_{iL'}(\underline{\alpha};\underline{L})$$

$$+ \Sigma_{k(\neq i)} \Sigma_{L'} G^\alpha_{iL,kL'} B^\alpha_{kL'}(\underline{\alpha};\underline{L}) + J^\alpha_{iL,o\underline{L}} \, \delta_{\alpha\underline{\alpha}} (1/k_\alpha\pi)^{1/2} \} \qquad (2.40)$$

$$+ \int_{\Delta\Omega} G^\alpha_0(\mathbf{r}-\mathbf{r}') \, \Sigma_{\alpha'} \, V^I_{\alpha\alpha'}(\mathbf{r}') \, f_{\alpha'}(\mathbf{r}') \, d^3r'$$

Notice that we have now affected the quantities $B^\alpha_{iL}(\underline{\alpha};\underline{L})$ by the indices $\underline{\alpha},\underline{L}$, marking the dependence on the inhomogeneous term $\delta_{\alpha\underline{\alpha}}\delta_{L\underline{L}}$. Therefore in Eq. (2.36)

$$B^\alpha_{iL} = \Sigma_{\underline{L}} \, i^{\underline{l}} \, B^\alpha_{iL}(\underline{\alpha};\underline{L}) \, Y_{\underline{L}}(\hat{\mathbf{k}}_{\underline{\alpha}}) \qquad (2.41)$$

Let us neglect, for the moment, the interstitial potential, i.e. let us put $V^I_{\alpha\alpha'}(\mathbf{r})=0$. Then the Eqs. (2.40), one for each i, determine the coefficients $B^\alpha_{iL}(\underline{\alpha};\underline{L})$, which through the relations (2.35) and (2.31), provide the functions $f_\alpha(\mathbf{r}_i)$ needed to calculate the transition matrix elements.

To interpret the $B^\alpha_{iL}(\underline{\alpha};\underline{L})$, we need to consider Eq. (2.21a) for $\mathbf{r} \notin \Sigma_k \Omega_k$ and use Eq. (2.18). Performing the surface integral and remembering the definition (2.35) we find

$$f_\alpha(\mathbf{r}) = N_\alpha \, e^{i\mathbf{k}_\alpha \cdot \mathbf{r}} \, \delta_{\alpha\underline{\alpha}}$$

$$- k_\alpha \Sigma_k \Sigma_{LL} \, i^{l+1} \, h^+_l(k_\alpha r_k) \, Y_L(\hat{\mathbf{r}}_k) \, B^\alpha_{kL}(\underline{\alpha};\underline{L}) \, Y_{\underline{L}}(\hat{\mathbf{k}}_{\underline{\alpha}})$$

$$+ \int_{\Delta\Omega} G^\alpha_0(\mathbf{r}-\mathbf{r}') \, \Sigma_{\alpha'} \, V^I_{\alpha\alpha'}(\mathbf{r}') \, f_{\alpha'}(\mathbf{r}') \, d^3r' \qquad (2.42)$$

Assuming again $V^I_{\alpha\alpha'}(\mathbf{r})=0$, this equation clearly shows the meaning of the $B^\alpha_{kL}(\underline{\alpha};\underline{L})$'s as scattering amplitudes into the channel α with angular momentum L emanating from site k in response to an excitation with angular momentum $\underline{L}$ into the channel $\underline{\alpha}$.

It is interesting to derive an explicit formula for the B^α_{iL}'s in the atomic case, which is obtained by suppressing the terms $k \neq i$ in Eqs. (2.40) and (2.42) and putting $i=o$.

From Eq. (2.40) in such a case we obtain, since[2] $J^\alpha_{oL,o\underline{L}} = \delta_{L\underline{L}} k_\alpha$,

$$\Sigma_{\alpha'L'} (T_a^{-1})^{\alpha\alpha'}{}_{LL'} B^{\alpha'}{}_{L'}(\underline{\alpha};\underline{L}) = -\delta_{LL}\delta_{\alpha\underline{\alpha}}(k_{\underline{\alpha}}/\pi)^{1/2} \qquad (2.43)$$

giving

$$-B^{\alpha}{}_L(\underline{\alpha};\underline{L}) = (T_a)^{\alpha\underline{\alpha}}{}_{L\underline{L}}(k_{\underline{\alpha}}/\pi)^{1/2} = \qquad (2.44)$$

$$= (k_{\underline{\alpha}}/\pi)^{1/2} \Sigma_{\alpha'L'} [W(j,f)]_{LL'}{}^{\alpha\alpha'}(k_{\alpha'})^{-1} [W(-ih^+,f)^{-1}]_{L'\underline{L}}{}^{\alpha'\underline{\alpha}}$$

This explains the definition in Eq. (2.37). The quantities $(T_a)^{\alpha\alpha'}{}_{LL'}$ are the natural generalization of the usual atomic T_a-matrices for non spherically symmetric potential in the multichannel case.

For the many center case the interpretation of the coefficients $B^{\alpha}{}_{iL}(\underline{\alpha};\underline{L})$ as scattering amplitudes is indeed confirmed by the physical meaning of the m.s. equations:

$$\Sigma_{\alpha'L'} (T_{ai}^{-1})^{\alpha\alpha'}{}_{LL'} B^{\alpha'}{}_{iL'}(\underline{\alpha};\underline{L}) + \qquad (2.45)$$

$$+ \sum_{k(\neq i)} \Sigma_{L'} G^{\alpha}{}_{iL,kL'} B^{\alpha}{}_{kL'}(\underline{\alpha};\underline{L}) = -J^{\alpha}{}_{iL,o\underline{L}}\delta_{\alpha\underline{\alpha}}(1/(k_{\underline{\alpha}}\pi))^{1/2}$$

which can also be written as

$$B^{\alpha}{}_{iL}(\underline{\alpha};\underline{L}) = -\Sigma_{\alpha'L'}(T_{ai})^{\alpha\alpha'}{}_{LL'} \sum_{k(\neq i)} \Sigma_{L''} G^{\alpha'}{}_{iL',kL''} B^{\alpha'}{}_{kL''}(\underline{\alpha};\underline{L})$$

$$- \Sigma_{L'}(T_{ai})^{\alpha\underline{\alpha}}{}_{LL'} J^{\underline{\alpha}}{}_{iL',o\underline{L}}(1/(k_{\underline{\alpha}}\pi))^{1/2} \qquad (2.46)$$

Since, from Eq. (2.38), $J^{\alpha}{}_{iL,o\underline{L}}$ is the exciting amplitude of the $\underline{L}$ angular momentum component of a plane wave impinging on the origin as seen from site i, Eq. (2.46) shows that $B^{\alpha}{}_{iL}(\underline{\alpha};\underline{L})$ is the sum of the scattering amplitude originated directly at site i by the exciting amplitude plus all the scattering amplitudes generated by the waves that are scattered by all other sites $k(\neq i)$ and propagate from site k to site i, where they are finally scattered into the final state.

If is interesting to look at the structure of the m.s. matrix Eq. (2.45):

$$S^{\alpha\alpha'}{}_{iL,kL'} = (T_{ai}^{-1})^{\alpha\alpha'}{}_{LL'}\delta_{ik} + (1-\delta_{ik})\delta_{\alpha\alpha'} G^{\alpha}{}_{iL,kL'}$$

$$= (K_{ai}^{-1})^{\alpha\alpha'}{}_{LL'}\delta_{ik} + (1-\delta_{ik})\delta_{\alpha\alpha'} N^{\alpha}{}_{iL,kL'} - i\delta_{\alpha\alpha'} J^{\alpha}{}_{iL,kL'}$$

$$= M^{\alpha\alpha'}{}_{iL,kL'} - i\Delta^{\alpha\alpha'}{}_{iL,kL'} \qquad (2.47)$$

where M and Δ are hermitian matrices (actually Δ is real symmetric). We have introduced the reactance atomic K_{ai}-matrix related to the T_{ai}-matrix by the usual relation

$$(T_{ai}^{-1})^{\alpha\alpha'}_{LL'} = k_\alpha \Sigma_{\alpha''L''} [W(-ih^+,f_i)]_{LL''}^{\alpha\alpha''} [W(j,f_i)^{-1}]_{L''L'}^{\alpha''\alpha'} =$$

$$= k_\alpha \Sigma_{\alpha''L''} [W(n,f_i)]_{LL''}^{\alpha\alpha''} [W(j,f_i)^{-1}]_{L''L'}^{\alpha''\alpha'}$$

$$- ik_\alpha \delta_{LL'} \delta_{\alpha\alpha'}$$

$$= (K_{ai}^{-1})^{\alpha\alpha'}_{LL'} - iI\, k_\alpha \qquad (2.48)$$

remembering that $-ih^+_l = n_l - ij_l$. The term $iI\, k_\alpha = i\,\delta_{\alpha\alpha'}\,\delta_{LL'}\,k_\alpha$ has been incorporated in Δ by lifting the restriction $i \neq k$ and using the relation $J^\alpha_{iL,iL'} = k_\alpha \delta_{LL'}$. In Eq. (2.47) we have used the decomposition (2.23).

By exploiting the sum rule[2]

$$\Sigma_L\, J^\alpha_{iL,oL}\, J^\alpha_{kL',oL} = k_\alpha J^\alpha_{iL,kL'} = k_\alpha\, \Delta^{\alpha\alpha'}_{iL,kL'}\, \delta_{\alpha\alpha'} \qquad (2.49)$$

it is now easy to derive a generalized optical theorem for the amplitudes $B^\alpha_{iL}(\underline{\alpha};\underline{L})$:

$$\Sigma_{\underline{\alpha L}}\, B^\alpha_{iL}(\underline{\alpha};\underline{L})\, [B^\alpha_{kL'}(\underline{\alpha};\underline{L})]^* = 1/\pi\, [(M-i\Delta)^{-1}\Delta(M+i\Delta)^{-1}]^{\alpha\alpha'}_{iL,kL'}$$

$$(2.50)$$

$$= 1/\pi\, \text{Im}\,[(M-i\Delta)^{-1}]^{\alpha\alpha'}_{iL,kL'} = 1/\pi\, \text{Im}\,\tau^{\alpha\alpha'}_{iL,kL'}$$

which we shall need in the following. For convenience we have put $(M-i\Delta)^{-1} = S^{-1} = \tau$, which is known as the scattering path operator.

The presence of an interstitial potential $V^I_{\alpha\alpha'}(r)$ merely modifies the quantities T_{ai}^{-1} and G in Eq. (2.45). However the general structure of the m.s. equations as well as the validity of the generalized optical theorem (2.50) remain unchanged. This is also true in presence of an outer sphere. We refer the reader to the already cited articles for details[1,2].

If we assume that the initial core state is localized at site i, we need the vave function $f_\alpha(r)$ inside the sphere Ω_i. From Eqs. (2.31) and (2.35) we obtain

$$f^i_\alpha(\mathbf{r}) = \Sigma_{\alpha'} \Sigma_{LL'} C_{iL'}{}^{\alpha'} f_{i;LL'}{}^{\alpha\alpha'}(r_i) Y_L(\hat{\mathbf{r}}_i)$$

$$= \Sigma_{\alpha'} \Sigma_{LL'} \Sigma_{\alpha''L''} \rho_i^{-2} [W(j,f_i)^{-1}]_{L'L''}{}^{\alpha'\alpha''} \quad (2.51)$$

$$B^{\alpha''}{}_{iL''} f_{i;LL'}{}^{\alpha\alpha'}(r_i) Y_L(\hat{\mathbf{r}}_i)$$

By defining the functions

$$\underline{f}_{LL'}{}^{\alpha\alpha''}(r) = \rho_i^{-2} \Sigma_{\alpha'L'} f_{LL'}{}^{\alpha\alpha'}(r) [W(j,f)^{-1}]_{L'L''}{}^{\alpha'\alpha''} \quad (2.52)$$

we can also write, making explicit the dependence on the incident wave vector k_α and using Eq. (2.41),

$$f_\alpha^i(\mathbf{r};\mathbf{k}_\alpha) = \Sigma_L \Sigma_{\alpha'L'} B^{\alpha'}{}_{iL'} \underline{f}_{LL'}{}^{\alpha\alpha'}(r_i) Y_L(\hat{\mathbf{r}}_i)$$

$$= \Sigma_L \Sigma_{\alpha'L'} \Sigma_{\underline{L}} B^{\alpha'}{}_{iL'}(\underline{\alpha};\underline{L}) i^{\underline{l}} Y_{\underline{L}}(\hat{\mathbf{k}}_\alpha) \underline{f}_{i;LL'}{}^{\alpha\alpha'}(r_i) Y_L(\hat{\mathbf{r}}_i) \quad (2.53)$$

To obtain the total cross section we have to sum over all possible photoelectron final states labelled by the index $\underline{\alpha}$. Since the wave functions f^i_α are normalized to one state per Rydberg we have, using the projection property $\mathcal{A}^2=\mathcal{A}$,

$$\sigma(\omega) = 4\pi^2 \alpha \hbar \omega \Sigma_{\underline{\alpha}}$$

$$\int d\hat{\mathbf{k}}_\alpha |(\Sigma_\alpha f_\alpha^i(\mathbf{r};\mathbf{k}_\alpha) \Psi_\alpha^{N-1} | \varepsilon \cdot \sum_{m=1}^N \mathbf{r}_m | N! \mathcal{A} \phi_c^i(\mathbf{r}) \Psi_G^{N-1})|^2$$

$$(2.54)$$

$$= 4\pi^2 \alpha \hbar \omega \Sigma_{\underline{\alpha}} \int d\hat{\mathbf{k}}_\alpha |(\Sigma_\alpha f_\alpha^i(\mathbf{r};\mathbf{k}_\alpha) | \varepsilon \cdot \mathbf{r} | \phi_c^i(\mathbf{r})) S_{\alpha 0}|^2$$

The last step follows from the orthogonality of f^i_α to all the initially occupied orbitals and the fact that we assume the arthogonality of $\phi_c^i(\mathbf{r})$ to all the orbitals appearing in the Ψ_α^{N-1}'s. $S_{\alpha 0} = (\Psi_\alpha^{N-1} | \Psi_G^{N-1})$ is the projection of Ψ_α^{N-1} onto the occupied configurations present in the initial state.

By introducing the expression (2.53) into Eq. (2.54), performing the angular integration over $\hat{\mathbf{k}}_\alpha$ and introducing the atomic matrix elements

$$M^{\alpha\alpha'}{}_{LL'} = (\underline{f}^{\alpha\alpha'}{}_{LL'}(r_i) Y_L(\mathbf{r}_i) | \varepsilon \cdot \mathbf{r} | \phi_c^i(\mathbf{r}_i)) \quad (2.55)$$

we can rewrite Eq. (2.54) as

$$\sigma(\omega) = 4\pi^2 \alpha \hbar \omega \sum_{\underline{\alpha}L} \sum_{\alpha'\alpha} \sum_{\beta'\beta} \sum_{L_fL'_f} \sum_{LL'} S_{\alpha 0} \; M^{\alpha\beta}_{L_fL} \; B^{\beta}_{iL} (\underline{\alpha};\underline{L})$$

$$[B^{\beta'}_{iL'} (\underline{\alpha};\underline{L}) \; M^{\alpha'\beta'}_{L'_fL'} \; S_{\alpha'0}]^*$$

(2.55a)

$$= 4\pi \alpha \hbar \omega \sum_{\alpha\alpha'} \sum_{\beta\beta'} \sum_{L_fL'_f} \sum_{LL'} S_{\alpha 0} \; M^{\alpha\beta}_{L_fL}$$

$$\{ \text{Im } \tau^{\beta\beta'}_{iL,iL'} \} [M^{\alpha'\beta'}_{L'_fL'} \; S_{\alpha'0}]^*$$

(2.55b)

using the generalized optical theorem Eq. (2.50).

From Eq. (2.53) it is immediate to write down an expression for the photoemission cross section for ejection of an electron into the state $\mathbf{k}_\alpha$ with energy $k_\alpha^2 = \hbar\omega - I_c - \Delta E_\alpha$

$$d\sigma(\omega)/d\hat{\mathbf{k}}_\alpha = 4\pi^2 \alpha \hbar \omega | (\Sigma_\alpha f_\alpha^i(\mathbf{r};\mathbf{k}_\alpha)| \epsilon \cdot \mathbf{r} | \phi_c^i(\mathbf{r})) S_{\alpha 0} |^2$$

(2.56)

$$= 4\pi^2 \alpha \hbar \omega | \sum_{\alpha L_f} \sum_{\alpha'L'} \sum_L B^{\alpha'}_{iL'}(\underline{\alpha};\underline{L}) \; i^L \; Y_L(\hat{\mathbf{k}}_\alpha) \; M^{\alpha\alpha'}_{L_fL'} \; S_{\alpha 0} |^2$$

In both cases the sum over L_f indicates the sum over the final angular momenta allowed by the dipole selection rule in Eq. (2.55). Notice that in Eq. (2.56) it is not possible to take advantage of the generalized optical theorem.

It is interesting to compare the expression (2.55) with the total cross section for electron molecule scattering. The general definition of scattering T-matrix in the multichannel case is derived by looking at the asymptotic behavior of the electron wave function

$$f_\alpha(\mathbf{r}) \underset{r \to \infty}{\sim} \sum_L 4\pi \; Y_L(\hat{\mathbf{k}}_\alpha) i^L [J^\alpha_L(\mathbf{r})\delta_{\alpha\underline{\alpha}} - ik_\alpha \sum_{L'} H^{+\alpha}_{L'}(\mathbf{r}) T^{\alpha\underline{\alpha}}_{L'\underline{L}}]$$

(2.57)

where $\mathbf{r}$ is referred to the center of the coordinates.

This expression has to be compared with Eq. (2.42), with $\mathbf{V}^I_{\alpha\alpha'} = 0$, after all coordinates $\mathbf{r}_k = \mathbf{r} - \mathbf{R}_k$ have been referred to the origin. To this purpose we use the reexpansion formula[2]

$$-ik_\alpha h^+_l(k_\alpha r_k) Y_L(\hat{\mathbf{r}}_k) = -i \sum_{L'} h^+_{l'}(k_\alpha r) Y_{L'}(\hat{\mathbf{r}}) \; J^\alpha_{oL',kL}$$

(2.58)

valid for $|\mathbf{r}_k - \mathbf{r}| = |\mathbf{R}_k| < |\mathbf{r}|$ since we look at $|\mathbf{r}| \to \infty$

Substituting this relation into Eq. (2.42) we obtain

$$f_\alpha(\mathbf{r}) \underset{r\to\infty}{\sim} \Sigma_L \, 4\pi \, Y_L(\hat{\mathbf{k}}_{\underline{\alpha}}) \, i^l \, [N_{\underline{\alpha}} \, J^\alpha_L(\mathbf{r}) \, \delta_{\alpha\underline{\alpha}}$$

$$-i/4\pi \, \Sigma_{kL} \, \Sigma_{L'} \, h^+_{l'}(k_\alpha r) \, Y_{L'}(\hat{\mathbf{r}}) \, J^\alpha_{oL',kL} \, B^\alpha_{kL}(\underline{\alpha}; \underline{L})]$$

$$= \Sigma_L \, 4\pi \, Y_L(\hat{\mathbf{k}}_{\underline{\alpha}}) \, i^l \, N_{\underline{\alpha}} \, [J^\alpha_L(\mathbf{r}) \, \delta_{\alpha\underline{\alpha}}$$

$$-i/(4\pi N_{\underline{\alpha}}) \, \Sigma_{L'} \, H^{+\alpha}_{L'}(\mathbf{r}) \, \Sigma_{kL} \, J^\alpha_{oL',kL} \, B^\alpha_{kL}(\underline{\alpha}; \underline{L})] \tag{2.59}$$

This gives for $T^{\alpha\underline{\alpha}}_{LL}$ the expression

$$T^{\alpha\underline{\alpha}}_{L\underline{L}} = \Sigma_{kL'} \, J^\alpha_{oL,kL'} \, B^\alpha_{kL'}(\underline{\alpha};\underline{L}) \, (\pi/k_\alpha^3)^{1/2} \tag{2.60}$$

The total scattering cross section into any channel α' starting from channel α is given by

$$\sigma^\alpha_{el}(E_\alpha) = 4\pi \, \Sigma_{\alpha'L'} \Sigma_L \, |T^{\alpha'\alpha}_{L'L}|^2 = 4\pi \, \Sigma_{\alpha'L'} \Sigma_L \, |T^{\alpha\alpha'}_{LL'}|^2 \tag{2.61}$$

using the detailed balance relation[4].

Using Eq. (2.60) we find

$$\Sigma_{\underline{\alpha}\underline{L}} |T^{\alpha\underline{\alpha}}_{L\underline{L}}|^2 = (TT^+)^{\alpha\alpha}_{LL} =$$

$$= \pi/k_\alpha^3 \, \Sigma_{\underline{\alpha}\underline{L}} \, \Sigma_{kL'} \, J^\alpha_{oL,kL'} \, B^\alpha_{kL'}(\underline{\alpha};\underline{L}) \, [\Sigma_{k'L''} \, J^\alpha_{oL,k'L''} \, B^\alpha_{k'L''}(\underline{\alpha};\underline{L})]^*$$

$$= \pi/k_\alpha^3 \, \Sigma_{kL'} \Sigma_{k'L''} \, J^\alpha_{oL,kL'} \, \text{Im} \, \tau^{\alpha\alpha}_{kL',kL''} \, J^\alpha_{oL,k'L''}$$

$$= 1/k_\alpha \, \text{Im} \, T^{\alpha\alpha}_{LL} \tag{2.62}$$

since $J^\alpha_{oL,kL'}$ is real and we have exploited the relation, derived from Eq. (2.45)

$$(k_\alpha\pi)^{1/2} \, B^\alpha_{kL}(\underline{\alpha};\underline{L}) = \Sigma_{k'L'} \, \tau^{\alpha\alpha}_{kL,k'L'} \, J^\alpha_{oL,k'L'} \tag{2.63}$$

Eq. (2.62) is nothing else that the optical theorem for the scattering T-matrix. As a consequence Eq. (2.61) takes the form

$$\sigma^\alpha_{el}(E_\alpha) = 4\pi/k_\alpha \, \Sigma_L \, \text{Im} \, T^{\alpha\alpha}_{LL}$$

$$= 4\pi/k_\alpha^3 \, \text{Im} \, \Sigma_{L''} \Sigma_{kL} \Sigma_{k'L'} \, J^\alpha_{oL'',kL} \, \tau^{\alpha\alpha}_{kL,k'L'} \, J^\alpha_{k'L',oL''} \tag{2.64}$$

exploiting the relation $J^{\alpha}_{oL,kL'} = J^{\alpha}_{kL',oL}$.

We shall discuss the relation of this expression with Eq. (2.55) in the next section.

3. THE GENERALIZED MULTIPLE SCATTERING EXPANSION

In the expressions (2.55), (2.64) the structural information is contained in the inverse $\tau = S^{-1}$ of the multiple scattering matrix Eq. (2.47) through the presence at the structure matrix elements $G^{\alpha}_{iL,kL'}$, in a rather involved way that intermingles dynamics as well as structure.

It turns out however that under certain circumstances, to be discussed shortly, one can expand the various cross sections in a convergent series the general term of which has a simple and direct physical meaning.

In fact, remembering the notation introduced in section 2, we have

$$\tau = S^{-1} = (T_a^{-1} + G)^{-1} = (I + T_a G)^{-1} T_a$$

so that if the spectral radius $\rho(T_a G)$ of the matrix $T_a G$ is less than one, where $\rho(A)$ is the maximum modulus of the eigenvalues of A, then

$$(I + T_a G)^{-1} = \sum_{n=0}^{\infty} (-1)^n (T_a G)^n \qquad (3.1)$$

the series on the right being absolutely convergent relative to some matrix norm. For short we shall henceforth define $G^{\alpha}_{iL,iL} \equiv 0$ to account for the factor $(1-\delta_{ik})$ in Eq. (2.47).

As a consequence the photoabsorption cross section Eq. (2.55) can be expanded in an absolutely convergent series

$$\sigma(\omega) = \sum_{n=0}^{\infty} \sigma_n(\omega) \qquad (3.2)$$

where

$$\sigma_0(\omega) = 4\pi\alpha\hbar\omega \sum_{\alpha\alpha'} \sum_{\beta\beta'} \sum_{L_f L'_f} \sum_{LL'} S_{\alpha 0} M^{\alpha\beta}_{L_f L} \operatorname{Im}(T_{ai})^{\beta\beta'}_{iL,iL'}$$
$$[M^{\alpha'\beta'}_{L'_f L'} S_{\alpha' 0}]^* \qquad (3.3)$$

is a smoothly varying atomic cross section and

$$\sigma_n(\omega) = 4\pi\alpha\hbar\omega \sum_{\alpha\alpha'} \sum_{\beta\beta'} \sum_{L_f L'_f} \sum_{LL'} S_{\alpha 0} M^{\alpha\beta}_{L_f L}$$
$$\text{Im}[(-1)^n (T_a G)^n T_a]^{\beta\beta'}_{iL,iL'} [M^{\alpha'\beta'}_{L'_f L'} S_{\alpha' 0}]^* \qquad (3.4)$$

represents the contribution to the photoabsorption cross section coming from process where the photoelectron, before being ejected at infinity, leaves the photoabsorbing atom, located at site i, with angular momentum L and channel state β, is scattered (n−1) times by the surrounding atoms and returns to site i with angular momentum L' and channel state β'. All these events are eventually to be multiplied by the corresponding amplitudes

$$S_{\alpha 0} M^{\alpha\beta}_{L_f L} \quad \text{and} \quad S_{\alpha'} M^{\alpha'\beta'}_{L'_f L'}$$

and summed together to give the n-th order contribution. It is clear that this term bears information on the n particle correlation and therefore is sensitive to the geometrical arrangement around the photoabsorbing atom.

In order to better illustrate these concepts let us treat some asymptotic cases. It is obvious that the condition $\rho(T_a G) < 1$ is satisfied at high photoelectron energy since

$$\lim_{k_\alpha \to \infty} |(T_a)^{\alpha\alpha'}_{LL'}| = 0.$$

In this regime one can safely write

$$(T_a)^{\alpha\alpha'}_{LL'} \simeq t^\alpha_{al}\delta_{LL'}\delta_{\alpha\alpha'} \quad \text{and} \quad M^{\alpha\alpha'}_{LL'} \simeq M^\alpha_L \delta_{LL'}\delta_{\alpha\alpha'} \qquad (3.5)$$

since the photoelectron is sensitive only to the atomic cores, which are spherically symmetric, and only the "incoming" channel $f_\alpha(r)$ in Eq. (2.14) is relevant, following the same argument leading to (2.15).

As a consequence the asymptotic cross section $\sigma_{as}(\omega)$ is given by

$$\sigma_{as}(\omega) = 4\pi\alpha\hbar\omega \sum_\alpha |S_{\alpha 0} M^\alpha_l|^2 \sum_m \text{Im}\, \tau^\alpha_{ilm,ilm} \qquad (3.6)$$

where, for simplicity, we have assumed a single l final state and $\tau^\alpha_{iL,kL'}$ is the inverse of S^α, the submatrix of S relative to the channel α:

$$[(\tau^\alpha)^{-1}]_{iL,kL'} = S^\alpha_{iL,kL'} = (t^\alpha_{ail})^{-1} \delta_{ik} \delta_{LL'} + G^\alpha_{iL,kL'} \qquad (3.7)$$

In other words the different channels decouple and they have identical m.s. structure, apart from the trivial dependence on the photoelectron propagation vector k_α and on the atomic scattering matrices t^α_{al}.

Eq. (3.6) is the form used by Rehr et al[6] to discuss the role of multielectron excitations in the EXAFS structure of the Br_2 molecule in the framework of the "sudden approximation".

The total cross section is therefore an incoherent sum of photoabsorption cross sections relative to different channels, so that we can limit ourselves to a single channel. On a theoretical basis, born out by experiments, one expects the predominance of a single channel in the sum (3.6) when the ground state of the system contains one single dominant configuration. In this case the biggest overlap factor among the $S_{\alpha,0}$'s is S_{00}, corresponding to the same relaxed configuration in the final state and to $\Delta E_\alpha = 0$ in Eq. (2.12). Depending on the systems, one has $0.7 < |S_{00}|^2 < 0.8$ so that one single channel accounts for 70~80 per cent of the spectrum. We shall see in a moment how to account for the rest in an approximate way.

In the energy region where Eq. (3.6) is valid we can also expand τ^α as

$$\tau^\alpha = (I + t^\alpha_a G^\alpha)^{-1} t^\alpha_a = \sum_{n=0}^{\infty} (-1)^n (t^\alpha_a G^\alpha)^n t^\alpha_a \qquad (3.8)$$

so that

$$\sigma_{as}(\omega) = 4\pi\alpha\hbar\omega \sum_\alpha |S_{\alpha 0} M^\alpha_1|^2 \sum_n \sum_m (-1)^n \, \text{Im}\, [(t^\alpha_a G^\alpha)^n t^\alpha_a]_{ilm,ilm}$$

$$= \sum_{n=0}^{\infty} \sum_\alpha \sigma^\alpha_{nas}(\omega) \qquad (3.9)$$

Analytic expressions for the m.s. terms, based on the Eq. (2.23) for the matrix elements $G_{iL,kL'}$, are available in the literature[7,8]. For our purpose it is sufficient to observe that each $G^\alpha_{iL,kL'}$ carries a factor $\exp\{ik_\alpha R_{ik}\}$ independent of L,L' contained in the Hankel function appearing in the definition (2.23), which can be better taken account of by defining the reduced matrix

$$\underline{G}^\alpha_{iL,kL'} = e^{-ik_\alpha R_{ik}} G^\alpha_{iL,kL'}$$

$$\underline{G}^\alpha_{iL,iL'} = 0 \qquad (3.10)$$

For the n-th order term in Eq. (3.9), we find

$$\sum_m [(t^\alpha_a G^\alpha)^n t^\alpha_a]_{ilm,ilm} = \sum_m \sum_{k_1 L_1} \cdots \sum_{k_{n-1} L_{n-1}} t^\alpha_{ilm} \underline{G}^\alpha_{ilm,k_1 L_1} t^\alpha_{k_1 l_1} \cdots$$

$$\cdots t^\alpha_{k_{n-1} l_{n-1}} \underline{G}^\alpha_{k_{n-1} L_{n-1}, ilm} t^\alpha_{ilm} \qquad (3.11)$$

The set $k_1 \ldots k_{n-1}$ defines a path p_n of order n that begins and ends at the central atom (located at site i), to which we can associate a total path length

$$R^{tot}_{p_n} = \sum_{m=1}^{n-1} R_{k_m k_{m+1}} \qquad (3.12)$$

Therefore, putting

$$A^1_n(k_\alpha, R^{p_n}_{ik}) \exp[i\phi^1_n(k_\alpha, R^{p_n}_{ik})] \qquad (3.13)$$

$$= \sum_m \sum_{L_1} \cdots \sum_{L_{n-1}} t^\alpha_{ilm} \underline{G}^\alpha_{ilm,k_1 L_1} \cdots \underline{G}^\alpha_{k_{n-1} L_{n-1}, ilm} t^\alpha_{ilm}$$

we can finally write

$$\sum_m [(t^\alpha_a G^\alpha)^n t^\alpha_a]_{ilm,ilm} =$$

$$= \sum_{p_n} A^1_n(k_\alpha, R^{p_n}_{ik}) \exp\{i[k_\alpha R^{tot}_{p_n} + \phi^1_n(k_\alpha, R^{p_n}_{ik})]\} \qquad (3.14)$$

so that the functional contribution of the n-th order m.s. term to the photoabsorption cross section in channel α is

$$\sum_{p_n} A^1_n(k_\alpha, R^{p_n}_{ik}) \sin[k_\alpha R^{tot}_{p_n} + \phi^1_n(k_\alpha, R^{p_n}_{ik})] \qquad (3.15)$$

This means that each path contributes an oscillatory signal in the cross section of period $2\pi/R^{tot}_{p_n}$ and amplitude $A^1_n(k_\alpha, R^{p_n}_{ik})$.

The quantities $A^1_n(k_\alpha, R^{p_n}_{ik})$ and $\phi^1_n(k_\alpha, R^{p_n}_{ik})$ are slowly varying functions of k_α, so that, indicating by $k_0=[\hbar\omega-I_c]^{1/2}$ the photoelectron wave vector of the primary channel, we can write approximatively in Eq. (3.14)

$$\Sigma_m [(t^\alpha_a G^\alpha)^n t^\alpha_a]_{ilm,ilm} =$$

$$\sum_{P_n} A^1_n(k_0, R^{P_n}_{ik}) \exp\{i[k_0 R^{tot}_{P_n} + \phi^1_n(k_0, R^{P_n}_{ik})]\} \quad (3.16)$$

$$\exp\{i(k_\alpha - k_0)[R^{tot}_{P_n} + (d/dk)\phi^1_n(k, R^{P_n}_{ik})|_{k=k_0}]\}$$

If we then define the complex number

$$B^1_n(k_0) e^{i\psi^1_n(k_0)} = |S_{00} M^{\alpha_0}_1|^{-2} \Sigma_\alpha |S_{\alpha 0}|^2 \quad (3.17)$$

$$\exp\{i(k_\alpha - k_0)[R^{tot}_{P_n} + (d/dk)\phi^1_n(k, R^{P_n}_{ik})|_{k=k_0}]\}$$

we can finally write

$$\sigma_{as}(\omega) = \Sigma_n \Sigma_\alpha \sigma^\alpha_n(\omega) = 4\pi\alpha\hbar\omega |S_{00} M^{\alpha_0}_1|^2 \Sigma_n \sum_{P_n} (-1)^n B^1_n(k_0)$$

$$A^1_n(k_0, R^{P_n}_{ik}) \sin[k_0 R^{tot}_{P_n} + \phi^1_n(k_0, R^{P_n}_{ik}) + \psi^1_n(k_0)] \quad (3.18)$$

This is the generalization of the result arrived at in Refs. 5,6. The modification needed when there are two or more configurations present in the ground state with comparable amplitudes, is straightforward. We easily find in this case

$$\sigma_{as}(\omega) = 4\pi\alpha\hbar\omega \Sigma_\beta |S_{\beta 0} M^\beta_1|^2 \Sigma_n \sum_{P_n} (-1)^n B^1_n(k_\beta) A^1_n(k_\beta, R^{P_n}_{ik})$$

$$\sin[k_\beta R^{tot}_{P_n} + \phi^1_n(k_\beta, R^{P_n}_{ik}) + \psi^1_n(k_\beta)] \quad (3.19)$$

where the Σ_β is over the corresponding relaxed configurations in the final state.

It should then be possible to discriminate in the experimental analysis between the various oscillatory signals appearing in the spectrum due to the presence of different main channels β.

However the formula (3.19) is only asymptotic and deviations from the sudden approximation (3.5) must be considered if one wants to exploit a larger energy range.

The general expansion to use in this case is given in Eq. (3.4). The lowest order term is n=2, since $G^{\alpha}_{iL,iL'}=0$. This is the usual EXAFS contribution given by

$$\sigma_2(\omega) = 4\pi\alpha\hbar\omega \Sigma_{\alpha\alpha'} S_{\alpha 0} M^{\alpha}_1 \Sigma_m \Sigma_{kL'} \sum_{\alpha_1\alpha_2} (T_{ai})_1^{\alpha\alpha_1} G^{\alpha_1}_{iL,kL'}$$

$$(T_{ak})_1^{\alpha_1\alpha_2} G^{\alpha_2}_{kL',iL} (T_{ai})_1^{\alpha_2\alpha'} [M^{\alpha'}_1 S_{\alpha'0}]^* \qquad (3.20)$$

where for simplicity we have assumed $(T_{ai})^{\alpha\alpha'}_{LL'}= (T_{ai})_1^{\alpha\alpha'} \delta_{LL'}$, and set $M^{\alpha\alpha'}_{LL'} \simeq M^{\alpha}_1 \delta_{\alpha\alpha'} \delta_{LL'}$, since terms proportional to $M^{\alpha\alpha'}_{LL'} (\alpha\neq\alpha')$ would be of higer order in this expansion.

The new feature now is given by the fact that at each scattering event the photoelectron can change its channel state, and consequently its propagation vector k_α. This fact can make difficult the detection of, say, a two channel in the EXAFS signal of fluctuating mixed valence compounds, especially for the first coordination shell whose atoms can participate to the relaxation effect of the photoabsorber.

However it is likely that there is no relaxation beyond the first shell so that one can write $(T_{ak})_1^{\alpha_1\alpha_2} \sim (T_{ak})_1^{\alpha_1} \delta_{\alpha_1\alpha_2}$ for atoms located in the second shell. Eq. (3.20) then implies that there are only two EXAFS signals, originating from this shell, each one with a definite propagation vector. Since for higher order shells the period of oscillation in k is shorter, it should be easier to detect the two signals. Recently interesting results concerning lattice relaxation in homogeneous and inhomogeneous mixed-valent materials have been obtained by the use of a two channel EXAFS analysis[9].

Equations like the one in (3.19) constitute the basis for a structural analysis of photoabsorption spectra. This analysis is in many way complicated by the need of taking configurational averages both dynamical (over the phonon spectrum) and structural, when it is the case (as in amorphous systems). The way to do this averaging processes is still a matter a research.

It is interesting at this point to compare the photoabsorption cross section Eq. (2.55b) which reduces to the following

$$\sigma(\omega) \simeq 4\pi\hbar\omega \Sigma_{\alpha\alpha'} S_{\alpha 0} M^{\alpha}_L \{\text{Im } \tau^{\alpha\alpha'}_{iL,iL}\} [M^{\alpha'}_L S_{\alpha'0}]^* \qquad (3.21)$$

if one takes the most important terms ($M^{\alpha\alpha'}_{LL'} \sim M^{\alpha}_L \delta_{\alpha\alpha'}$, where L represents the l channel selected by the dipole matrix element with initial core electron angular momentum l-1), with the expression (2.64) for electron- molecule (i.e. cluster of atoms total cross section, which we rewrite here for convenience

$$\sigma^{\alpha}_{el}(E_\alpha) = $$
$$= 4\pi/k_\alpha^3 \, \text{Im} \, \Sigma_{L''} \Sigma_{kL} \Sigma_{k'L'} \, J^{\alpha}_{oL'',kL} \, \tau^{\alpha\alpha}_{kL,k'L'} \, J^{\alpha}_{k'L',oL''} \qquad (3.22)$$

The greater structural and angular momentum selectivity of the photoabsorption cross section is apparent. In Eq. (3.21) only paths beginning and ending at the photoabsorbing site with the same angular momentum are possible. No such selection rule exists in Eq. (3.22). Moreover in the greatest majority of cases, when only one single configuration is dominant in the ground state, only the primary channel α_0 matters, the effect of the remaining channels resulting into a smoothing action on the primary transition. Therefore, as a structural probe, photoabsorption has to be preferred to electron collisions.

It is also interesting to compare Eq. (3.21) with the photoemission cross section Eq. (2.56), which under the same assumptions reduces to

$$\frac{d\sigma}{d\hat{\mathbf{k}}_\alpha} \simeq 4\pi^2 \hbar \omega \, | \Sigma_\alpha \Sigma_L \, B^{\alpha}_{iL}(\underline{\alpha}; \underline{L}) \, i^L \, Y_L(\hat{\mathbf{k}}_\alpha) \, M^{\alpha}_L \, S_{\alpha 0} |^2 \qquad (3.23)$$

By using the solution (2.63) for B^{α}_{iL}, together with the definition (2.23) for $J^{\alpha}_{oL,kL'}$ and the relation (2.38), we find

$$\Sigma_L B^{\alpha}_{iL}(\underline{\alpha}; \underline{L}) \, i^L \, Y_L(\hat{\mathbf{k}}_\alpha)$$
$$= (k_\alpha/\pi)^{1/2} \, \Sigma_{kL'} \, \tau^{\alpha\underline{\alpha}}_{iL,kL'} \, i^{L'} \, Y_{L'}(\hat{\mathbf{k}}_\alpha) \, e^{i\mathbf{k}_\alpha \cdot \mathbf{R}_{ko}}$$
$$= (k_\alpha/\pi)^{1/2} \, \Sigma_{kL'} \, [(I + T_a G)^{-1} T_a]^{\alpha\underline{\alpha}}_{iL,kL'} \, i^{L'} \, Y_{L'}(\hat{\mathbf{k}}_\alpha) \, e^{i\mathbf{k}_\alpha \cdot \mathbf{R}_{ko}}$$
$$\qquad (3.24)$$

At "high" photoemission energies, again

$$(I + T_a G)^{-1} T_a \simeq [I - T_a G + (T_a G)^2 + \ldots] T_a$$

retaining only terms up to the second.

Within this approximation and putting for simplicity $(T_{ai})^{\alpha\alpha'}{}_{LL'} \simeq (T_{ai})_1{}^{\alpha\alpha'} \delta_{LL'}$ we derive

$$\Sigma_L \, B^\alpha{}_{iL}(\underline{\alpha};\underline{L}) \, i^l \, Y_L(\hat{\mathbf{k}}_\alpha) = (k_\alpha/\pi)^{1/2} \, \Sigma_{\alpha'} \, (T_{ai})_1{}^{\alpha\alpha'}$$

$$\{ i^l \, Y_L(\hat{\mathbf{k}}_\alpha) \, \delta_{ik} \, \delta_{\underline{\alpha}\alpha'} - \Sigma_{kL'} \, G^{\alpha'}{}_{iL,kL'} \, (T_{ak})_1{}^{\alpha'\underline{\alpha}} \, i^{l'} \, Y_{L'}(\hat{\mathbf{k}}_\alpha)$$

$$+ \Sigma_{\alpha''} \, \Sigma_{kL'} \, \Sigma_{jL''} \, G^\alpha{}_{iL,jL''} \, (T_{aj})_{1''}{}^{\alpha'\alpha''} \, G^{\alpha''}{}_{jL'',kL'}$$

$$(T_{ak})_1{}^{\alpha''\underline{\alpha}} \, i^{l'} \, Y_{L'}(\hat{\mathbf{k}}_\alpha) + \ldots \} e^{i\mathbf{k}_\alpha \cdot \mathbf{R}_{ko}} \qquad (3.25)$$

which has to be inserted in Eq. (3.23).

As can be seen from this equation, now there are contributions coming from paths beginning at the photoabsorbing site and ending anywhere in the system, as it is obvious since the photoelectron is detected outside, in free space.

The structural analysis is more complicated than in the photoabsorption case, but can still be done and is giving its fruits[10]. The expression (3.25) incorporates the multichannel structure which can help analysing photodiffraction experiments with more completeness.

Of practical importance in the structural analysis is an accurate approximation to the exact, but computationally cumbersome, expression (2.23).

The following approximation

$$G^\alpha{}_{iL,kL'} = -4\pi \, k_\alpha \, i^{l-l'} Y_L(\hat{\mathbf{R}}_{ik}) Y_{L'}(\hat{\mathbf{R}}_{ik}) G(\rho^\alpha{}_{ik}; \alpha_{ll'}, \beta_{ll'}) \qquad (3.26)$$

where

$$g(\rho; \alpha, \beta) = [1 + \alpha/(2\rho)^2]^{1/2} \, J_0(\beta/\rho) \, 1/\rho$$
$$\exp\{i\rho[1+(\alpha/(2\rho)^2]\} \qquad (3.27)$$

with

$$\alpha_{ll'} = 2[l(l+1) + l'(l'+1)];$$
$$\beta_{ll'} = [l(l+1)l'(l'+1)]^{1/2}; \qquad \rho^\alpha{}_{ik} = k_\alpha R_{ik} \qquad (3.28)$$

gives rather accurate results for m.s. paths of low order (n=2,3,4) when compared with the exact expressions. In Eq. (3.27) $J_0(\rho)$ is the Bessel function of order zero[8,11].

The nice feature of Eq. (3.26) is the proportionality to $Y_L Y_{L'}$ which allows to close intermediate angular momentum summations through the addition theorem for spherical harmonics

$$(2l+1)/(4\pi)\, P_l(\hat{R}_1 \cdot \hat{R}_2) = \sum_m Y_{lm}(\hat{R}_1) Y_{lm}(\hat{R}_2) \qquad (3.29)$$

For example the second term in Eq. (3.25), putting the origin o at site i, becomes

$$-\sum_{kL'} G^{\alpha'}{}_{iL,kL'} (T_{ak})_{l'}{}^{\alpha'\underline{\alpha}} i^{l'} Y_{L'}(\hat{k}_{\underline{\alpha}}) e^{i k_{\underline{\alpha}} \cdot R_{ki}}$$

$$= 4\pi\, k_{\alpha'} i^l Y_L(\hat{R}_{ik}) \sum_{kL'} Y_{L'}(\hat{R}_{ik})$$

$$g(\rho^{\alpha'}{}_{ik}; \alpha_{11'}, \beta_{11'}) (T_{ak})_{l'}{}^{\alpha'\underline{\alpha}} Y_{L'}(\hat{k}_{\underline{\alpha}}) e^{i k_{\underline{\alpha}} \cdot R_{ki}}$$

$$= \sum_k i^l Y_L(\hat{R}_{ik}) \sum_{l'} (2l'+1) P_{l'}(\hat{k}_{\underline{\alpha}} \cdot \hat{R}_{ik}) \qquad (3.30)$$

$$g(\rho^{\alpha'}{}_{ik}; \alpha_{11'}, \beta_{11'}) k_{\alpha'} (T_{ak})_{l'}{}^{\alpha'\underline{\alpha}} e^{i k_{\underline{\alpha}} \cdot R_{ki}}$$

$$= \sum_k i^l Y_L(\hat{R}_{ik}) f_{eff}(\rho^{\alpha'}{}_{ik}; \hat{k}_{\underline{\alpha}} \cdot \hat{R}_{ik}) e^{i k_{\underline{\alpha}} \cdot R_{ki}}$$

where $f_{eff}(\rho^{\alpha'}{}_{ik}; \hat{k}_{\underline{\alpha}} \cdot \hat{R}_{ik})$ is an effective scattering amplitude off the atom located at site k, calculated at the angle arcos $(\hat{k}_{\underline{\alpha}} \cdot \hat{R}_{ik})$ between the vector joining the photoabsorbing site i with the scattering atom and the direction $\hat{k}_{\underline{\alpha}}$ of escape of the photoejected electron. A similar form is valid for the third term if one introduces an effective scattering amplitude off atoms located at site k and j.[12]

The expression (3.26) can also be efficiently used for computing m.s. terms like those in Eq. (3.13) for photoabsorption. We refer the interested reader to Refs. 8,11.

Until now we have simply assumed that $\rho(T_a G) < 1$ and given an argument $(|T_{ai}|^{\alpha\alpha'}{}_{LL'}| \to 0$ for $k_\alpha \to \infty)$ to show that there exists an energy regime for which this relation holds.

However, by simply considering the behavior of $\rho(T_a G)$ as a function of $\hbar\omega$ (hence of the various k_α), one can predict some general features of photoabsorption spectra.

In fact the spectral radius $(T_a G)$ is a continuous function of $\hbar\omega$ and, as already observed, goes to zero for $\hbar\omega \to \infty$. At the other extreme however, i.e. near threshold $(\hbar\omega \sim I_c)$, it is reasonable to assume that $\rho(T_a G) \to \infty$, due to the singularity of the Hankel functions $h^+_1(k_\alpha R_{ij})$ appearing in the definition (2.23) of the matrix elements of G (the product $k_\alpha h^+_1(k_\alpha R_{ij})$ goes like k_α^{-1}). Consequently $\rho(T_a G)$ must cross at least once the value $\rho=1$ in the range $I_c < \hbar\omega < \infty$. Moreover, the nearer to 1 is its value, the slower is the convergence of the m.s. series.

On the basis of this simple consideration we can therefore conclude that there are at least three regimes in a photoabsorption spectrum: a full multiple scattering regime (FMS) ($\rho(T_a G) \geqslant 1$), where a great number of m.s. paths of high order contribute significantly to shape up the photoabsorption spectrum or even an infinite number of them, depending on whether the m.s. series converges or not; an intermediate multiple scattering regime (IMS) where only a few m.s. paths of low order are relevent (typically $n \leqslant 4$) so that interatomic configurational correlations of this order are accessible; a single scattering (SS) regime where only the lowest order term of the m.s. series (n=2) is detectable and provides information on the atomic pair correlation function.

The energy extent and even the sequential order, as a function of increasing photon energy, of the regimes described above are obviously system dependent. Usually the FMS regime precedes the IMS which, in turn, merges into the SS region. This is the normal situation; however there are exceptions to this. In copper K-edge spectrum, for example, in the first ~50 eV above the absorption edge the EXAFS like $\sigma_2(\omega)$ term alone is capable of reproducing the experimental spectrum and the exact band calculation. However a substantial discrepancy shows up in the energy range 50÷200 eV, where clearly m.s. contributions of order higher that two are present[11].

This behavior can be understood on the basis of the peculiarity of the relevant atomic phase shifts that are small (modulo π, by Levinson theorem) at low energy and must cross $\pi/2$ (again modulo π) before going to zero at high energy. At the crossing $|t_{a1}|=|\sin\delta_1|\sim 1$, so that the coupling of the photoelectron with matter becomes again substantial.

Summarizing, since the magnitude of $\rho(T_a G)$ depends on the interplay between the atomic T-matrices and the structure factors G, both ingredients must be considered in discussing a photoabsorption spectrum. The bearance of the multichannel structure of T_a on the magnitude of ρ is still an interesting subject open to research.

Experimental analysis based on the preceeding considerations is confirming that structural information can indeed be obtained from the SS and IMS energy region of the spectrum[13,14]. In the FMS region the presence of many scattering paths in a limited energy range (usually 2-5 Rydbergs) makes it impossible to derive any detailed information whatsover on the various paths. However it is an empirical experimental fact that clusters of similar atoms (in the sense that they have similar scattering power, i.e. atomic phase shifts, like atoms in neighboring or corresponding positions along the periodic table) with the same geometrical arrangement give quite similar features, like fingerprints in photoabsorption spectra.

This is quite evident in molecules where these particular features have been named "cage" or "shape resonances"[15]. They afford a kind of global information about both the structure and the type of atoms participating in the resonance.

These resonances are the cluster analogues of the scattering or photoabsorption resonances which are well known in the single atom scattering case. They have been mainly asssociated with the presence of some effective repulsive potential that creates a sort of cage that traps the final state electron in a quasi-bound state decaying away with a lifetime $\tau = \hbar \Gamma_\gamma^{-1}$ connected with the tunneling probability through the barrier. In reality this is only a partial, model view of the potential resonance theory[16]. Be as it may be, these resonances, which show up as more or less sharp maxima in the cross section, are associated with a singularity of the reactance matrix k related to the atomic t-matrix by the relation (see Eq. (2.48) for the general case)

$$t_l = (1/k_0) e^{i\delta_l} \sin \delta_l ; \quad k^{-1}_l = \cotg \delta_l = t_l^{-1} + ik_0$$

$$t_l = k_l / (1 - ik_0 k_l) \tag{3.31}$$

where we have indicated by k_0 the wave vector of the electron and introduced the potential phase shift δ_l.

Since the cross section is proportional to Im t_l (see Eq. 2.55b) we find

$$\text{Im } t_1 = (1/k_0)\sin^2\delta_1 = k_0/(k_0^2 + k_1^{-2}) \qquad (3.32)$$

so that at a maximum

$$k_1^{-1} = k_0 \cotg\delta_1 = 0 \rightarrow \delta_1 = \pi/2 \text{ (modulo } \pi)$$

This implies that at a resonance k_1 is singular.

It is easy to convince oneself that quite similarly, in the cluster case, resonances are associated with singularities of the cluster K_c-matrix which can be shown to be given by

$$(K_c)^{\alpha\alpha'}{}_{LL'} = \Sigma_{kL'} \Sigma_{jL''} J^\alpha{}_{oL,kL'} (M^{-1})^{\alpha\alpha'}{}_{kL',jL''} J^\alpha{}_{jL'',oL'} \qquad (3.33)$$

where the matrix M has been defined in Eq. (2.47), and to be related to the cluster T_c matrix Eq. (2.64) by the usual relation, analogous to (3.31)

$$(T_c)^{\alpha\alpha'}{}_{LL'} = [(I - i \underline{k} K_c)^{-1} K_c]^{\alpha\alpha'}{}_{LL'} \qquad (3.34)$$

where we have introduced the diagonal matrix $\underline{k} = k_\alpha \delta_{\alpha\alpha'} \delta_{LL'}$.

The matrix K_c is hermitian, so that its real eigenvalues λ_m can be identified with the tangent of the eigenphase shifts: $\lambda_m = \tan \delta_{\lambda m}$.

Therefore in the electron molecules scattering, as in the atomic case, resonances occur whenever some eigenvalue λ_m goes to infinity ($\delta_{\lambda m} \rightarrow \pi/2$), i.e. whenever

$$\text{Det } \|M\| = \text{Det } \|(K_{ai})^{\alpha\alpha'}{}_{LL'} \delta_{ik} + (1-\delta_{ik})\delta_{\alpha\alpha'} N^\alpha{}_{iL,kL'}\| = 0 \qquad (3.35)$$

due to (3.33). Similarly for the photoabsorption case Eq. (2.55b) where the cross section is proportional to

$$\text{Im } \tau = \text{Im } (M - i\Delta)^{-1} = \text{Im}(I - iM^{-1}\Delta)^{-1} M^{-1} .$$

The sharpness of the resonance depends on how fast, as a function of energy, the eigenphase δ_m increases through an odd multiple of $\pi/2$.

Eq. (3.35) is the natural generalization to the multichannel case of the resonance condition already discussed in Ref. 16 for the one channel case. It gives the wanted, global relation between scattering power of the constituent atoms and their geometrical organization in the molecule or cluster. In the one channel case, under the assumption that the relevant atomic phase shifts are non resonating and actually depend smoothly on the energy, it leads to the rule $k_r R=$ constant, where k_r denotes the resonance wavevector and R the average coordination bond length, in molecules or clusters with identical angular geometrical arrangement but different bond length scale. This follows from the fact that the structure matrix elements $N_{iL,kL'}$ depend on energy only through the combination kR. We refer for applications and more details to Refs. 16,17.

As a final remark, we note that the condition Det $\|M\|=0$ does not entail necessarily the other condition $\rho(T_a G)>1$. Stated differently, at a resonance the m.s. series might even converge, although one is always in the FMS regime where $\rho \simeq 1$.

4. THE ONE CHANNEL APPROXIMATION AND THE OPTICAL POTENTIAL

The multichannel m.s. theory approach to the description of photoabsorption and photoemission processes in condensed matter is a relatively recent development that makes use of concepts already known in atomic or molecular physics. This approach is substantially equivalent to the configuration interaction method (Fano, Davis, Feldkamp)[18]. For the relation of this latter approach to some aspects of the many-body calculational approach, see Chang and Fano[19], although in the general case this relation can be quite involved.

As a general trend, however, the calculation of the EXAFS signal in photoabsorption and photoelectron diffraction processes in condensed matter or molecular physics has been traditionally based on an effective one particle approach, that is one particle moving in an effective (real or complex) potential. Multielectron excitations effects are added on top, so to say[6].

In this section we want to illustrate the relation of this one particle approach to the general theory of section 2 and show how the multiple scattering approach provides the unifying scheme in which to frame the different ways of solving the one particle problem.

The first reduction procedure one can think of is to eliminate in the set of equations (2.13), supplemented by the boundary conditions (2.14), all the "inelastic channels", i.e. those with $\alpha \neq \alpha'$ and such that $\Delta E_\alpha \neq 0$, in favour of the "elastic" one. This elimination is in principle possible and leads to an effective Schrödinger equation with a complex potential, which in fact describes exactly the effect of the eliminated channels. This potential is known as optical potential. The contribution of the inelastic channels to the total absorption cross section is neglected altogether. Actually, since the optical potential is quite complicated, approximate forms based on ad hoc theoretical considerations are used in practical calculations, where quite often the imaginary part is neglected.

As a further approximation one reduces the potential to a muffin-tin form, although this is done only for computational convenience. In order to keep the discussion and the notation simple we shall assume this form for the potential, so that the relations (3.5) apply. The necessary generalization of the following considerations for non muffin-tin potentials is left to the reader.

The problem is therefore reduced to the calculation of the quantity

$$\sigma(\omega) = 4\pi^2 \alpha \hbar \omega \sum_f |(\phi_f|\varepsilon \cdot \mathbf{r}|\phi_i)|^2 \delta(\hbar\omega - E_f + E_i) \quad (4.1)$$

where now $|\phi_f)$ and $|\phi_i)$ refer to one particle eigenstates with energies E_f and E_i respectively, of the effective one-electron Hamiltonian. For the final state

$$(k^2 - H)\phi_f = (\Delta + k^2 - V(\mathbf{r}))\phi_f = 0 \quad (4.2)$$

where $V(\mathbf{r}) = \sum_k V_k(\mathbf{r})$ is a collection of muffin-tin potentials and $k^2 = \hbar\omega - I_c$ is the photoelectron energy.

Three methods have been used to calculate the quantity in Eq. (4.1):

a) the scattering method, where one calculates the time reversed scattering wave function ϕ^-_k for ϕ_f, with energy k^2 and normalized to one state per Rydberg. Then

$$\sigma_{sc}(\omega) = 4\pi^2 \alpha \hbar \omega |(\phi^-_k(\mathbf{r})|\varepsilon \cdot \mathbf{r}|\phi_i(\mathbf{r}))|^2 \quad (4.3)$$

with

$$(\nabla^2 + k^2 - V(\mathbf{r}))\, \phi^-_k = 0$$

$$\phi^-_k \underset{r\to\infty}{\simeq} (1/4\pi)(k/\pi)^{1/2}\, [e^{-i\mathbf{k}\cdot\mathbf{r}} + f^*(\mathbf{k'},\mathbf{k})(e^{-ikr}/r)]$$

b) the Green's function method, whereby one transforms Eq.(4.1) as $(T = \boldsymbol{\varepsilon}\cdot\mathbf{r})$

$$\sigma_{GF}(\omega) = 4\pi^2 \alpha \hbar \omega\, (1/\pi)\, \mathrm{Im}(\phi_i | T^+ (k^2-H)^{-1} T | \phi_i)$$

(4.4)

$$= 4\pi\alpha\hbar\omega\, \mathrm{Im} \int dr^3 dr'^3\, \phi_i(\mathbf{r})\, \boldsymbol{\varepsilon}\cdot\mathbf{r}\, G^-(\mathbf{r},\mathbf{r'})\, \boldsymbol{\varepsilon}\cdot\mathbf{r'}\, \phi_i(\mathbf{r'})$$

where $(k^2 - H)G^- = I$ or in the coordinate representation

$$[\nabla^2 + k^2 - V(\mathbf{r})]\, G^-(\mathbf{r},\mathbf{r'}) = \delta(\mathbf{r}-\mathbf{r'}) \tag{4.5}$$

G^- being the Green's function operator, with incoming wave boundary conditions.

c) the band structure approach for periodic systems, whereby the scattering states are replaced by Bloch states $\phi_\mathbf{q}^n(\mathbf{r})$, where $\mathbf{q}$ indicates the wave vector in the reduced Brillouin zone (BZ) and n is the band index. Then Eq.(4.1) becomes

$$\sigma_{BS}(\omega) = 4\pi^2\alpha\hbar\omega \sum_n v/(2\pi)^3$$

$$\int_{BZ} d^3q\, \delta(k^2 - \varepsilon_n(\mathbf{q}))\, |(\phi_\mathbf{q}^n(\mathbf{r})|\boldsymbol{\varepsilon}\cdot\mathbf{r}|\phi_i(\mathbf{r}))|^2$$

(4.6)

where $\varepsilon_n(\mathbf{q})$ gives the dispersion low for the band of index n and v is the volume of the unit primitive cell.

It is not at all immediate that the three expressions can be cast into the same final form for identical systems. We shall show that this is possible in the framework of the m.s. theory.

We can obtain the result of the scattering approach by performing the necessary index reduction in Eq.(2.55a). We obtain, assuming $i \equiv o$

$$\sigma_{BS}(\omega) = 4\pi^2\alpha\hbar\omega \sum_{L_f L'_f} M_{L_f} \sum_L B_{oL_f}(\mathbf{L})\, B^*_{oL'_f}(\mathbf{L})\, M^*_{L'_f} \tag{4.7}$$

where

$$M_L = (R^\circ_L(\mathbf{r}) | \varepsilon \cdot \mathbf{r} | \phi_o(\mathbf{r})) \tag{4.8}$$

$$R^\circ_L(\mathbf{r}) = R^\circ_l(r)\, Y_L(\hat{\mathbf{r}}) \tag{4.9}$$

and $R^\circ_l(r)$ is that solution of the radial Schrödinger equation that matches smoothly to $j_l(kr)\,\mathrm{ctg}\,\delta_l - n_l(kr)$ at the radius of the muffin-tin sphere of the absorbing atom and behaves like $r^l(1 + ...)$ at the origin. It can be shown to be identical with the reduced form of the function $\underline{f}^{\alpha\alpha}{}_{LL'}$ introduced in Eq.(2.52).

If the potential $V(\mathbf{r})$ is real, by using the optical theorem (2.19) we obtain the alternative form (2.55b)

$$\sigma_{sc}(\omega) = 4\pi\alpha\hbar\omega \sum_{L_f L'_f} M_{L_f}\, \mathrm{Im}\, \tau_{oL_f, oL'_f}\, M_{L'_f} \tag{4.10}$$

where we have dropped the star on $M_{L'}$, since in this case $R_L(\mathbf{r})$ is real. This form will be useful for comparison with the Green's function approach.

The solution of Eq.(4.5) for a collection of muffin-tin potentials is given by (but see Ref. 21 for a more complete definition)

$$G^+(\mathbf{r},\mathbf{r}') = - \sum_{LL'} R_L(\mathbf{r}_o)\, \tau_{oL,oL'}\, R_{L'}(\mathbf{r}'_o) - \sum_L R_L(\mathbf{r}_o) S_L(\mathbf{r}'_o)$$
$$\text{for } \mathbf{r},\mathbf{r}' \in \Omega_o \tag{4.11a}$$

$$G^+(\mathbf{r},\mathbf{r}') = - \sum_{LL'} R_L(\mathbf{r}_i)\, \tau_{iL,kL'}\, R_{L'}(\mathbf{r}'_k)$$
$$\text{for } \mathbf{r} \in \Omega_i,\ \mathbf{r}' \in \Omega_k \tag{4.11b}$$

where $S_L(\mathbf{r}) = S_l(r) Y_L(\hat{\mathbf{r}})$ and $S_l(r)$ is that solution of the radial Schrödinger equation that matches smoothly to $j_l(kr)$ at ρ_o and is singular at the origin. We need only the expression (4.11a), since $\phi_o(\mathbf{r})$, being a core state, is localized at site o. Its insertion in Eq.(4.4) gives, since $G^- = (G^+)^*$,

$$\sigma_{GF}(\omega) = 4\pi\alpha\hbar\omega \sum_{L_f L'_f} \mathrm{Im}\{ M_{L_f}\, \tau_{oL_f, oL'_f}\, M_{L'_f} + M_{L_f}\, \underline{M}_{L'_f} \} \tag{4.12}$$

where

$$\underline{M}_L = (S_L(\mathbf{r}) | \varepsilon \cdot \mathbf{r} | \phi_o(\mathbf{r})) \tag{4.13}$$

For real potentials, since M_L and $\underline{M}_L$ are real, we recover Eq. (4.10).

Finally, in an infinite regular lattice, where for simplicity we assume all sites to be equivalent, the KKR method writes the Bloch function as

$$\phi^n_q(\mathbf{r}) = \Sigma_L \, \alpha^n_L(\mathbf{q}) \, R_L(\mathbf{r}) \tag{4.14}$$

with the same definition of $R_L(\mathbf{r})$ as before.

The coefficients $\alpha^n_L(\mathbf{q})$ satisfy the homogeneous equations

$$\Sigma_{L'} (t_1^{-1} \delta_{LL'} - G_{LL'}(\mathbf{q})) \, \alpha^n_L(\mathbf{q}) = 0 \tag{4.15}$$

where $t_1 = e^{i\delta_1} \sin\delta_1$ is the usual l wawe atomic t-matrix, common to all sites, and

$$G_{LL'}(\mathbf{q}) = (1/N) \Sigma_{ik} \, e^{-\mathbf{q} \cdot (\mathbf{R}_i - \mathbf{R}_k)} \, G_{iL,kL'}$$

$$= \Sigma_{k(\neq o)} e^{-\mathbf{q} \cdot (\mathbf{R}_o - \mathbf{R}_k)} \, G_{oL,kL'} \tag{4.16}$$

since now the second term is independent of the initial site o.

A non trivial solution of Eq.(4.15) demands that

$$\mathrm{Det} \, \| t^{-1}(\varepsilon) - G(\mathbf{q};\varepsilon) \| = 0 \tag{4.17}$$

which determines the band dispersion $k^2 = \varepsilon = \varepsilon_n(\mathbf{q})$. Correspondently Eqs. (4.15) provide $\alpha^n_L(\mathbf{q})$. Using the expression (4.14) for the final state wave function, the Eq. (4.6) gives

$$\sigma_{BS}(\omega) = 4\pi^2 \alpha \hbar \omega \, \Sigma_{L_f L'_f} M_{L_f} M_{L'_f} \, \Sigma_n \, v/(2\pi)^3$$

$$\int_{BZ} d^3q \, \delta(k^2 - \varepsilon_n(\mathbf{q})) \, \alpha^n_{L_f}(\mathbf{q}) [\alpha^n_{L'_f}(\mathbf{q})]^* \tag{4.18}$$

Now this expression is nothing else that

$$\sigma_{BS}(\omega) = 4\pi \alpha \hbar \omega$$

$$\mathrm{Im} \int dr^3 \, dr'^3 \, \phi_o(\mathbf{r}) \, \boldsymbol{\varepsilon} \cdot \mathbf{r} \, G_{BS}(\mathbf{r},\mathbf{r}') \, \boldsymbol{\varepsilon} \cdot \mathbf{r}' \, \phi_o(\mathbf{r}') \tag{4.19}$$

where

$$\text{Im } G_{BS}(\mathbf{r},\mathbf{r}') = \text{Im } \Sigma_q \Sigma_n \frac{[\phi^n_q(\mathbf{r})]^* \phi^n_q(\mathbf{r}')}{k^2 - \varepsilon_n(\mathbf{q})} =$$

$$= \pi \Sigma_q \Sigma_n [\phi^n_q(\mathbf{r})]^* \phi^n_q(\mathbf{r}') \, \delta(k^2 - \varepsilon_n(\mathbf{q})) \qquad (4.20)$$

But the function $G_{BS}(\mathbf{r},\mathbf{r}')$ is a solution of the Schrödinger equation

$$(\nabla + k^2 - V(\mathbf{r})) \, G_{BS}(\mathbf{r},\mathbf{r}') = \delta(\mathbf{r} - \mathbf{r}') \qquad (4.21)$$

which satisfies periodic boundary conditions

$$G_{BS}(\mathbf{r}+\mathbf{R}_k, \mathbf{r}'+\mathbf{R}_k) = G_{BS}(\mathbf{r},\mathbf{r}') \qquad (4.22)$$

due to the property of the Bloch states

$$\phi^n_q(\mathbf{r}+\mathbf{R}_k) = e^{i\mathbf{q}\cdot\mathbf{R}_k} \phi^n_q(\mathbf{r})$$

Such a solution is provided by the function defined in Eq. (4.11), where now $\tau_{iL,kL}$ depends only on the difference $\mathbf{R}_i - \mathbf{R}_j$ due to the periodicity of the lattice. When inserted in Eq. (4.19) this solution provides the usual result (4.10), since $V(\mathbf{r})$ is assumed to be real in band structure calculations.

The equivalence of the three approaches, just proved, reconciles the apparently different point of view of the chemist, who usually thinks in terms of wave function amplitude, with the physicist attitude, who is inclined to think in terms of density of unoccupied states. In fact Im $G(\mathbf{r},\mathbf{r}')$, for $\mathbf{r},\mathbf{r}' \in \Omega_o$, is proportional to the local projected density of states, of which a particular l character is selected when performing the weighted trace in Eq. (4.4). This equivalence is not surprising, since the presence of a potential modifies at the same time the amplitude of the wave function and the density of the available states.

When the potential is complex, there is no more equivalence between the scattering and the Green's function approach. In fact the generalized optical theorem Eq. (2.50) does not hold in this case. One must then resort to theoretical considerations to know which method to use.

The imaginary part of the complex optical potential

describes the reduction of the wavefunction amplitude of the elastic channel due to transitions to all the other channels.

As it is known in scattering theory[4], the imaginary part of the forward scattering amplitude is greater than the integral of its modulus, the difference giving the flux of particles scattered in the inelastic channels. So for the electron-molecule scattering the form (2.64) is still to be used for the total cross section, elastic plus inelastic.

In the photoabsorption process we add an electron to the ground state of the (Z+1)-equivalent atom, therefore we neeed to describe the propagation of the added electron in the presence of all the other electrons of the system. The amplitude of this propagation is the probability amplitude that the added electron remains in the original state in which it has been added to the system. Its imaginary part, as in the scattering case, gives the total probability of scattering in and out the initial state.

This propagation is described by the one particle Green's function $G(\mathbf{r},\mathbf{r}';E)$ which obeys an effective one particle Schrödinger equation, better known as Dyson equation,

$$(\nabla^2 + E - V_c(\mathbf{r})) G(\mathbf{r},\mathbf{r}';E) - \int d^3r'' \Sigma(\mathbf{r},\mathbf{r}'';E) G(\mathbf{r}'',\mathbf{r}';E) = \delta(\mathbf{r}-\mathbf{r}') \qquad (4.23)$$

where $\Sigma(\mathbf{r},\mathbf{r}';E)$ is an energy dependent, complex and in general non local, effective exchange and correlation potential, whereas $V_c(\mathbf{r})$ is the usual Coulomb or Hartree potential. Therefore in calculating the photoabsorption cross section we have to replace Eq. (4.5) with Eq. (4.23) and use formula (4.12).

Much work has gone into approximating the self-energy Σ in a way suitable for numerical applications. Hedin and Lundqvist[22], by incorporating the Sham-Kohn[23] density-functional formalism for excited states within the single-plasmon pole approximation of the electron-gas dielectric function, have produced a useful, theoretically sound, local approximation to Σ given by

$$V_{xc}(\mathbf{r}) \simeq \Sigma_h(p(\mathbf{r}), E - V_c(\mathbf{r}); \rho(\mathbf{r})) \qquad (4.24)$$

Here Σ_h is the self-energy of an electron in an homogeneous electron gas with momentum p(**r**), energy E - V_c(**r**) and density ρ(**r**), the local density of the actual physical system.

The local momentum p(**r**) is defined as

$$p^2(\mathbf{r}) = k^2 + k_F^2(\mathbf{r}) - \mu_F \qquad (4.25)$$

where k^2 is the photoelectron energy, $k_F^2(\mathbf{r}) = [3\pi^2 \rho(\mathbf{r})]^{1/3}$ is the local Fermi momentum and μ_F is the Fermi energy of the system as a whole. For molecules it should be the first ionization energy.

Since E - V_c(**r**) ~ p^2(**r**) we can write with Lee and Beni[24]

$$V_{xc}(\mathbf{r}) \simeq \Sigma_h(p(\mathbf{r}), p^2(\mathbf{r}); \rho(\mathbf{r})) \qquad (4.26)$$

To calculate Σ_h, one uses Eq.s (25.14) and (25.15) of Ref. 25

$$\text{Re } \Sigma_h(\mathbf{p}, \omega) = -\int \frac{d^3q}{(2\pi)^3} \frac{4\pi}{q^2} \frac{f(\mathbf{p}+\mathbf{q})}{\varepsilon(\mathbf{q},(\mathbf{p}+\mathbf{q})^2 - \omega)}$$

$$- \omega_p \int \frac{d^3q}{(2\pi)^3} \frac{1}{2\omega_1(\mathbf{q})} \frac{1}{\omega_1(\mathbf{q}) - \omega + (\mathbf{p}+\mathbf{q})^2}$$

$$\text{Im } \Sigma_h(\mathbf{p}, \omega) = \frac{\pi \omega_p^2}{2} \int \frac{d^3q}{(2\pi)^3} \frac{4\pi}{q^2} \frac{1}{\omega_1(\mathbf{q})}$$

$$\{ f(\mathbf{p}+\mathbf{q}) \delta[(\mathbf{p}+\mathbf{q})^2 - \omega_1(\mathbf{q}) - \omega]$$

$$- [1-f(\mathbf{p}+\mathbf{q})] \delta[(\mathbf{p}+\mathbf{q})^2 + \omega_1(\mathbf{q}) - \omega] \}$$

where the dielectric function is approximated by

$$[\varepsilon(\mathbf{p}, \omega)]^{-1} = 1 + \frac{\omega_p^2}{[\omega^2 - \omega_1^2(\mathbf{q})]}$$

$$\omega_1^2(\mathbf{q}) = \omega_p^2 + \varepsilon_F^2 [(4/3)(q/k_F)^2 + (q/k_F)^4]$$

ω_p is the plasmon frequency and $f(\mathbf{k})$ is the Fermi distribution function. A useful analytical approximation to these equations is given in ref. 6 where other approximate forms of effective potentials are discussed, like the X-α and the Dirac-Hara potentials[26].

5. CONCLUSIONS

The multichannel multiple scattering theory outlined in section 2 provides a simple, natural scheme in which to study two main problems that are still a subject of active research: the evolution from the adiabatic to the sudden regime and the interplay between excitation dynamics and structure.

In fact the nature of the crossover from sudden to adiabatic behavior is an interesting theoretical question which is not yet well understood. We refer to ref. 27 for a review discussion on this point, mainly based on the articles of Fano and Cooper[28] and Lee and Beni[24]. A more quantitative attempt is contained in the work by Chou et al[6] (but se also references therein) and in ref. 29.

In our formulation we see from Eq.(2.47) that the driving terms that control the crossover are the off-diagonal (i.e. interchannel) $(K_{ai})^{\alpha\alpha'}{}_{LL'}$ matrix elements of the atomic reactance matrices. It is not the purpose of the present notes to develop this aspect of the theory which is still a matter of investigation.

Another aspect which is clarified by the present theory is that of the relation between excitation dynamics and geometrical and electronic structure of the ground state. It is not surprising, looking at the structure of the m.s. matrix Eq. (2.47) and the resonance condition Eq. (3.35) that the general shape of a photoabsorption spectrum is determined mainly by the geometrical structure of the ground state and by the configurations present in it.

Certainly many applications are needed to establish the relative role of the various factors that contribute to a photoabsorption or to a photoemission spectrum. What was missing was a unifying interpretative scheme which we think is now provided by the multichannel multiple scattering theory.

After the completion of this notes I became aware of the

fact that the questions treated here had been addressed by Bardyszewski and Hedin in a different scheme provided by a novel perturbation theory approach, with applications to photoemission and X-ray spectroscopy[30]. Their conclusions are qualitatively similar to those presented here although further study is needed to established the relation between the two approaches.

ACKNOWLEDGEMENTS

It is a pleasure to acknowledge the stimulating collaboration of Dr. M. Benfatto during which a large part of the results presented here have been obtained.

I also want to thank Mrs. L. Invidia for her patience and valuable suggestions in editing the manuscript.

REFERENCES

1) C.R. Natoli, M.Benfatto, C. Brouder and M.Ruiz-Lopez, to be submitted to Phys. Rev. A.
2) The derivation of the multichannel multiple scattering (m.s.) equations follows very closely the method used to derive the m.s. equations for general non muffin-tin potentials in C.R. Natoli, M. Benfatto and S. Doniach, Phys. Rev. A34, 4682 (1986) to which we also refer for the derivation of the various reexpansion formulae used in the text.
3) The problem of exchange is well known among theoretical molecular physicists and quantum chemists. They have devised a wealth of methods to cope with it. For a review see N.F. Lane, Rev. Mod. Phys. 52, 29 (1980) and F.A. Gianturco and A. Jane, Phys. Rep. 143, 347 (1986) and references therein.
4) N.F. Mott and H.S.W. Massey, The Theory of Atomic Collisions (Clarendon, Oxford, 1965) 3rd Ed.
5) J.J. Rehr, E.A. Stern, R.L. Martin and E.R. Davidson, Phys. Rev. B17, 560 (1978).
6) S.H. Chou, J.J. Rehr, E.A .Stern and E.R. Davidson, Phys; Rev. B35, 2604 (1987).

7) S.J. Gurman, N. Binsted, and I. Ross, J. Phys. C: Solid State Phys. 19, 1845 (1986).
8) M. Benfatto, C.R. Natoli and M. Ruiz-Lopez, to be submitted to Phys. Rev. B.
9) G. Krill, J. de Phys. C8-907 (1986); N. Wetta, G. Krill, P. Haen, F. Lapierre, M.F. Ravet, L. Godart and F. Holtzeberg, ibidem p. 965.
10) J.J. Burton and D.A. Shirley, Phys. Rev. B32, 1982 (1985); M. Sagurton, E.L. Bullock, R. Saiki, A. Kaduwela, C.R. Brundle, C.S. Fadley and J.J. Rehr, Phys. Rev. B33, 2207 (1986); C.S. Fadley in Progress in Surface Science, S. Davison Editor, Vol. 16, pag. 275 (Pergamon, N.Y. 1984).
11) J.J. Rehr, R.C. Albers, C.R. Natoli and E.A. Stern, Phys. Rev. B34, 4350 (1986)
12) J.J. Rehr, J. Mustre de Leon, C.R. Natoli and C.S. Fadley, J. de Phys. 47, Coll. C8, 213 (1986)
13) M. Benfatto, C.R. Natoli, A. Bianconi, J. Garcia, A. Marcelli, M. Fanfoni and I. Davoli, Phys. Rev. B34, 5774 (1986);
14) A. Bianconi, A. Di Cicco, N.V. Pavel, M. Benfatto, A. Marcelli, C.R. Natoli, P. Pianetta and J. Woicik, to appear in Phys. Rev. B; October 1987. M. Ruiz-Lopez, M. Loos, J. Goulon, M. Benfatto and C.R. Natoli to be submitted to Chem. Phys.
15) J.L. Dehmer, J. Chem. Phys. 56, 4496 (1972); J.L. Dehmer and D. Dill, Phys. Rev. Lett., 35, 213 (1975).
16) C.R. Natoli in "EXAFS and Near Edge Structure", A. Bianconi, L. Incoccia and S. Stipcich Editors, Springer Series in Chemical Physics 27, 43 (1983).
17) C.R. Natoli, EXAFS and Near Edge Structure III, Vol. 2 of Springer Proceedings of Physics edited by K.O. Hodgson, B. Hedman, J.E. Penner-Hahn (Springer, Berlin, 1984) pag. 38; J. Stohr, J.L. Gland, W. Eberhardt, D. Outka, R.J. Madix, F. Sette, R.J. Koestner and U. Döbler, Phys. Rev. Lett. 51, 2414 (1983).
18) U. Fano, Phys. Rev. 124, 1866 (1961); L.C. Davis and L.A. Feldkamp, Phys. Rev. B15, 2961 (1977); L.C. Davis and L.A. Feldkamp, Phys. Rev. B23, 6239 (1981).
19) T.N. Chang and U. Fano, Phys. Rev. A13, 263 (1976).
20) C.R. Natoli and M. Benfatto, J. de Phys. 47, Coll. C8, 11 (1986).

21) J.S. Faulkner and G.M. Stocks, Phys. Rev. B21, 3232 (1980).
22) L. Hedin and B.I. Lundqvist, J. Phys. C: Solid State Phys, 4, 2064 (1971).
23) L.J. Sham and W. Kohn, Phys. Rev. 145, 561 (1966).
24) P.A. Lee and G. Beni, Phys. Rev. B15, 2862 (1977).
25) L. Hedin and S. Lundqvist, Solid State Phys. 23, 1 (1969).
26) For the X-α potential see J.C. Slater, The Self-Consistent Field for Molecules and Solids, Quantum Theory of Molecules and Solids (Mc Graw-Hill, N.Y., 1974) and K. Schwarz, Phys. Rev. B5, 2466 (1972); for the Dirac-Hara potential see S. Hara, J. Phys. Soc. Jpn 26, 376 (1967).
27) T.M. Hayes and J.B. Boyce, Solid State Phys. 37, 173 (1982).
28) U. Fano and J.W. Cooper, Rev. Mod. Phys. 40, 441 (1968).
29) J. Stohr, R. Jaeger and J.J. Rehr, Phys. Rev. Lett. 51, 821 (1983).
30) W. Bardyszewski and L. Hedin, Physica Scripta 32, 439 (1985); Proc. of the X-87 Intern. Conf. Paris 14-18 September (1987), to be published on J. de Phys. Colloques.

TRENDS OF EXAFS AND SEXAFS IN SOLID STATE PHYSICS

A.Fontaine

LURE Bat. 209d CNRS, CEA, MEN, F91405 Orsay

1 INTRODUCTION

Within recent years, the X-ray absorption spectroscopy (XAS) has become a widely used technique to determine the local atomic structure in material science as well as in physics, chemistry, biophysics, and geophysics[1]. The absorption cross section of a selected atom when embedded in condensed matter shows near edge features (XANES=X-ray Absorption Near Edge Structure) and oscillations above the edge (EXAFS=Extended X-ray Absorption Fine Structure) which contain informations about the oxidation state, the symmetry of the local environment, the partial density of unoccupied states, the coordination number and the distances of the neighbors. Almost no attempt was successful to clarify this spectroscopy until the beginning of the 70's even if famous scientists showed interests in that field[2,3]. A couple of reasons may be given. First no tractable theory was available to give a frame to analyze the Krönig oscillations (as it was termed before the 70's) existing above the K or L edges of any element present in a solid, liquid, or even in a molecule. D.Sayers' thesis achieved in E. Stern's group[4] showed for the first time that a subsequent Fourier transform yields a pseudo-radial distribution.

The second reason which precluded any development of this technique before the 70's, was experimental. No white X-ray source of reasonable intensity was available to collect data systematically. The advent of synchrotron radiation (SR) is certainly one of the major break-through of these last fifteen years. New experimental SR-oriented techniques arose which drew benefits of the dramatic gain in intensity (three or four orders of magnitude compare to a rotating anode). In addition, the broad energy band pass, the linear polarization and the collimation are tremendous extra advantages which allow angular dependent X-ray absorption spectroscopy.

The first SR X-ray sources started using the storage rings ADONE (Frascati-Italy) ACO (Orsay-France) and SPEAR (Stanford-California). Then it has been possible to reduce the data collection time per spectrum to 10 minutes instead of a few days using the bremsstralhung of a sealed tube, with in addition a significant gain in signal to noise ratio. Let us point out that the combination of a dispersive focusing optics with a position sensitive detector based on a cooled selfscanned photodiode array, which has the unique capability to work under high flux conditions, allows to measure an X-ray absorption spectrum in 2.7ms [5].

2 SYNCHROTON RADIATION[6,7]

When a charged particle (electron or positron) is circulating in a storage ring its trajectory is bent when it enters the magnetic gap. A full turn is realized when all dipoles are considered. Additional elements are implemented to keep a good optics for the electron (or e^+) beam (quadrupoles, hexapoles, octopoles,...), to inject, and to restore the lost energy to the charged particles (RF cavity). A bent trajectory is due to the acceleration induced by the magnetic field in the dipole, along the horizontal radial direction. Consequently this accelerated charged particle emits electromagnetic waves within the classical cone in its own frame. Since the electron speed in the storage ring is close to the light velocity, the cones of emission are fold in the forward direction in the laboratory frame. The photon beam is emitted tangent to the trajectory with a divergence limited to $\gamma^{-1} = m_o c^2 / E(GeV)$ where $m_o c^2$ is the energy of the electron at rest (0.511 MeV) and E its energy in the storage ring. The main characteristics of this X-ray source are the:
1-High intensity
2-Large energy band pass
3-Collimated beam
4-Polarized beam since the electric field is parallel to the acceleration
5-Pulsed structure, which must be mentioned even if only few works

have used this quality up to now. The pulsed structure deals with the obligation to confine the charges into separated bunches in order to allow injection in the storage ring and the energy restoration by the RF cavity.

The pulsed structure has been successfully used to achieve time-resolved X-ray absorption spectroscopy on the iron K-edge of carbon monoxide-myoglobin recombination after laser photolysis of 1mM MbCO solution. A time resolution of 300 microseconds[8] was achieved by synchronizing a neodynium:yttrium-aluminium-garnet pulsed laser with the bursts of X-rays emitted from the Cornell High Energy Synchrotron Source.

The characteristics of the synchrotron emission form a dipole can be improved or shifted by using spatially periodic magnetic devices inserted in the straight sections of the ring. The electron is thus obliged to follow an oscillating trajectory whose lateral amplitude is more or less large. If the magnitude of the overall deflection ψ_0 is smaller than γ^{-1}, the electron emission is kept in phase from each bending of the trajectory. This is the undulator regime. The X-ray emission is made of lines whose wavelengths are:

$$\lambda_n = \{\lambda_u / (2\gamma^2 n)\} \times \{1 + (\psi_0 \gamma)^2\}$$

where λu is the magnetic period. The width of the harmonics n varies according to $1/(nN)$ where N is the number of magnetic periods.

If $\psi_0 \gg \gamma^{-1}$ each bending emits out as an independent source. This the wiggler regime which is very alike the dipole radiation except that the characteristics wavelength is often shifted to a shorter value due to the smaller radius of curvature of the trajectory, and the intensity is improved, up to a factor hundred according to the wavelength considered and the number of poles.

Table 1

		σx(mm)	σx'(mrad)	σz (mm)	σz'(mrad)
→ESRF	Bending magnet	0.16		0.13	0.008
	Wiggler K=10 ($\psi_0 = 10\gamma^{-1}$)	0.07	0.09	0.05	0.013
	Undulator K=1 ($\psi_0 = \gamma^{-1}$)	0.41	0.015	0.084	0.007
→DCI(LURE)	B.M.	2.72		1.06	0.06
→NSLS	B.M.	0.25		0.10	0.01

The next table summarizes the RMS dimensions from the three generic sources found in a storage ring. The full dimensions are obtained easily multiplying by the classical factor 2.35 of the Gaussian distribution.

These divergences of the electron bunches have to be convoluted by the divergence of the synchrotron emission which is given by [$\gamma^{-1}=1/(1957\ E)$], E being in GeV. One can see the very promising improvement of the newly SR-dedicated storage ring. These very small divergences are rather impressive but it is worthy to recall that the acceptances of the optics (perfect crystal of Silicium) are of the order of 5 arcseconds ($\simeq 25\mu rad$) which is 10 times smaller than $2.35\gamma^{-1}$.

3 X-RAY ABSORPTION SPECTROSCOPY : BASIC PRINCIPLES

The interaction between an X-ray photon and matter is widely dominated by the absorption. Elastic and inelastic scattering cross-sections are negligible if the photon energy is smaller than 10^5 eV, which is always the case for this spectroscopy. The interaction with the nucleii requires higher energy too.

Because of the photoelectric effect the energy of the impinging photon is used to promote a bound electron of an atom to the continuum. The difference between the photon energy $h\omega$ and the electron binding energy E_i (<0) gives rise to the kinetic energy of the photoelectron $E_f = E_i + h\omega$.

Within the dipole approximation the Fermi Golden Rule yields the probability of the absorption process:

$$\sigma(\omega) = \Sigma_{i,f}\ |\langle i|\mathbf{E}\cdot\mathbf{r}|f\rangle|^2\ \delta(E_f - E_i - h\omega).$$

The photon energy is in the KeV range (0.1 up to 30) since the initial states $\langle i|$ to be considered herein are the core levels, typically 1s, or $2p^{3/2}$, $2p^{1/2}$, 2s for K or L_{III}, L_{II}, L_I edges respectively. The other electrons much less bonded contribute only to a small background which is weakly and smoothly energy-dependent and thence can be easily subtracted.

Among the three contributions in the matrix element $\langle i|\mathbf{E}\cdot\mathbf{r}|f\rangle$ the non-trivial part to evaluate in the ab-initio computation is the final state $|f\rangle$. Two cases are usually considered according to the kinetic energy of the photon electron.

3-a Characteristic times and hole-induced modifications of the final states

If the kinetic energy of the photon electron is **small**, the **relaxation** of the electronic levels can take place immediately after the hole creation in the inner shell. This hole creation is a short process ($\tau_1 \sim 10^{-19}$s) compared to the following steps involved. These steps can be schematically separated as the electronic level relaxation ($\tau_2 \sim 10-15$s<=>5eV), the decay to erase the hole ($\tau_3 \sim \tau_2$), the lifetime of the photoelectron within the absorbing atom ($\tau_4 \sim \tau_2/20$<=>$E_f=100$eV; $\tau_4 \sim 5\tau_2$<=>1eV) and the overall lifetime of the photoelectron within the condensed matter (τ_5) which is controlled by the mean free path ($E_f=100$eV$\rightarrow \lambda \sim 5$Å$\rightarrow \tau_5 \sim 10^{-16}$s, $E_f=10$eV$\rightarrow \lambda \sim 20$Å$\rightarrow \tau_5 \sim 4.10^{-15}$s). These given τs' values[9] are certainly wrong in most cases but are helpful to get an idea of the order of magnitude of the characteristic times of this spectroscopy compared to other ones. All these times are shorter than the characteristic phonon time $\tau_6 \sim 10^{-12}$s ; in contrast of the electronic levels of the absorbing atom there is no hole-induced change in position of the atomic neighbors during the photoelectric absorption.

In addition the hole is more or less screened according to the nature of the absorbing ion (existing f, or d, or p, s electrons) or/and the nature of the dipole-allowed transition. In contrast of XPS and EXAFS where the kinetic energy of the photon electron is **large** (**sudden approximation** τ_4 and $\tau_5 < \tau_2$ and τ_3) this **adiabatic regime** ($\tau_2 < \tau_4$) gives us final states which can be shuffled compared to the order existing in the ground state. The importance of this phenomenum in the Near Edge core hole spectroscopy is discussed extensively in the Fuggle and Sawatzky's lectures. This effect is recognized to be of particular importance in rare-earth compounds where the narrow f band is closed to the Fermi level[10]. Generally speaking all the "quasi intra-atomic" final states of the inner shell spectroscopy is sensitive to the hole creation and there is no straightforward interpretation of the edge features.

In addition as well-known the electronic levels close to the Fermi levels are sensitive to the local environment, creating ionic bonds or covalent ligands with the orbitals of the neighbors. The X-ray absorption spectroscopy probes the local projected density of states. Using the scattering language specifically for the low kinetic energy range (the Xanes regime) the photoelectron has a large mean free-path and thence is allowed to be scattered more than once by the nearest or next nearest neighbors. This is very important since it gives access to correlation function of order higher than 2. The multiple scattering theory describes the motion of the final state photoelectron in an effective, multicenter, energy-dependent potential. A clear presentation of this part of the spectrum is developed by C.R.Natoli in this volume.

A practical way to separate Xanes from the Exafs domain is to compare the wavelength of the photoelectron to the distance between the absorber and the backscatterer. A short wavelength means Exafs whereas long wavelengths call for multiple scattering. A convincing demonstration of the splitting of the X-ray absorption spectrum in two domains, XANES and EXAFS according to the kinetic energy of the photoelectron has been given in manganese compounds[11]. Such an homologous discrimination between the "quasi intra-atomic" final states and the solid state-induced final states is not worthy to seek in the Near Edge Spectrum but specific examples can be found where one of the two aspects is clearly dominant. In any case the present paper will include none of these two aspects which are covered in this school by the real experts.

Thence the following theoretical discussion will be limited to the Exafs formalism needed by the experimentalist even if many fruitful informations from the Xanes spectra will be used as complements of Exafs data in the sections dedicated to the experimental results.

3-b <u>Exafs formalism</u>

Two different ways of calculating the final state in the Exafs regime has been developed:
i/ **the band structure approach** requires a triple periodicity and then is restricted to ordered solids whereas
ii/ **the scattering formalism** takes |f> to be a solution of the excited potential and improves it systematically including the single backscattering from neighboring atoms, the spatial variation of the scattered wavelets, (and multiple scattering events as needed to account for the low energy range, Xanes). Unlike the band structure approach calculation, the scattering approach is of general applicability; it covers disordered condensed matter, clusters and molecules as well[12].

Of particular importance is the fact that, except in unusual circumstances, multiple scattering does not contribute much in the high energy part of the spectrum ≳50-100 eV. Then the Exafs data extends over a large domain. Compared to other diffraction techniques (neutrons and X-rays) this is a strength. The fact that the initial part of the spectrum is not relevant of the same mathematical analysis limits the power of this technique to determine accurately long distances but gives other benefits. The Exafs modulations (figure 1) $\chi(E)$ are extracted from the X-ray absorption cross-section using:

$$\sigma(E) = \sigma_0(E) [1+\chi(E)]$$

where $\sigma_0(E)$ is the free atom-like cross section and E the photoelectron kinetic energy. E is related to the energy of the impinging photon by $E=\hbar\omega-E_0$ where E_0 is the origin of the kinetic energy of the photoelectron. E0 is close to the threshold energy, but does not have to coincide with this value. For a free electron metal E_0 should be close to the bottom of the conduction band as has been shown for pure aluminum and magnesium[13].

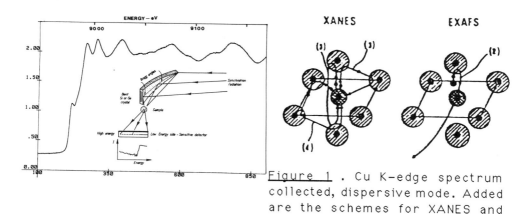

Figure 1. Cu K-edge spectrum collected, dispersive mode. Added are the schemes for XANES and EXAFS.

Thus in the frame of the plane wave approximation if multiple scattering effects and multiple electronic-absorption process are neglected, the final state is merely a superimposition of the outgoing spherical wave and the wavelets backscattered by the neighbors. This gives rise to interferences which are constructive or destructive according to the optical path. One approximates the solids as localized potential wells with flat plateau inbetween. This is the muffin-tin approximation. Refinements of the model should consider possible anisotropy of the potential to account for the ligand field. But in most practical case it is not a strong point as commented later.

Thence the commonly and practical model gives:

$$\chi(k) = -\frac{1}{k}\sum_j \frac{3\cos^2(R_j,E)}{R_j^2} \exp-2\sigma_j^2 k^2 \exp(-2R_j/\lambda(k)) |f_j(\pi,k)| \sin(\Psi(R_j,k))$$

$$\Psi(R_j,k) = 2kR_j + 2\delta_1(k) + \arg(f_j(\pi,k)) = 2kR_j + \Phi_j(k)$$

where the momentum k is given by $E = hk^2/2m$, the position of the backscattering atom: R_j, $\lambda(k)$ the electron mean free path in the material. δ_1 is the phase shift experienced by the photoelectron during its travel in the potential of the absorbing atom.

$f_j(\pi,k)$ is the backscattering atom which is a complex number: both its amplitude and its phase allow the chemical identification of the backscattering atom.

The first exponential ($\exp-2\sigma_j^2 k^2$) is the usual Debye-Waller damping which accounts for the thermal vibration in the harmonic approximation.

Since SR is highly polarized in the orbit plane the dot product of the dipole hamiltonian generates the square of the cosine of the angle between R_j and E the electric field. The amplitude of the outgoing wave is thus proportional to $\cos(R_j \cdot E)/R_j$. If the probed atom is in a cubic symmetry $\Sigma_j 3\cos^2(R_j \cdot E)$ is usually replaced by $\Sigma_j N_j$ where the index j labels now the successive neighboring shells instead of the individual atoms. In order to keep the practical shell description of the local ordering, an effective number $N_j^* = \Sigma_i 3\cos^2(R_i \cdot E)$ of atoms is often used, the subscribe i running for all the atoms within the shell j. Surfaces are obvious anisotropic system. Thence one this polarization-dependent spectroscopy is a very powerful tool, able to determine what site is involved in adsorbate-surface interactions as illustrated below.

This polarization dependence is even more attractive when one deals with the Xanes regime where the large electron mean free path at low energy allows multiple scattering events or longer paths to direct backscattering. In this domain close to the edge it can be profitable to think in band structure, and then XAS probes the empty orbitals. For example, the CO molecule has been shown to sit perpendicular to the

Cu (100) surface [14] looking at the angular dependence of the π*, and σ* resonances at the oxygen K-edge using XAS or Electron Energy Loss Spectroscopy. The first one is maximum when the electric field is parallel to the surface whereas the second one is minimum. The reverse is progressively reached when the sample is turn by 90°.

3-c Phase-shifts and backscattering amplitudes

An ab-initio fitting routine needs the knowledge of phase-shifts and backscattering amplitudes to perform the usual comparison between the experimental data and the predicted spectrum yielding as output the structural informations which are sought. The analysis of the first generations proceeds with theoretical values calculated within the mentioned-above approximation: one electron process giving rise to a plane wave travelling in the matter discribed as a muffin-tin potential, keeping only single-scattered wavelets. It means that this theoretical calculations ignored the chemistry of the solids. Fortunately the energetic photoelectron takes most of its phase-shifts in the central part of the atomic potential and consequently is not too much sensitive to the actual chemistry. This is less true for the amplitudes the theoretical calculations of which incorporates more phase shifts.

This is not a real problem. It was soon recognized that the best way to know accurately the phase-shifts and backscattering amplitudes is to measure them in a model compound, the structure of which are well-known and chemically close to that of the unknown sample under investigation. The simplicity and consistency of this approach alleviates three major sources of errors:

1-The plane wave approximation is not correct and spherical wave corrections are energy-dependent. Using model compounds with similar bond lengths cancel this approximation.

2-The unaccuracy in the choice of the origin of the kinetic energy results in unaccuracy in the phase-shifts ($\delta\Psi \sim (\delta E/k)*R_j$). If for metals one can get a crude idea to locate the extrapolated bottom of the parabolic-like partial density of states, it is clearly impossible for most of the materials. Thence, locating E_0 somewhere close to the edge of the model compound and keeping that energy scale during the analysis of the data of related samples under investigation outweigh again the magnitude of the E_0-related uncertainty.

3-The muffin-tin approximation ignores the chemistry of the real system. If the ligand field of the model compound is close to the investigated sample the chemical aspects including the local anisotropy is included in the phase-shifts and backscattering amplitudes extracted and transferred from the model compound.

Thence the use of model compound EXAFS data is essential for reliable structural parameter determinations. But the constraint of the chemical closeness of the model compounds does not tightly bound the experimentalist. Experimental values of phase-shifts and backscattering amplitudes of model compounds can be used for backscattering and absorbing atoms with theoretical corrections if the atomic numbers Z differ by $\leqslant\pm2$ from those in a more convenient and reliable model compound. These corrections are usually small, and any systematic errors in the theoretical values are even smaller due to cancellation. As a general idea Ψ is 80% due to the optical path, 15% due to the $2\delta_1$ absorber contribution, and only a small 5% for the backscatterer. A loss of precision in this last term does not kill the data analysis.

3-d Available structural informations

Once more let us recall that Exafs is a selective tool to investigate the local atomic order around the atom which has been chosen according to its binding energy of the core electrons. The promotion of the photoelectron creates a spherical electronic wave emitted by that specific atom pinned up by the choice of the photon energy. In other words one deals with low energy electron diffraction in a spherical symmetry with the source of electron inside the sample itself.

The **bond lengths** which intervene in the sine term via the optical paths are clearly evidenced by a **Fourier transform**. Excellent determination of phase shifts from a standard the structure of which is close to the examined sample is very often available. Then bond lengths in crystalline systems or molecules with high symmetry are determined currently with a **0.01Å accuracy**.

Chemical identification of the neighbors proceeds with a more sophisticated analysis. But both the energy-dependence of the amplitude and even better the phase of the signal are specific of one atomic specy. Differentiation between two possible types of neighbors is found easy if the atomic numbers differ by more than a few units. The phase of the backscattering of amplitudes of two atoms (Al, Zn for example[15] see section5-b) can be in phase opposition over almost the full range of energy. In that case Exafs is very sensitive to **coordination numbers** if both atoms are two possible neighbors.

The **Debye-Waller** factor yields a limited number of papers. As anyone of our community of experimentalists would have guessed, the metallic copper foil, which is our canonical sample, has been investigated in a rather sophisticated manner by Greegor and Lytle[16] and it was probably the first one which addressed a fundamental comparison between theories and experimental data. It had been probably the only one for a long time. But recently a smart paper investigates the anisotropy of the thermal disorder of surface atoms, a monolayer of cobalt, deposited on copper(111). This study will be briefly reported in the Sexafs section.

Of particular importance is the **limited value of the mean free path** of the photoelectron. Then the Exafs signal contains only informations on the closest neighbors. It can be seen as a bad story since only limited informations are available but on the other hand the available informations are extracted clearly since they are not merged into a too large set. On the same line of reasoning one must point out again the large extension of the signal in the k-space which can easily reach $16 \text{\AA}^{-1}$, a large figure compared to other diffraction techniques.

4 VARIOUS SCHEMES OF DATA COLLECTION

In most cases the data collection proceeds by the **classical transmission** mode. A monochromatized beam intensity is measured before and behind the absorbing sample, using ion chambers. Both values I_0 and I_1 are recorded versus the photon energy which is scanned **step by step**. $LOG(I_0/I_1)$ yields the cross-section. For each point of the spectrum, a dead time is dedicated to allow the mechanical movement and stabilization of both mechanics and electronics since the constant time should be large enough to yield a noise-reduced of currents in the typical range $10^{-12} \rightarrow 10^{-8}$ amps. In total each data point needs about one second, half for the dead time and half for the effective data collection. It results in a very good signal to noise ratio since each measurement I_0 and I_1 is made out of 10^9 photons in the usual case.

An alternate scheme combines a dispersive optics *-which reflects a polychromatic X-ray beam and focuses it on the sample-* with a cooled photodiode array *-which is a position sensitive detector able to work under high flux conditions-*. The triangle-shaped crystal is bent in order to change continuously the Bragg angle along the footprint of the beam on the crystal. The output of such an X-ray optics is a polychromatic beam, but with a correlation between the energy and the direction of propagation. This correlation transforms in the detector into a energy-position correlation. Hence after tuning the average incident angle to the Bragg angle of the energy of the edge, one can collect all the data at once, drawing benefit from the superior qualities of the silicon-based detector. The advantages **of the dispersive mode**[5] over the step by step scan comes from the lack of mechanical movement and the collection of the full set of data at once which gives an extreme stability to the energy scale which subsequently gives an extreme sensitivity for small change in a series of closely related samples. In addition the sample can be very small ($<1mm^2$) which makes for example high pressure experiments[17] with a diamond anvil cell feasible. The fast and simultaneous acquisition opens the field of kinetic experiment at the 100 millisecond-1second time-scale[18]. Obviously this scheme cancels any possibility of detection of the X-ray absorption signal via the intensity of the decay process and thus does not permit investigation of ultra-dilute samples.

The detection of **the decay process** offers an alternative way to measure the probabilty of the creation of the photoelectron. The two main routes of the system to relax to the fundamental state (fluorescence and Auger emissions) are independent of the energy of the photon used to create the core hole. Thus the intensities of the decay processes image only the energy-dependent probability to create the hole. They are of importance for dilute systems which can be typified by biophysical samples and adsorbates on surfaces as well.

The **fluorescence yield** (FY) technique has been preferred generally for thick and dilute samples, since in the X-ray range the absorption length of the fluorescent photon is of the order of a few µm. A tremendous advantage comes from the lack of background if one uses an energy-selective detector. But such a detector, generally covers a limited solid angle and then is unappropriated to collect photons emitted within 4π. The game to play is then to cover a solid angle as large as possible (a large fraction of 2π since only one side of the space is available). The outmost realization[19] uses Si-photodiodes operated in the photovoltaic mode at -120°C. Four large Silicon sheets (developed as multistrip detector for the high energy physics) cover the walls of the cryostat. This realization offers several advantages over other fluorescence detectors which are mainly ion chambers of special design: high efficiency, perfect linearity of the response, vacuum compatibility for soft X-ray measurements.

Owe to the low mean free path of the electrons (between 5 to 500 A according to their kinetic energy) the **electron yield** detection is a surface sensitive method which has been regarded generally only possible in ultra high vacuum. If the electron detector is tuned to the narrow energy window of the Auger electrons **(AY)**, the Sexafs signal is coming from the upper part of the sample. If one wants to investigate a deeper part of the sample one can include in the signal secondary electrons which are created by the Auger's ones after "collisions" due to the limited mean free path of the electrons in matter. In case of total electron yield **(TY)** the signal comes from the first 300Å typically. Often partial electron yield is preferred **(PY)** since, in most cases one deals with a monolayer or less of adsorbates on a surface of another materials. Thence, there is no more ambiguity, the signal is coming from the true upper part of the sample since the X-ray energy is tuned to the characteristic edge of the adsorbate.

Kordesch and Hoffman[20] have used recently an helium atmosphere to convert the energetic Auger electrons (5-6 keV.) emitted from an X-ray illuminated sample. Then these bunches of secondary electrons are collected on a biased electrode. We shall illustrate this development which is currently used at LURE[21] to study thin films, ion-implanted materials··· The advantages are related to the simplicity of the technology involved as well as the capability to investigate the top of materials in its real world.

Belonging to the same class of tools to study thin bidimensionnal samples is **REFLExafs** which allows the investigation of a buried interface for example. A section will be devoted to thin films or thin layers with the numerous experimental schemes known nowadays.

Other Exafs-like techniques

Numerous facsimile techniques have emerged in that time whose goals are specifically geared to match or approximate those of EXAFS and more particularly SEXAFS. I would like to point out the angle-resolved X-ray photoelectron diffraction which received different acronyms. Let us use the simpler one, XPD, even if some tiny differences can be found to discriminate one from another. In XPD[22] the oscillatory behavior of the differential cross-section $d\sigma/d\Omega = |<i|E.r|f(k)>|^2 \delta(E_f(k)-E_i-\hbar\omega)$ where **k** is the momentum of the photoelectron. When $d\sigma/d\Omega$ is integrated over all possible directions of the final photoelectron state one must obtain the total cross section and consequently the formula discussed in the previous section.

Five different Exafs-like phenomena[23] have been described in the literature which use electron as the primary particle to interact with matter. They give rise to these rude acronyms: SEELFS Surface Extended Energy Loss Fine Structure, EXELFS Extended Energy Loss Fine Structure, EAFPS Extended Appearance Potential Fine Structure, EXBIFS Extended X-ray Bremstrahlung Isochromat Fine Structure,

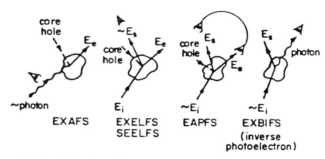

Figure 2. Scattering diagrams for Exafs-like phenomena.

EXFAS Extended Fine Structure in Auger Spectra. I borrow once more to the Ed Stern's paper the schematic scattering diagrams to give a short presentation of these two first Exafs-like phenomena which have been used for real investigations in contrast with the three other ones which remain themselves objects of studies.

In EXAFS an incoming photon is absorbed, ejecting a photoelectron from a core level into |f> with a kinetic energy E_e leaving a core hole in the atom. The measure deals with the number of absorbed photons versus their energy (symbol~). For both SEELFS and EXELFS an incoming electron of **fixed energy** E_{inc} interacts with an atom pulling out a core electron into a final state |f> with energy E_e, again leaving a core hole behind. The number of primary electrons which undergo this inelastic scattering is measured and varies according to the value of E_e

5 ILLUSTRATIONS OF SPECIFIC FIELDS OF APPLICATIONS

The variety of applications is now too large to allow any extensive report. Thence the choice of the specific fields is somehow arbitrary.The two driving forces of the science discussed in the following sections will be the simplicity of the problems which were tackled and my personal proximity to the subject. The reader should find a practical way to find informations on his/her own field by reading through the proceedings of the four International Conferences hold to cover this spectroscopy. Thence I apologize to the readers for hiding fundamentals results in catalysis, in disordered condensed matter, in biophysics···

5-a Elastic core effect around a solute in solid solutions

According to the Vegard's law, the average lattice parameter "**a**" of a substitutional solid solution varies linearly with concentration. The variation of the parameter is a macroscopic effect due to the introduction of solute atoms which are point defects dispersed at random into an elastic matrix. This solute atoms cause a local displacement of the first neighbor atoms of the matrix. This positive or negative core effect provokes an overall change of the lattice parameter which is measured from the shift of the Bragg peaks. If this last measurement is easy to achieved, the measurement of the elastic core effect had been always out of the realm of direct investigations until the advent of Exafs. The displacement of the matrix atoms neighbor of an isolated solute atom has been measured for three aluminum solid solutions and yielded the change δb in b which is the radius of the first shell. Table 2 summarizes the experimental values and compares them to δb values calculated from the changes in lattice parameter δa as measured from the shifts of the Bragg reflections and describing the matrix as a pure elastic medium as proposed by Eshelby[24].

Table 2

Solid solution	δb (Exafs)	δb(Exafs)/δb(Elastic med.)
Al-Mg	0.08 Å	2.0
Al-Cu	-0.125 Å	3.1
Al-Zn	-0.02 Å	2.9

The large discrepancy (factor 2 or 3) between the experimental and calculated values should be assigned to the charge screening of the solute atoms by the conduction electrons of the matrix to compensate for the difference in valency between the solute and the matrix atoms. This extra localized charge relaxes as Friedel's oscillations spread over several shells around the impurity. Thus an additional force generated by this charge-induced electric field acts on matrix ions has to be included in the relationship between δb and δa. This consideration was the basic motivation to undertake investigations in isovalent binary alloys. We studied Cu-Ag where as Cu-Au was measured independently. The two solute Ag and Au have similar atomic radii. They give similar δb values which are only 50% larger than the δb value expected from the relative change of lattice parameters treating the matrix as a continuous medium. This is not a definitive answer about the importance of Friedel's oscillations, but it is a reasonable clue to encourage further investigations in order to get idea about magnitude and phase of this oscillating charge relaxation at the next neighbor position. These measurements were carried out at the very beginning of the Exafs development. The improvements of the fluorescence signal measurement should extend the number of dilute solid solutions which can be studied.

In the transmission mode optimized signals of dilute systems are obtained when the atomic number of the dilute specy is higher than the average Z number of the matrix. Thus biological samples offer good systems. This favorable combination was also found in the aluminum-based alloys which was our first domain of interest.

5-b Exafs Sensitivity to clustering

This property was checked in many systems including catalysts which explain the interest of the major oil companies to the X-ray absorption spectroscopy. The capability to find surface segregation in small particles of binary alloys totally miscible in bulk and the findings of substrate-induced shapes for these particles were the basic elements which initiates dozens of studies on catalysts.[25]

The Al-Zn system is chosen here for its capabilty to illustrate one positive aspect of Exafs, namely the Exafs capability to distinguish between backscattering atoms because of their phase-shifts which can differ by π. Chemical signatures provided by the complex backscattering factor is one of the key-advantage of this structural technique over electron X-ray scattering. In fact this phase shift-related advantage appears erratically in neutron scattering when the experimentalist focus on specific systems such as CuTi where lengths of diffusion are of opposite signs, or when it is possible to use isotopes like Ni ones, to cancel the nuclear cross-section. For neutron scattering this opportunity is limited to dedicated systems. It is more widely encountered in Exafs and will be illustrated herein with the Al-Zn systems.

The metastable Al-Zn solid solution decomposes into Guinier-Preston zones spread within the depleted matrix, when the initial concentration is within the misciblity gap. Hence Zn atoms are embedded in Zn-enriched fcc lattice, when the Al-6.8at.%Zn solid solution is annealed for 50 days at room temperature.

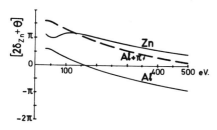

Figure 3a. Phase shifts of the atomic pairs Φ_{Zn-Al} Φ_{Zn-Zn}

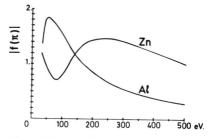

3b. Al and Zn backscattering amplitudes.

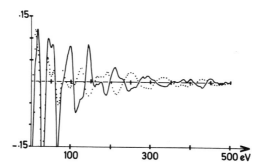

Figure 4. Zn spectra for solute atom dispersed at random (solid line) for Zn in well-formed G.P. zones appearing at the end of the annealing process. dotted line)

The GP zone composition reaches 80% as determined by Exafs[15], thence the first shell around Zn is mostly made out of zinc atoms which yield the Exafs spectra of fig 4. The dotted line is the Exafs spectrum of the stable Al-0.83at.%Zn solid solution. In that case Zn atoms are surrounded by almost 12 Al atoms. The phase opposition between the two spectra is clearly evidenced. It is related to the difference close to π between the two arguments of the complex backscattering amplitudes of aluminum and zinc (fig3a).

Hence a progressive substitution within the first shell around Zn central atoms of aluminum neighbors by zinc neighbors should first decrease the amplitude of the Exafs signal. Indeed " a beat" in the K-space is observed which is a point of zero amplitude which separates the low k domain whose phase-shifts is imposed by the aluminum atoms and the high k range which exhibits oscillations with the phase-shifts of zinc. This beat node shifts from the higher energy range to the lower energy part of the spectrum according to the increasing number of zinc inside the first shell which is the result of clustering. Clustering trends appear because of increase in initial concentration or/and annealing of metastable solution. The first peak of the Fourier transform of the as-quenched metastable Al-4.4at.%Zn solid solution is well-fitted by mixing 10.5 Al atoms with 1.5 Zn atoms in the first shell (fig.5 , curve 2). The curve 3 (dotted line) shows a good fit of the Fourier transform of the as-quenched metastable Al-6.8at.%Zn solid solution. The curve 4 (crosses) exhibits the surprising splitted peak which is the theoretical FT of an Exafs-simulated spectrum mixing 7 Al atoms an 5 Zn atoms in the first shell. This was experimentally observed when clustering is advanced (the metastable Al-4.4at.%Zn solid solution annealed at room temperature for 12 days or the metastable Al-6.8at.%Zn solid solution annealed at room temperature for 90 minutes). More generally all the investigated supersaturated alloys whose composition ranges from 4.4 at.% to 15 at.% decompose via a common sequence of Exafs spectra with different time scales obviously[26].

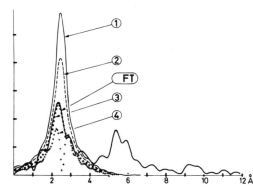

Figure 5. Fourier transforms of theoretical spectra compared to the Al-6.8at.%Zn one.
1- 12 Al at 2.83Å
2- 10.5 Al at 2.83Å and 1.5 Zn at 2.81Å
3- 8.75 Al at 2.83Å and 3.25 Zn at 2.81Å
4- 7 Al at 2.83Å and 5 Zn at 2.81Å

In this particular case where the system is well-conditioned to yield a significant signal, namely a heavy atom in a light matrix plus a π difference in the phase-shift, Exafs reveals a powerful ability to determine the coordination number with an accuracy of the order of 0.25 atom. In addition the first shell radii are currently measured with a 0.01Å precision. Such pairings suited for the Exafs point of view were a clear benefit in Sexafs investigation of Ni or Co on Silicium.[27]

5-c Pseudobinary alloys and virtual crystal approximation

Covalent pseudobinary alloys ($Ga_xIn_{1-x}As$) with the zincblende-typed structure have been investigated to confront the approximation of the virtual crystal to the contradictory notion of invariant covalent radii introduced by L.Pauling. For anion-cation (b* distance, anion-anion, cation-cation distance determination, three spectra were available[28,29] in addition of Bragg diffraction giving the average lattice parameter with respect to x. Such a study has been extended to ionically bonded compounds with the rocksalt structure ($K_xRb_{1-x}Br$ and $RbBr_xI_{1-x}$). For clarity we present the results for the first covalent alloy, but similar conclusions are drawn when the substitution takes place within the anion sublattice.

1- The first neighbor distribution is rather simple and does not differ from a pure binary system when one looks from one of the cations. It consists of 4 anion neighbors at a well-defined distance close to that one of the pure binary compound. But a small difference exists which is composition-dependent. Arsenic is at the center of 5 possible tetraedra according to the number of Ga (or In) at the apex of this geometrical unit with defined Ga-As and In-As distances. It is assumed that the cations sit exactly on the sites of the fcc sublattice of the zinc blende structure whereas As is off-center in the intermediate tetrahedra where at least one of the cation differs from the others.

2- The second neighbor distributions are more complex. Close anions are bounded via a cation which can be one or the other. The anion-anion distribution is bimodal with As-Ga-As and As-In-As distances close to those in pure GaAs and InAs, respectively. The cation-cation distribution consists of a single broadened peak at the virtual crystal distance.

3- For $Zn(Se_x,Te_{1-x})$ which is more ionic and which has anion substitution rather than cation substitution, the situation is almost identical. For the rocksalt-typed ionic compounds the results follow a similar scheme but a little more complex because there are 6 neighbors in the first shell instead of 4.

4- The main difference between covalent and ionic alloys is that the composition dependence of the cation-anion distance is larger for the ionic case. For covalent alloys the total variation δb^* **a of b*** is about 20% of that given by the virtual crystal geometry dbv. In the ionic case it is about 20%. (db_V is merely the difference between the two cation-anion distances). This difference has been reasonably explained considering compressibility[28]. It cancels in similar metallic solutions.

5-d $\chi(fcc)$-Fe and $\alpha(bcc)$-Fe clusters in two Cu-based matrix as evidenced by Xanes[30]

Iron clusters were precipitated from two supersaturated copper-based matrix by annealing $Cu_{98}Fe_2$ and $Cu_{67}Au_{30}Fe_3$ for 120 minutes at 620°C, and $Cu_{69}Au_{30}Fe_1$ for 2 days at 540°C. Melting and annealing processes were done in a quartz tube under H_2-atmosphere. The

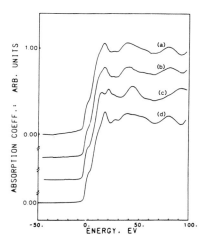

Figure 6a : Fe K-edge XANES spectra :
(a) $Cu_{67}Au_{30}Fe_3$ (Fluorescence),
(b) $Cu_{69}Au_{30}Fe_1$ (F),
(c) $Cu_{98}Fe_2$ (F),
(d) pure Fe (Transmission).

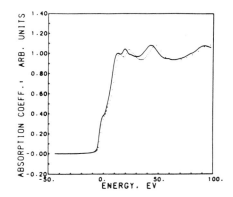

Figure 6b : Solid line : Fe K-edge XANES for $Cu_{98}Fe_2$ (Fluorescence).
Dotted line : Ni K-edge XANES for pure Ni (Transmission).

equilibrium solubility of Fe in Cu which is 3 at.% at 1060°C drops sharply with decreasing temperature and is only 0.5at.% at 600°C.

The Xanes and the Exafs spectra for the CuAuFe alloys are quite similar to the spectra of α-Fe which seems to reflect the clustering of ferromagnetic bcc Fe (fig.6a). On the other hand the Xanes and the Exafs spectra for $Cu_{98}Fe_2$ closely resembles those of fcc Ni, strongly indicating the environment of antiferromagnetic fcc γ-iron (fig.6b). This is in agreement with recent ^{57}Fe-Mössbauer measurements made at 4.2K: iron in atoms in $Cu_{69}Au_{30}Fe_1$ and $Cu_{67}Au_{30}Fe_3$ are dominantly (>80%) ferromagnetic, whereas iron atoms in $Cu_{98}Fe_2$ manifest antiferromagnetic interactions. When the lattice parameter changes within the same structure (fcc for the Cu-based the matrix), the change is known to result in energy shifts of features in the corresponding Xanes. However such a drastic difference in electronic structure, as shown herein by Fe Xanes spectra is not expected to result from a mere change of lattice parameter. Fourier transforms of the Exafs spectra confirm the conclusions yielded by Xanes

6 TIME-RESOLVED SPECTROSCOPY:
In-Situ OBSERVATION of ELECTROCHEMICAL INCLUSION of METALLIC CLUSTERS within a CONDUCTING POLYMER

Conducting polymers represent an important class of materials in the fields of catalysis and energy storage, from both a fundamental and technological point of view. TOURILLON et al.[31] reported catalytic properties of poly-3-methylthiophene (PMeT) loaded with Cu and Pt aggregates. The size of the aggregates and the polymer-cluster interactions are obviously important to control the catalytic efficiency. An appealing feature of polythiophene is its good stability against moisture and oxygen[32] and its very high purity since it is synthesized electrochemically without any catalyst.

The growth and the interaction of polymer-supported metal clusters was determined by (EXAFS), and (XANES) via informations on oxidation states, coordination geometry, and bonding angles[33]. The dispersive scheme offers the possibility of **in-situ investigations** of the electrochemical inclusion of copper into PMeT using specifically:
-i) the small size of the x-ray beam to probe the PMeT grafted on a Pt wire spatially .
-ii) the fast data acquisition to follow the time-dependent evolution of the copper clusters in the polymer.

6-a XANES approach of the electrochemical inclusion

Information on the copper ionic valency is contained in the near-edge spectra as shown in fig. 7 for Cu^{2+}, Cu^{1+}, and metallic copper. Two sets of spectra were taken to characterize the electrochemical steps which lead to
(i) the complexing of Cu^{1+} ions by the polymer
(ii) the formation of metallic-copper aggregates in PMeT.

The evolution of the copper K-edge spectrum was followed versus the cathodic polarization time. The kinetics of the inclusion processes were determined from the time dependence of Cu^{2+}, Cu^{1+}, and Cu^0 concentrations. These concentrations were obtained by deconvolution of the near-edge spectrum. These experiments have been achieved by use of a triangular-shaped bent Si(311) dispersive crystal in order to get superior energy resolution[34].

The grafting of poly-3 methylthiophene (PMeT) on the Pt wire involves the oxidation of the monomer, 3-methylthiophene, 0.5M in CH_3CN + 0.5M $N(C_4H_9)_4SO_3CF_3$ at 1.35 V/SCE (saturated calomel electrode). The polymer is formed directly in its doped conducting state. This modified electrode is put in a 3 mm-thick electrochemical cell composed of a Teflon ring covered by two Kapton windows. An aqueous copper solution (H_2O + 50 mM $CuCl_2$, pH = 6) is then added into the cell. The grafted electrode and another Pt wire are used to develop the copper inclusions. Since the sample attenuates the X-ray beam by a factor of about 200, the collection time for each spectrum was chosen as long as 3.6 s. Data acquisitions were spaced 7.2 s apart, twice the acquisition time of each spectrum.

Fig. 8 shows a series of spectra of the near-edge region of the Cu K-edge. A rapid shift of the copper edge towards lower energy (by 8.5 eV) appears, in addition to a small bump in the rise of the absorption just at the edge. It clearly shows that the Cu^{2+} ions transform into Cu^{1+} ions for the first step of the reaction. This was checked by the formation of a red complex when the material was treated with a bipyridine solution. Conversely, a blue complex is observed with the Cu^{2+} ion. Since Cu^{1+} ions are unstable in aqueous solution, they must be stabilized by the dopant or the polymer.

If the cathodic polarization is continuously applied, the changes in the XANES spectrum are consistent with the evolution of the system to reduced species. The bump in the rise of the absorption is shifted slightly from a value consistent with Cu^{1+} to the Cu^0 value. Also, emerge the first two oscillations, of the metallic spectrum which are separated by 9 eV. Fifteen minutes after starting the polarization, characteristic oscillations of bulk metallic copper can be clearly seen (fig. 8). These inclusion process are fully reversible. Reversing the potential to +1 V leads to Cu^0 cluster dissolution into Cu^{1+} ions, followed by reappearance of Cu^{2+} ions.

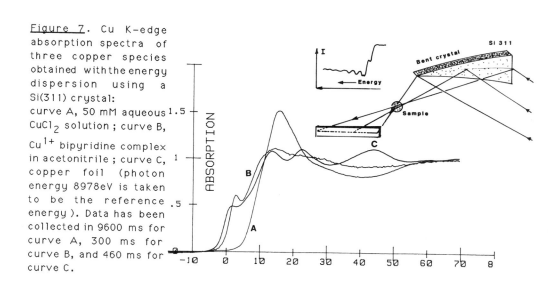

Figure 7. Cu K-edge absorption spectra of three copper species obtained with the energy dispersion using a Si(311) crystal: curve A, 50 mM aqueous $CuCl_2$ solution; curve B, Cu^{1+} bipyridine complex in acetonitrile; curve C, copper foil (photon energy 8978eV is taken to be the reference energy). Data has been collected in 9600 ms for curve A, 300 ms for curve B, and 460 ms for curve C.

As shown in fig. 7, three different kinetic domains must be considered:

1) The first one is a fast $Cu^{2+} \rightarrow Cu^{1+}$ transformation where Cu^{1+} ions are stabilized by the polymer backbone and form complexes with the $SO_3CF_3^-$ ions. This assumption is consistent:

(i) with the well-known stabilization of $Cu^{1+}SO_3CF_3^-$ by benzene[35] (giving rise to a crystallized compound), and

(ii) with our EXAFS data, which show that Cu^{1+} ions are surrounded by oxygen which can come only from $SO_3CF_3^-$ [36]. The time constant of this rapid fixation is 27 s.

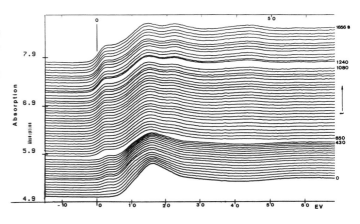

Figure 8. In-situ measurements of the evolution of the Cu K-edge when PMeT is cathodically polarized in an $H_2O-CuCl_2$ 50 mM electrolytic medium.

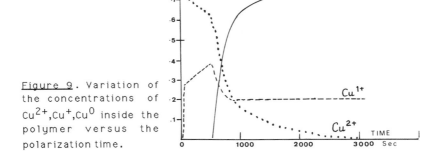

Figure 9. Variation of the concentrations of Cu^{2+}, Cu^+, Cu^0 inside the polymer versus the polarization time.

2) The second kinetic domain is characterized by a longer time constant [600s (Fig. 9)], where the Cu^{1+} ion concentration increases form 25 % to 40 %. Since the doping level of PMeT is equal to 25 %, no more $SO_3CF_3^-$ ions are available for this process. Thus the newly synthesized Cu^{1+} ions should come from a direct interaction with the polymer backbone.

3) The last step is dominated by the metallic-copper cluster formation. We assume that in the initial process, all the accessible sulfur sites of PMeT are saturated. Then, in the absence of stabilizing agent in aqueous solution, the monovalent copper ions undergo disproportionation to produce Cu^{2+} ions and metallic copper. Additional Cu^{2+} is then drained from the solution, resulting in an increase in the absolute copper content.

6-b Time-dependent Exafs investigation

The EXAFS data have been collected using a 23 cm-long triangular-shaped Si(111) crystal. Tuned to the copper K-edge the energy band pass can be as wide as 800 eV when the bending radius is 7 meters. This is large enough to collect a full EXAFS spectrum.

A selected set of the k^3-weighted Fourier transforms using a 350 eV-wide window are shown in Fig. 10. One can observe as a first evolution (fig.10 a→b) a shift of the principal peak towards the short distances, as well as a broadening (Δt = 160s). After 168s this peak is splitted into two components (fig.10c) whose intensity ratio inverts with the polarization time. The fast evolution of the system is illustrated in fig. 11 which corresponds to two consecutive spectra separated by only 8 sec. Later a shoulder appears at long distance on the main peak which becomes more and more pronounced. After 280 sec, a single peak is observed which shifts continuously towards longer distances.

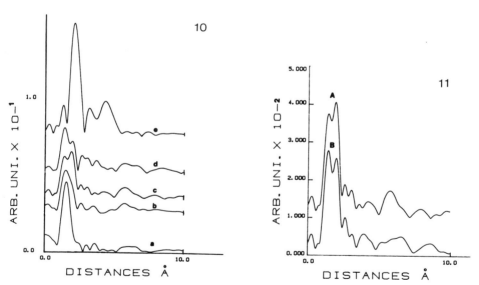

Figure 10. Main steps of evolution of the k^3-Fourier transforms versus polarization time. a : 0s, b : 160s, c : 168s, d : 200s, c : 460s.

Figure 11. k^3 Fourier transforms of two consecutive spectra separated by 8 sec. Note the inversion of the intensity ratio for the two peaks of the first shell. (A:160s, B:168s)

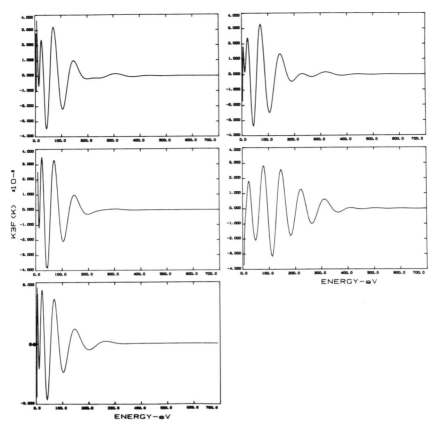

Figure 12. Fourier backtransformed copper spectra. Steps of Fig.10. Note that the splitted peak (c) results in a node of a beat at 220 eV., showing an exact compensation between the oxygen and sulphur contributions to the complex backscattering amplitude.

Then the intensity of the FT peak increases showing that the atoms of the first shell have either a larger backscattering amplitude or are in increasing number. At the end of the process, the characteristic FT of metallic copper is obtained (fig.10e) 1. Fig. 12 (a,b,c,d) shows the filtered backtransformed spectra of the first shell. These curves exhibit a continuous decrease of the amplitude of the oscillations with the appearance of a beat node at about 250 eV (Fig. 12 c) directly related to the splitting of the Fourier transform. This beat node evidences that two different atoms with a π difference in their phase shifts contribute to the EXAFS oscillations. A direct explanation involves the O and S atoms in the first shell. This is consistent with the

EXAFS characteristics drawn from two samples used as standards: the octahedrally hydrated Cu^{2+} solution and Cu_2S which exhibit a π difference in their phase shifts.

As mentioned, the first chemical step corresponds to the formation of Cu^{1+} ions complexed by O atoms given by the $(SO_3CF_3)^-$ dopant and the second step dealing with the fixation of new synthesized Cu^{1+} ions by the S atoms of the backbone of the polymer. These conclusions account for:

1) the broadening of the peak observed after ~80 sec which results from the presence of Cu^{2+}-O, Cu^{1+}-O and Cu^{1+}-S bonds.

2) the formation of metallic copper clusters in the matrix which appears as the last step. The intensity ratio of the 3th and 4th shell is inverted when compared to the bulk f.c.c. metallic copper : this implies that the early stage of metallic clustering involves a (111) platelet growth since the distances inside the (111) plane are observed preferentially to the distances between those planes.

The mechanisms and the kinetics of the electrochemical inclusion of copper particles inside an organic conducting polymer yield a complete illustration of the typical experiments achievable with the energy dispersive spectrometer.

7 THIN FILMS

7-a Si_3N_4/GaAs Buried Interfaces investigated by Reflexafs[37]

Although plasma deposited silicon nitride is widely used to passivate or to encapsulate GaAs(100) wafers, little is known about the interfacial chemistry between GaAs and the insulating layer. X-ray absorption spectroscopy using the total reflection scheme (REFLExafs) is suited to investigate the very first 25A of the semiconductor below the 175 Å-thick Si_3N_4 passive layer.

The X-ray index of refraction is just below one ($n = 1-\delta-i\beta$) which is well-understood if one refers to the Drude model. The electrons act like harmonic oscillators which are excited by an external electric field whose frequency is far above each natural frequencies. Consequently each oscillator (except eventually the considered core level) responds with a small negative amplitude. Then total external reflection is possible when the glancing angle (measured referred to the surface) is smaller than the critical angle: $\theta_c = \sqrt{2\delta}$.

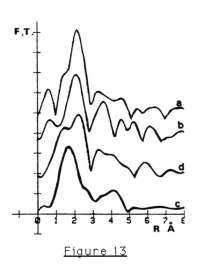

Figure 13

The imaginary part β which can be obtained from δ by the Kramers-Krönig transform is just proportional to the cross-section σ: $β=σλ/4π$. If the surface is set close but below θc, then the incident wave and the reflected wave form standing waves with antinodes of the electric field on the surface. The evanescent wave which penetrates only 25-30 A, undergoes then the photoelectric effect and the reflected beam carries the Exafs signal when all the photon energies close to the edge are considered. Since δ is roughly proportional to the electronic density, X-rays are able to penetrate the topmost surface layer and retrieve information from a buried interface provided the upper film is of low electronic density compared to the substrate.

Three samples elaborated at Thomson's laboratory, consist of 2-inch GaAs wafers optically flat, each of them coated by a 175A-thick Si_3N_4 layer fabricated by Plasma Enhanced Chemical Vapor Deposition for samples 1 and 2 and Reactive Sputtering for sample3. Before Si_3N_4 deposition sample 2 was exposed to a nitrogen plasma. In addition we ground a small piece of a GaAs crystal to obtain a standard measured in transmission mode [5].

Figure13 reports the Fourier transforms of the Ga K-edge spectra which are averages of about 30 spectra each of them being collected in dispersive mode in 13.6 seconds. Similar data have been recorded for the As K-edge. The soft PECVD method used for sample 1(b) keeps the GaAs structure unaltered as checked by comparison to the GaAs powder (a). Sample 2 (c) shows the shorter distance GaN than the GaAs (a) one as expected. The RS sample contains both GaN and GaAs-type environment as evidenced by the splitted FT first peak. In agreement with XPS measurements these data are reflecting a trend to form rough interfaces of the island type in case of RS whereas PECVD keeps well-defined interfaces.

7-b Fluorescence Yield under Grazing Incidence

If one uses a step by step scan instead of the dispersive optics which allows to collect all the data at once, one should have been able to get strictly equivalent data measuring the fluorescent yield of the Ga Kα line. This method gives superior results even for monolayer if the sample is illuminated below the critical angle to limit the probed depth to the very first top. Heald et al.[38] got the Exafs signal from a gold-coated silica mirror looking at the LIII-edge. In half an hour they collected a signal of extremely good quality from the equivalent of one single layer. Since they were working at room temperature the gold atoms form small clusters. Their signal compares well to that of a bulk foil measured in transmission. If the decay process of interest is the non radiative one, i.e. the electron yield, one has to take care to ground correctly the sample (especially insulators) in order to avoid charging effect.

7-c Electron Yield Under Helium Atmosphere[20,21]

The detector is a small vessel which can be pumped down to 10^{-6} Torr with
  1-a kapton window in its front part
  2-a turnable sample holder
  3-a large biased collector with a slot milled out just in the central part in order to permit the beam to hit the sample which is the second plate of this ion chamber
  4-inlets to allow pumping and to introduce He gas at controlled atmosphere.

When the photoelectrons or the high energetic electrons yielded by the non-radiative decay process cross this He 3mm-thick slice, they are converted into bunches of low energy electrons which are collected on the biased electrode. These ionization current variations versus the photon energy scanned step by step give the Exafs signal of the top of the sample whose thickness is defined by the mean free path of the electrons which is always small compared to the X-ray attenuation length.

At LURE [21] we undertook to check the effective thickness probed by such a scheme using several gold layers deposited on a Ni substrate. The thicknesses were adjusted between 50 and 2000Å. With a 70° glancing angle (again referring to the surface as usual in the X-ray field) and +45 Volt-biased collector, the signal has been proven to come from the first ~300Å. By reversing the polarity of the collector and collecting the He ions, the probed thickness can decrease down to ~20Å. In this configuration only electrons whose energy is larger than ~30 eV contributes to the signal. Since this energy coincides with the minimum of the electronic mean free path, one samples the top layers essentially.

As an example we show in figure 14 the structural evolutions of a bulky polycrystalline nickel after phosphor implantation. The curve a refers to the Fourier transform of the spectrum of the initial fcc Ni sample looking at the K-edge.

$2.5 \ 10^{17}$ P atoms/cm² have been implanted in such a sample at 40 keV and low temperature. The top, within the first 20 A, is amorphous as evidenced by the curve c which has been fitted with only 4 Ni first neighbors (instead of 12 in fcc) at the regular distance (2.47A±0.01), a large Debye-Waller factor ($\Delta\sigma \sim 0.008A$), and the complete disappearance of the 2nd, 3rd, and 4th shells. Curve b is the Fourier transform of the Exafs signal given by the electron collection and then the response of 300A. Curve fitting leads to 8 Ni first neighbors at 2.46A. Hence this scheme is depth-selective. Good efficiency in electron collection permits to cover a 2π solid angle. In the near future it should be possible to combine a chamber where fluorescent yield or electron yield are put into operation at convenience.

One development is aiming to add the energy resolution to the fluorescent photon detection by implementing a UHV-compatible proportional counter.

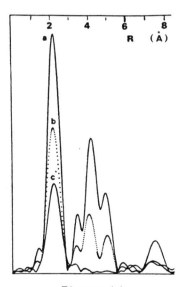

Figure 14

7-d Amorphization by Solid State Reaction

Co/Sn MULTILAYERS

Since its discovery in 1983, amorphization[39] by solid state reaction in multilayers has attracted as a new technique of preparation of non crystalline states. Cobalt/Tin multilayers are amorphized via a two stage process[40]. The first stage is fast and then needed the energy dispersive scheme to collect time-dependent data. The sample were prepared by alternate evaporation in an UHV chamber of pure Co and Sn (both 99.999%) onto a kapton substrate at 77K. 150 bilayers of Co(16Å) and Sn(48Å) were deposited. The samples were kept at 77K until transfer in an He cryostat at LURE for immediate measurements.

A total of 100 spectra (60 can be stored in a fast buffer before transfer to the computer) have been recorded, most of them close to room temperature, where diffusion is known to occur from resistivity measurements. A Co foil was also measured for calibration41.

Xanes evolution from 77K to room temperature is shown in figure 15 with pure Cobalt (b) and amorphous CoSn (c) as references. Fourier transforms are shown on figure 16.

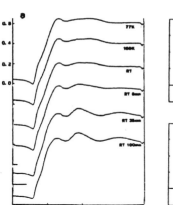
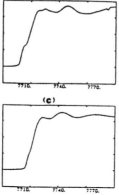
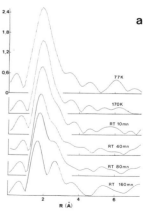

Figure 15a . XANES spectra of Co in Co/Sn multilayers from 77 K to 300K. b : XANES of pure metallic Co and c : XANES of amorphous $Co_{53}Sn_{47}$.

Figure 16a . Fourier transforms of EXAFS spectra or CO in Co/Sn multilayers form 77 K to 300 K.
b : Fourier transform of metallic Co at RT.

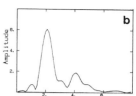

The first striking feature of these spectra is that the initial stage does not look like metallic hcp Co. The Exafs main peak is much broader and lower than the corresponding peak in Co which is a clue of the disorder which occurs at the first stage of the diffusion.

Generally an increase in temperature leads to an enhanced Debye-Waller factor and then the Exafs peaks get broader. On the contrary figure 16 shows a narrowing of the main peak from 77K to 170K. This structural reorganization has been confirmed by the small nonreversibility of resistivity as a function of temperature, significant of structural relaxation in disordered materials. At high temperature diffusion is more effective for structural change. The progressive appearance of the second peak reveals than more and more Sn atoms belong to the immediate vicinity of Co atoms. Annealed for 100 minutes at room temperature leads to a state which is very alike the $Co_{57}Sn_{43}$ amorphous spectrum.

Ce/Ni MULTILAYERS

This is a second illustration of solid state amorphization by diffusion in metallic multilayers. It is given because it combines the high sensitivity of the local environment-induced change in the Ce valence well-evidenced with the LIII Xanes spectrum, and the Exafs spectra recorded at both edges: namely the Ni-K edge and the Ce-LIII edge[42]. Samples were made by alternate evaporation in ultra-high vacuum chamber of ultrapure Ni and Ce layers onto a Kapton polyimide substrate stuck to a cooled (77K) copper plate. 130 bilayers of Ni (30Å) and Ce (35Å) were deposited. Again, the samples were kept permanently in liquid nitrogen until they were transferred into a helium cryostat at the energy dispersive X-ray absorption spectrometer at LURE.

To study the dynamics of the solid state at constant temperature, 3 spectra were taken at 250K, (2 spectra between 250K and 263 K), and 4 spectra at 263K, and finally 6 spectra between 263K and room temperature (292K). Two samples were measured at each edge, and reproducibility was excellent. A complete spectra is collected in one second, so samples do not evolve.

The Cerium valence depends only on the local environment in this binary system following the Jaccarino-Walker model which was successful to describe the valence change in Ce(Cu$_{1-x}$Ni$_x$)$_5$ alloys[43].

To account for the Ce valence change in the multilayers two families of profiles are derived, according to the process of diffusion chosen. Various models varying in sophistication exist. For the present study only two simple one-parameter models have been tried, namely the "homogeneous model" and the "Fick model".

Briefly, in the first model, a homogeneous amorphous layer of uniform and fixed composition is assumed to be formed at the interface between pure Ce and pure Ni. Once one of the species has been completely transformed into the amorphous phase, the other specy is absorbed to form an homogeneous amorphous alloy at varying concentration. The alternate Fick model deals with the classical equation $\partial C/\partial t = D\, \partial^2 C/\partial z^2$ where the different letters have the classical meaning. Starting from the step-like profile, the concentration can be easily obtained versus the dimensionless parameter $\Theta = 4\pi^2 Dt/z^2$.

Even if Exafs by itself is not sensitive enough to yield the exact profile, it is shown that it can be used to discriminate between two different models of diffusion. Clearly the homogeneous model gives good fit whereas the Fick model yields too small amplitudes. Inherently the Fick model deals with a broad distribution of many environments, thence Exafs averages over many spectra, leading to interference effects which flatten the signal. In the homogeneous model the Ce$_{40}$Ni$_{60}$ amorphous phase and pure Ni are in phase, so the presence of pure Nickel enhances the signal up to its experimental value.

The main point of this paper is to demonstrate that even for thin layers, it is possible to discriminate between two different models, showing that the homogeneous amorphous interlayer model propounded by Johnson[44] in early papers on solid reaction supports very well this set of data. In addition this experiment shows the capability of the dispersive scheme to achieve time-dependent experiments in solid state physics.

7-e Organic films on platinum electrodes

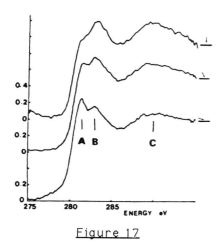

Figure 17

Using polarization- dependent Xanes, thin films down to a monolayer of poly 3 methylthiophene (PMeT) were investigated at the carbon K-edge collecting the Auger electrons (AY) with a CMA in the UHV chamber at UV 14 at Brookhaven. These five membered heterocycles can be regarded as being derived from benzene or from a long polyacetylen-like chains stabilized by the heteroatom. They were electrochemically deposited on a glass slice coated with platinum. In addition changes in the unoccupied antibonding π^* are clearly evidenced when the doping anion makes this polymer conductive.

This confirms the narrowing of the band gap related to the appearance of a metallic-like behavior in this organic conducting polymer [39].

In figure 17 we show the polarization dependence of the Carbon K-edge for a 20A-thin undoped film. Peak A, which is associated with transitions into the empty π^* level is stronger than peaks B and C when photons hit the sample at grazing incidence. At the opposite peaks B and C which are assigned to the (C-S)σ^* and the (C-C)σ^* respectively are stronger at normal incidence whereas the π^* intensity follows the reverse. This implies that the film is ordered with the σ bonds parallel to the surface and the π bonds normal to it, in average. The first layer is thus π bonded to the surface, i.e. the heterocycles lie flat onto the platinum electrode. Upon doping, the polarization dependence is still observed but becomes weaker. The incomplete extinction of π^* and σ^* at normal and grazing incidence respectively is in connection of a disorder induced by anions intercalation during doping. Increasing the PMeT film thickness leads to a lack of the substrate-induced orientation.

8 SURFACE EXAFS

As pointed out, last year in a good and generous humoristic style by P.Citrin1 in the 8TH review paper about Sexafs, dedicated to the full set of 48 papers (preprinted by the time of July86) including 7 theoretical papers and 19 experimental ones whose findings were viewed as non-definitive, only a limited number of systems (32) have been investigated. They are reproduced in the following table, which lists papers concerned by more or less a monolayer of adsorbates deposited on surfaces of single crystals almost exclusively. (This table is borrowed from P.Citrin's paper which gathers a lot of informations which form the backbone of this section. If the reader is eager to know more I would warmly recommend a direct access to P.Citrin's paper).

8-a Chronological summary of "important" developments in Sexafs following P. Citrin Int. Conf. Fontevraud 1986

Thence if one uses the more straightforward gauge for evaluating the utility of an experimental method which is the number of published papers, one would find this technique rather overrated. I feel exactly

Year	Description	System	Coverage	Ref
1976	SEXAFS experiment suggested using Auger electron detection			2
				3
1977	SEXAFS feasibility demonstrated-sensitivity Auger detection	I/Ag(111)	⩾1ML	4
1978	Bond length, absolute amplitude determined	I/Ag(111)	~1/3ML	5
	Total yield detection			6
	Partial yield (~2 eV.) applier to low Z elements Al_2O_3		~30Å	7
1979	PY (~325 eV) higher sensitivity	O/GaAs(110)	~1ML	8
	sensitivity to surf. conc. using TY demonstr.	O/Al(111)	~1ML	9
1980	Adsorption sites determ. using relative amp.	I/Cu(100)	~1/4ML	10
1981	Higher neighbor dist. used for sites determ.	S/Ni(100)	~1/2ML	11
1982	Adsorp. sites identified on semiconductors	I/Si(111)	<1ML	12
	Metal adsorbates studied, non react.& react.	Ag,Pd/Si(111)	1,3ML	13
	Adsorption coverage dependence studied	O/Ni(100)	~.25 →.5ML	14
	anamalous adsorption site found on metal	Te/Cu(111)	~1/3ML	15

(continued)

1983	adsorption on amorphous substrate studied	Cl/a-Ge	<1ML	16
	reactive chemisorption pathway stuied	Ni/Si(111)	~1/2ML →5ML	17
1984	Similar multiple surface bond lengths	O/Cu(110)	~1/2ML	18
	fluorescence yield detect. glancing geometry	Au/SiOé	>1ML	19
1985	Multishell analysis applied to surfaces	I/Ni(100)	>1ML	20
	Structure of clean surface studied	a-Si	~1ML	21
	Molecular adsorbates studied on metals	HCOO/Cu(100)	>1/2ML	22
		HCOO/Cu(110)	≤1/2ML	23
	FY det. applied to conventionnal geometry	S/Ni(100)	≤0.1ML	24
1986	Adsorbate-induced reconstruction of metal identified	O/Ni(110)	~1/3ML	25
	Anisotropy of Debye-Waller factor studied	Co/Cu(111)	~1ML	26
	Anharmonic Debye—Waller effect on bond length studied.	Cl/Ag(111)	~2/3ML	27

the opposite. Had the tunneling microscope not be created in the very recent years, I would think that Sexafs is at the top of the breakthroughs for surface investigations. The limited production of results is linked to the accumulations of the difficulties of surfaces investigations in connection with the poor flexibility of beam time allocation. The emergence of experimental ports fully dedicated to Sexafs with beam always available when the surface is ready (Bessy, NSLS, DCI-Orsay) changed the evolution. On the top of that, these experiments are eager of photons since only a monolayer interacts with the beam.

It has been the merit of P.Lee[2] and U.Landman and D.L.Adams[3] to suggest this surface experiment in 1976.

8-b Available informations

In all respects the informations derived from the Sexafs spectroscopy do not differ from those mentioned above for the bulk. But because of the inherent nature of the surface, the polarization-dependence of the signal is of prime importance. And the crystallographic orientation of the surface is essential in the interaction adsorbate-substrate. Thence it turns out that the structural questions differ from what is searched in the bulk. Sexafs is ideally suited for determining bond lengths and positions of adsorbates on single crystal surfaces. Additional informations related to the substrate, such as the nature of the reconstruction of the surface induced by an adsorbate, or its structure prior to being covered by an adsorbate, i.e. clean, have recently been obtained and will be discussed later in this section[25].

The two most important factors involved in limiting the information content of the data are the accessible range of data above the absorption edge and the required signal/noise ratio.

The data length is determined on one hand by physical restrictions which are the interference with other absorbing edges and photopeaks (both particularly serious for low Z elements) and on the other hand by the spurious Bragg reflection due to the single crystal nature of the sample studied (more important for photons of high energy whose choice comes from the Z number of the absorbing element).

The proper statistics is often not reached within the lifetime of the surface system under investigation which does not remain the controlled submonolayer on a clean surface for a long time. Actually this experiment calls for more powerful sources but with very good harmonics rejection by the optics.

The last element which controls the quality of the data is the choice of the detection scheme: Auger electron yield, total electron yield, partial electron yield, and fluorescence x-ray yield. To our knowledge only one attempt has been made to compare these different schemes for a single system. One of the reason certainly deals with the fact that the ultravacuum-compatible proportional counters used are still in their infancy even if they are very promising[28].

Besides the difficulties inherent to the surface study one has to take great care to avoid any production of electrons or fluorescent photons coming from any part of the sample holder or the vacuum vessel hitted by the incoming photons. It is not simple since one wants to work both at small and large glancing angles and with homogeneous samples which subsequently are often small.

As a partial conclusion of this presentation of the experiment let us point out that Sexafs data are polarization-dependent which is actually an advantage, compared to bulk Exafs data, but this is often balanced out by the shorter data length and higher noise level. Data accumulated by averaging many scans over long periods of time should be checked for time-dependent non-reproducibilities due to beam motion, monochromator drift, or sample degradation. It is always needed to compare different fractions, e.g., the first and second halves, of the averaged data. It happens that many hours of work are simply discarded.

8-c Accomplishments

Site determination: The direct and local nature of Sexafs as a probe to determine bond lengths and adsorption sites has allowed to solve lively problems. It is obvious that interpretations of Sexafs data can also lead to conflicts with other experiments or theory. Let us sample of what we feel we can call resolved problems since there is a convergence of interpretations in some way :

—Oxygen on Al(111)[9,29], on Ni(100)[14], on Cu(110)[18,30], on Ni(110)[25,31], on Cu(100)[32]

—Cl on Ag(100)[33,34,27], Cl on Ge(111)[35]

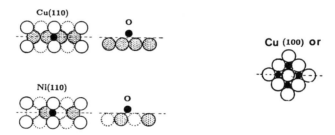

Figure 18. Chemisorption geometries found by SEXAFS. Adsorbate atoms are filled circles, substrate surface-layer atoms are open, and substrate second layer are shaded. The sawtooth 2x1 reconstruction includes a second row along [001] missing in the second layer(110).

The following drawings schematize the adsorbing sites of oxygen on the two 3d metals. Of particular interest are the findings of the Ni(110)2x1-O[25,31] and Cu(110)2x1-O[30] structures in which the "sawtooth" and "missing row" reconstructions, respectively, were identified.

Bond lengths clearly depend on coordination number, and so too are the surface bond lengths. Using the arguments and empirical numbers introduced by L.Pauling[36] trends to surface bond ionicity have been discussed.

Thermal disorder of surface atoms[26]: The study of the surface Debye-Waller factor shows evidence of the importance of the correlation between the motion of the central atoms and its neighbors and points out the inherent anisotropy of σ_j^2 introduced by the surface. Both experimental and theoretical studies of the DW factor have been achieved for bulk metals and a monolayer of Cobalt adsorbed on the (111) face of Copper. This system has been chosen

Table 3

	Cu(100)	Cu(110)	Ni(100)	Ni(110)
Oxygen	1.94Å (4)[32]	1.84(2)[18]	1.96 (4)[14]	1.85(2)[25]

because LEED and Auger[37] spectroscopy have shown that Cobalt grows layer by layer at room temperature. Sexafs investigation confirms clearly the two dimensional character of the monolayer of cobalt[38]. A second argument deals with the Debye temperatures of the two metals which are both just above room temperature. In that case the magnitude of the damping of the Exafs oscillations is expected to be large, but it is nevertheless reasonable to keep the harmonic oscillation up to this temperature. The third reason comes from the good knowledge of their elastic properties. Additionally the (111) dense face provides a good contrast between the adsorbate-adsorbate bonds and the adsorbate-substrate bonds. When the polarization is parallel to the surface the ratio between the two contributions is 9:1.5 while it is reverse 1:9 when it is at 75° from the surface. Thence it is a system where the two contributions are well-separated.

The experimental spectra collected in normal incidence are shown in figure19. The full time of data acquisition was about one hour for each spectra extending up to 600eV above the edge.

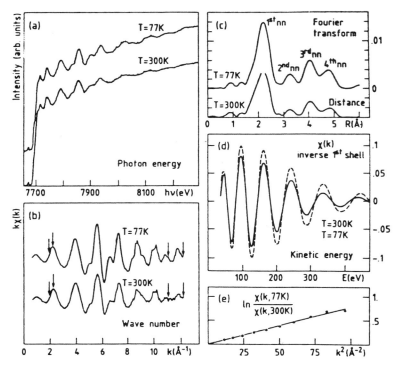

Figure 19. from ref.26 (a) Experimental absorption spectra of a monolayer of Co on Cu(111) at the K-edge of Cobalt at both 77 and 300K. The polarization of the light is parallel to the surface. (b) Exafs modulation {k χ(k)} normalized to the height of the edge jump. The arrows show the limits of the cosine window used for the Fourier transforms of the spectra. (c) Fourier transforms of the spectra. The peaks corresponding to the first, second, third, and fourth-nearest neighbors clearly appear. (d) Inverse Fourier transform of the first neighbor peak. (e) Logarithm of the ratio of the amplitudes at 77K and 300K as a function of k^2 for the first shell of neighbors. Points are the experimental data and the continuous line is the linear regression corresponding to these data.

Experimentally the DW factor is deduced from the temperature dependance of the damping of the Exafs oscillations measures at the Cobalt K-edge. Again the use of the polarization of the synchrotron radiation allows to distinguish between bonds involving two surface atoms or one surface atom only and then to determine the anisotropy of the surface atom vibrations.

The linearity of LOG{ $\chi(T=77K)/\chi(T=300K)$} is well established in the range 10 to 100 Å^{-2} of k^2 and justifies the use of the harmonic approximation. The slope gives the variation of the DW factor between 77 and 300K and then $\Delta\sigma^2_{//} = 3.9 \pm 0.3\ 10^{-3}$ Å^{+2} and $\Delta\sigma^2_L = 4.9 \pm 0.5\ 10^{-3}$ Å^{+2}. This is a significant anisotropy which contains a surface effect: $\Delta\sigma^2_L/\Delta\sigma^2_{//} \sim 1.25$. In order to check the origin of this anisotropy the local phonon densities of modes for motions perpendicular and parallel to the surface have been calculated in the harmonic approximation considering central and angular forces between first

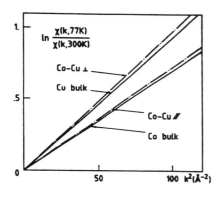

Figure 20. Logarithm of the ratio of the amplitudes at 77K and 300K as a faction of k^2 for the first shell of neighbors:
———bulk cobalt and copper,
— . — . ——— one monolayer of cobalt on the (111) copper surface for both orientation with respect to the polarization.

and second nearest neighbors. For motions parallel to the surface the density of modes for the surface layer is very similar to that one of the deeper layers, but for motion perpendicular to the surface there is a well-defined increase of the low frequency modes which are surface-related. $\Delta\sigma_j^2$ as measured by Exafs is related to the projected density of modes along the bonding direction. The calculated values $\Delta\sigma_j^2$ compare very well with the experimental data.

Table 4. Theoretical and experimental values of $\Delta\sigma_j^2 \times 10^3$

	Co (bulk)	Cu (bulk)	Co/Co ads→ads	Co/Cu ads↘sub
Theory	2.9	4.7	3.1	5.4
Exp	3.8±0.3	4.7±0.5	3.9±0.3	4.9±0.5
	$\Delta\sigma_{Co}^2$	$\Delta\sigma_{Cu}^2$	$\Delta\sigma_{//}^2$	$\Delta\sigma_L^2$

9 CONCLUSION

This must be regarded as a limited review of the wide variety of the different topics covered by X-ray absorption spectroscopy. The chosen examples bear the virtue to be presented with short explanations. More sophisticated systems have been solved by Exafs. Despite Surface Exafs is the major advent provided by SR, only a limited number of papers dealing with this technique have been already published. Thin films are even a more virgin field. But the generalized development of fluorescent detection and yield detection with samples set at grazing incidence should alleviate the difficulties related to UHV. The new SR sources such as NSLS (Brookhaven), BESSY (Berlin), SUPERACO (LURE-Orsay) are of great interest because of their tremendous brightness compared to the original ones whose designs were not tuned up to light production. The Chasman-Green lattice of general use for this new storage rings, in addition of insertion devices as it is planned for the 6 Gev machine at Grenoble is a subsequent jump of several orders of magnitude for the flux, the collimation. The time-dependent spectroscopy, as well as extremely dilute systems should draw a real benefit of such sources.

I – REFERENCES of Sections 1 to 7

1- IVth International Conference EXAFS and Near Edge Structure, July 1986 Fontevraud E.Dartyge, A.Fontaine, A.Jucha, J. Physique **47**, C8,sup12, edited by P.Lagarde, D.Raoux, J.Petiau, p.907, (1986)

IIIrd Int. Conf July 84 Stanford Proceedings Edited by K.O.Hodgson, B.Hedman, J.E.Penner-Hahn, , Springer Proc. in Physics 2

IInd Int. Conf Sept 82 Frascati, Italy, 13-17 September 1982, Proceedings edited by A. BIANCONI, L. INCONNIA and S. STIPICICH, Springer Series in Chemical Physics Vol. 27 (Springer, New-York), 1983)

2- De Broglie CR Acad. Sci. **157**,924(1913)
3- R. De L. Kronig, A. Physik **70**, 317 (1931)
4- D.E. Sayers, E.A. Stern, F.W. Lytle, PRL,(1971), **27**, 1024
5- E. Dartyge, C.Depautex, J.M. Dubuisson, A. Fontaine, A. Jucha, P. Leboucher, G. Tourillon, NIM **A246**, (1986), 452
6- ESRF: Foundation phase report (red book) Feb 1987, ESRF BP 220 F38043 Grenoble
7- J.D. Jackson : Classical electrodynamics, Chapt 14, Radiation by moving charges (John Wiley,1975)
8- D.Mills, A.Lewis, A. Harootunian, J.Huang, B.Smith Science, **223**,813(1984).
9- C Noguera PhD Thesis, Orsay, (1980)
10- G.Krill J. Physique **47**,C8,sup12, edited by P.Lagarde, D.Raoux, J.Petiau,p.907, (1986)
11- M. Benfatto, C.R. Natoli, A.Bianconi, J.Garcia, A. Marcelli, M.Fanfoni, I.Davoli, Phys. Rev B, (1986),
12- J.E. Muller, W.L. Schaich, Phys. Rev.**B27**, (1983), 6489
13- D.Raoux, A. Fontaine, P. Lagarde, A.Sadoc, Phys. Rev. **B24**, (1981), 5547
14- F. Sette, J. Stohr, Proc of the IIIrdInt.Conf. of Exafs and Near Edge structure, Ed K.O.Hodgson, B.Hedman, J.E.Penner-Hahn, Stanford July 1984, Spinger Proc. in Physics 2,p250
15- J. Mimault, A. Fontaine, P. Lagarde, D. Raoux, A. Sadoc, D. Spanjaard, J.Phys. **F11**, (1981), 1211
16- R.B.Greegor, F.W.Lytle Phys.Rev.**B20**,4902,(1979)
17- J.P.Itié, M. Jean-Louis, E.Dartyge, A.Fontaine, A.Jucha, J. Physique **47**,C8,sup12, edited by P.Lagarde, D.Raoux, J.Petiau, (1986) p.897
18- G.Tourillon, E.Dartyge, A.Fontaine, A.Jucha, PRL**57**, 5, (1986) p.506
19- A.Retournard, M.Loos, I.Ascone, J.Goulon, M.Lemonnier J. Physique **47**, C8, sup12, edited by P.Lagarde, D.Raoux, J.Petiau, p.143, (1986)
20- M.E. Kordesch, R.W. Hoffman, Phys. Rev. **B29**, (1984), 491
21- G. Tourillon, E. Dartyge, A. Fontaine, M. Lemonnier, F. Bartol, Physics Letters A,**121**,5,(1987)p.251
22- C.S.Fadley Progress in surface science S.Davison Ed, 16,p275 Pergamon NY 1984, Physica Scripta **35**,(1987),

23- E.A.Stern J. Physique **47**,C8,sup12, edited by P.Lagarde, D.Raoux, J.Petiau,p.3, (1986)
24- J.D.Eshelby in Solid State Physics edited by F.Seitz, H.Ehrenreich, D.Turnbull (Academic, NY 1956)**3**,p79
25- G.H.Via, G.Meitzner, J.H.Sinfelt, R.B.Greegor, F.W.Lytle Proceedings Edited by K.O.Hodgson, B.Hedman, J.E.Penner-Hahn, p176, Springer Proc. in Physics 2
26- E. Dartyge, A. Fontaine, J.Mimault Proceedings edited by A. BIANCONI, L. INCONNIA and S. STIPICICH, Springer Series in Chemical Physics Vol. 27 (Springer, New-York), 1983)p.80.
27- F.Comin Proceedings Edited by K.O.Hodgson, B.Hedman, J.E.Penner-Hahn,p.238, Springer Proc. in Physics 2
28- J.C.Mikkelsen, J.B.Boyce,Phys.Rev.**B28**,(1983)p.7130
29- A.Balzarotti, A.Kiesel, N.Motta, M.Zimnal- Starnawska,M.T.Czyzyk, M.Podgorny Phys.Rev.**B30**,(1984)p.2295
30- J.I.Budnick, M.H.Choi, G.H.Hayes, D.M.Pease, Z.Tan, E.Klein, B.Illerhaus J. Physique **47**, C8,sup12, edited by P.Lagarde, D.Raoux, J.Petiau,p.1037, (1986)
31- G. Tourillon, E. Dartyge, H. Dexpert, A. Fontaine, A. Jucha, P. Lagarde and D.E. Sayers J. Electroanal. Chem. **178**, 357 (1984)
32- G. Tourillon, in Handbook on Conjugated Electrically Conducting Polymers,, edited by T. SKOTHEIM (Marcelle DEKKER, New-York, 1985), Vol. 1, Chap. 9, p. 294
33- A. Bianconi, M. Dell'Arricia P.J. Durham and J.P. Pendry, Phys. Rev. **B 26**, 6502 (1982)
34- H. Tolentino, E. Dartyge, A. Fontaine, G. Tourillon J.Appl. Crist. in press(1987)
35- T. Spee and A. Macjor, J. Am. Chem. Soc. **103**, 6901 (1981)
36- E.Dartyge,A.Fontaine,G.Tourillon, A.Jucha, J.Physique **47**, C8, sup12, edited by P.Lagarde, D.Raoux, J.Petiau, p.607, (1986)
37- E.Dartyge,A.Fontaine,G.Tourillon, A.Jucha, J.F.Peray, R.Joubard, P.Alnot, J.Physique **47**,C8,sup12, edited by P.Lagarde, D.Raoux, J.Petiau,p.415, (1986)
38- S.Heald, E.Keller, E.A.Stern, Phys.Let.A**103**,(1984),155
39- R.B. Schwartz, W.L. Johnson, **PRL, 51**, (1983), 415.
40- P. Guilmin, P. Guyot, G. Marchal, Phys. Lett. **109A**, (1985), 174.
41- C. Brouder, G. Krill, E. Dartyge, A. Fontaine, G. Marchal, P. Guilmin, J.Physique **47**,C8,sup12, edited by P.Lagarde, D.Raoux, J.Petiau, p.1065, (1986)

42- C. Brouder, G. Krill, E. Dartyge, A. Fontaine, G. Tourillon, G. Marchal, P. Guilmin, submitted to Phys.Rev.B
43- D. Gignoux, F. Givord, R. Lemaire, H. Launois, F. Sayetat, J Physique **43**, 173(1982)
44- W.L. Johnson, B. Dolgin, M. Van Rossum, in Glass current Issues, Edited by A.F. Wright and J. Dupay), p172(1985) NATO ASI Series, E-92 Martinus Nijhoff, Boston.
45- G. Tourillon, A. Fontaine, R. Garrett, M. Sagurton, G.P. Williams, Phys.Rev.B, Rapid Com.,**36**, 5, 15 August 1987

II- REFERENCES of the Sexafs section

1- P.H. Citrin J. Physique **47**, C8,sup12, edited by P. Lagarde, D. Raoux, J. Petiau, p437, (1986).
2- P.A. Lee, Phys.Rev.B, **13**,5261 (1976)
3- U. Landman and D.L. Adams, Proc. Nat. Acad. Sci. USA **73**, 2550, (1976)
4- P.H. Citrin, Bull.Am. Phys. Soc.**22**,359 (1977).
   P.H. Citrin, P. Eisenberger, R.C. Hewitt, J. Vac. Sci. Tech. **15**, 449 (1978), NIM **152**,330, (1978)
5- P.H. Citrin, P. Eisenberger, R.C. Hewitt, PRL**41**,309(1978)
6- G. Martens, P. Rabe, N. Schwenther, J.Phys.C**11**,3125(1978)
7- J. Stöhr, D. Denley, P. Perfetti, Phys.Rev.B**18**,4132,(1978)
8- J. Stöhr, R.S. Bauer, J.C. McMenamin, L.I. Johansson, and S. Brennan, J.Vac.Sci. Tech.**16**,1195(1979)
9- L.I. Johansson, J. Stöhr, PRL,**43**,1882(1979)
   L.I. Johansson, J. Stöhr, S. Brennan, Appl. Surf. Sci. **6**,419(1980)
   J. Stöhr L.I. Johansson, S. Brennan, M. Hetch, J.N. Miller, Phys. Rev. B**23**,2102(1981)
10- P.H. Citrin, P. Eisenberger, R.C. Hewitt, PRL**45**,1948(1980), **PRL47**, 1567 (1981)
11- S. Brennan, J. Stöhr, R. Jaeger, Phys.Rev.B**24**,4871 (1981)
12- P.H. Citrin, P. Eisenberger, J.E. Rowe,**PRL48**,802(1982)
13- J. Stöhr, R. Jaeger J.Vac.Sci. Tech.**21**,619(1982)
14- J. Stöhr, R. Jaeger, T. Kendelewicz **PRL49**,142,(1982)
15- F. Comin, P.H. Citrin, P. Eisenberger, J.E. Rowe, Phys.Rev.B**26**,7060 (1982)
16- P.A. Bennett, F. Comin, J.E. Rowe, P.H. Citrin, G. Margaritondo, N. Stoffel, Bull.Am.Soc.**28**,532(1983)

17- F.Comin, J.E.Rowe, P.H.Citrin, **PRL51**,2402(1983)
18- U.Döbler, K.Baberschke, J.Haase, A.Puschmann, **PRL52**, 1437(1984)
19- S.M.Heald, E.Keller, E.A.Stern, Phys.Lett.**103A**,155(1984)
20- R.G.Jones, S.Ainsworth, M.D.Crapper, C.Somerton, D.P.Woodruff, R.S.Brooks, J.C.Campuzano, D.A.King, G.M.Gamble, Surf.Sci.**152**,443(1985)
21- F.Comin, L.Incoccia, P.Lagarde, G.Rossi, P.H.Citrin, **PRL54**, 1437, 122 (1985)
22- J.Stöhr, D.A.Outka, R.J.Madix, U.Döbler **PRL54**,1256(1985)
D.A.Outka, R.J.Madix, J.Stöhr, Surf.Sci.164,235(1985)
23- A.Puschmann, J.Haase, M.D.Crapper, C.E.Riley, D.P.Woodruff, **PRL54**, 2250 (1985)
M.D.Crapper, C.E.Riley, D.P.Woodruff, A.Puschmann, J.Haase, Surf.Sci.**171**,1(1986)
24- J.Stöhr, E.B.Kollin, D.A.Fischer, J.B.Hastings, F.Zaera, F.Sette, **PRL55**,1468(1985)
25- K.Baberschke, U.Döbler, L.Wenzel, D.Arvanitis, A.Baratoff, K.H.Rieder, Phys.Rev.**B33**,5910(1986)
26- P.Roubin, D.Chandesris, G.Rossi, J.Lecante, M.C.Desjonquères, G.Tréglia, **PRL56**,1272(1986)
27- G.Lamble, D.A.King, Phil.Trans.R.Soc.Lond.**A318**,203(1986)
28- D.Arvanitis, U.Döbler, L.Wenzel, K.Baberschke, J.Stohr, J. Physique **47**,C8,sup12, edited by P.Lagarde, D.Raoux, J.Petiau, p173, (1986).
29- D.Norman, S.Brennan, R.Jaeger,J.Stöhr, Surf.Sci.**105**,L297 (1981)
30- M.Bader, J.Haase, K.H.Franck, C.Ocal, A.Puschmann, J. Physique **47**,C8,sup12, edited by P.Lagarde, D.Raoux, J.Petiau, p491, (1986).
31- U.Döbler, L.Wenzel, D.Arvanitis, K.Baberschke J. Physique **47**,C8,sup12, edited by P.Lagarde, D.Raoux, J.Petiau, p473, (1986).
32- U.Döbler, K.Baberschke, J.Stöhr, D.A.Outka, Phys. Rev. **B31**, 2532 (1985)
33- P.H.Citrin, D.R.Hamann, L.F.Mattheiss, J.E.Rowe, PRL49, 1712 (1982)
PRL50,1824(1983)
34- G.Lamble, D.J.Holmes, D.A.King, D.Norman J. Physique **47**, C8, sup12, edited by P.Lagarde, D.Raoux, J.Petiau, p509, (1986).
35- P.H.Citrin, J.E.Rowe, P.Eisenberger Phys.Rev.**B28**,2299(1983)
36- L.Pauling The nature of the chemical bond, Cornell Univ.Press NY(1960)pp231,260
37- L.Gonzalez, R.Miranda, M.Salmeron, J.A.Verges, F.Yndurain, Phys.Rev.**B 24**,3245(1981)
38- D.Chandesris, P.Roubin, G.Rossi, J.Lecante Surf. Sci. **169**, 57 (1986)

Figure captions

Figure 1: Cu K-edge spectrum collected in dispersive mode. Added are the schematic model for final states in the XANES and EXAFS domains.

Figure 2: reports the Fourier transforms of the Ga K-edge spectra which are averages of about 30 spectra each of them being collected in dispersive mode in 13.6 seconds. Similar data have been recorded for the As K-edge. The soft PECVD method used for sample 1(b) keeps the GaAs stucture unaltered as checked by comparison to the GaAs powder (a). Sample 2 (c) shows the shorter distance GaN than the GaAs (a) one as expected. The RS sample contains both GaN and GaAs-type expected. The RS sample contains both GaN and GaAs-type environment as evidenced by the splitted FT first peak. In agreement with XPS measurements these data are reflecting a trend to form rough interfaces of the island type in case of RS whereas PECVD keeps well-defined interfaces.

As an example we show in figure 3 the structural evolutions of a bulky polycrystalline nickel after phosphor implantation. The curve a refers to the Fourier transform of the spectrum of the initial cfc Ni sample looking at the K-edge.

In figure 4 we show the polarization dependence of the Carbon K-edge for a 20A-thin undoped film. Peak A, which is associated with transitions into the empty π^* level is stronger than peaks B and C when photons hit the sample at grazing incidence. At the opposite peaks B and C which are assigned to the (C-S)σ^* and the (C-C)σ^* respectively are stronger at normal incidence whereas the π^* intensity follows the reverse. This implies that the film is ordered with the σ bonds parallel to the surface and the π bonds normal to it, in average. The first layer is thus π bonded to the surface, i.e. the heterocycles lie flat onto the platinum electrode. Upon doping, the polarization

Fig.5 a: XANES spectra of Co in Co/Sn mutilayers from 77K to 300K. b: XANES of pure metallic Co and c:XANES of amorphous Co53Sn47.

Fig.6 a: Fourier transforms of EXAFS spectra of Co in Co/Sn multilayers from 77K to 300K
   b: Fourier transform of metallic Co at RT.

PAX (PHOTOELECTRON AND X-RAY EMISSION) SPECTROSCOPY: BASIC PRINCIPLES AND CHEMICAL EFFECTS

David S. Urch

Chemistry Department, Queen Mary College
(University of London)
Mile End Road, London E1 4NS, UK

1. INTRODUCTION

When an atom is subject to X-ray irradiation or to electron bombardment, photoelectrons can be ejected leaving an ion. If the ion has a vacancy in a core orbital then relaxation follows, either by X-ray emission, in which an electron from a less tightly bound orbital 'jumps' to the inner orbital and the energy released is observed as an X-ray photon, or by the formation of an Auger electron whose kinetic energy derives from a similar transition and which leaves a doubly ionised atom. These processes together with related transitions that can occur if the initial ionising event has caused electronic excitation or even further ionisation are summarised in figure 1.

A most important special case exists if a monoenergetic source is used to ionise the atom for it is then possible to equate the kinetic energy of the ejected photoelectron [$E(k)$] and the binding energy with which that electron had been bound in the atom [$E(b)$] with the incident energy ($E = h\nu$, where ν is the frequency of monochromatic radiation), as follows,

$$E = h\nu = E(b) + E(k) + \phi \tag{1}$$

where ϕ is a small correction to cover recoil energy, charging effects and work function terms in solids, etc. The measurement of the kinetic energies of photoelectrons produced in this way is the basis of Photoelectron Spectroscopy (PS).

The ions produced in photoelectron spectroscopy can relax either by Auger electron, or by X-ray, emission. In the former process the energy of the ejected electron $E(A)$ is,

$$E(A) = E(i) - E(x) - E(y), \tag{2}$$

where $E(i)$ is the ionisation energy of the initial core hole state and where $E(x)$ and $E(y)$ are the ionisation energies of the two final hole states x and y, (<u>not</u> the same as the ionisation energies of those states in the initial atom - atomic number Z - although to assume that one of the energies can be approximated in this way whilst the energy of the other state may be taken from the atom with atomic number $Z + 1$, gives a

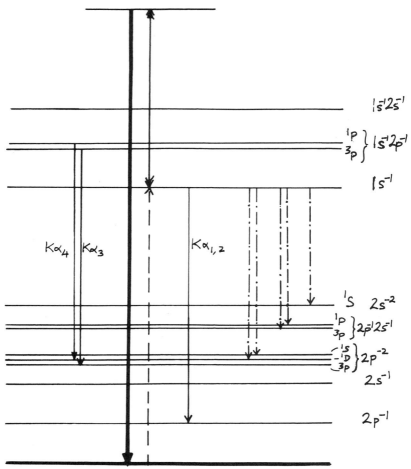

Figure 1 Energy level diagram showing possible relaxation processes following inner shell ionisation for an atom such as neon. Heavy vertical line indicates the initial photoionisation energy, the dashed line the ionisation energy of the 1s electron - and the double headed arrow its kinetic energy. X-ray emission processes are single headed arrows. Auger transitions are shown, as dash dot lines.

convenient estimate for E(A)). The other relaxation process open to core ionised atoms is that of direct emission of radiation, in which case the energy of the X-ray E(X) is, E(X) = E(i) - E(f) where E(f) is the ionisation energy of the final state. As both E(i) and E(f) can be investigated directly by PS, a close relationship exists between X-ray emission and photoelectron spectroscopies. Furthermore transitions are also observed in X-ray Emission Spectroscopy (XES) which occur in doubly-ionised species: this indicates a connection with the final states of the Auger process.

It is the nature and importance of the relationships between Photo-electron, Auger and X-ray spectroscopies that will be discussed in these lectures and it will be convenient to refer to them collectively as PAX (Photoelectron, Auger and X-ray). This first lecture will concentrate on effects that can be observed in atoms in different chemical states or environments, the second lecture will show how PAX spectroscopy can be used to study electronic structure.

2. PAX SPECTROSCOPY: EXPERIMENTAL PROCEDURES

PE Spectroscopy

The fundamental requirement is a source of monochromatic radiation which, in the ordinary laboratory situation, may be either light from a noble gas resonance discharge lamp or from an X-ray tube. Not all elements are suitable as X-ray targets, either because of their physical nature or because the intense characteristic lines which they emit are too broad. Thus magnesium and aluminium have been singled out and are now the most widely used elements as sources of radiation in X-ray photo-electron (XP) spectroscopy.[2,3] When noble gases are used e.g. HeI(21.2eV) ultra-violet photoelectron (UP) spectroscopy results.[4,5] The essential features of UP and XP spectrometers are shown in figs. 2 and 3. The radiation from the X-ray sources passes through a thin filter made of the same material as the anode. This has the effect of removing most of the high energy bremsstrahlung and some of the low energy photons as well, leaving the $K\alpha_{1,2}$ lines as the most intense radiation (but the satellites $K_{3,4}$ with about 10% $K\alpha_{1,2}$ intensity and about 10eV higher in energy are still present). A more effective way of producing clean photon source with a narrow line width is to allow the X-rays to pass through a curved crystal monochromator. The total reduction in intensity is large but great improvements are found in peak to background ratio and in providing spectra free of "$K\alpha_{3,4}$ satellites".

The advent of synchrotron radiation has made the distinction between UPS and XPS irrelevant in principle as it is now possible to choose radiation of any desired frequency, but only of course if one has access to a synchrotron!

Once a suitable source of radiation has been chosen, it can be used to initiate photoemission from a sample. The photoelectrons are then sorted according to their kinetic energy, either by electrostatic or magnetic deflection, and detected, usually by a channeltron or similar device.

As the photoelectrons produced in the usual types of XPS have energies no greater than 1.5 keV, their escape depths will be of the order of a few nanometers. Conditions must therefore be provided within the spectrometer which will prevent the formation of even a few mono-layers of contamination on the surface of the sample during the experiment. This requires a vacuum regime of 10^{-9} torr, or better, even for routine

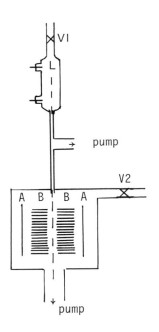

Figure 2 A simple ultra-violet photoelectron spectrometer. L is the gas discharge lamp to which a noble gas is admitted through valve V1. Light from the lamp, — — — passes through the thin vertical capillary to the lower chamber. The gas to be studied is admitted to this chamber through valve V2. Photoelectrons produced in the centre of the chamber are collected on the cylindrical anode A, having first been collimated by the discs B. A retarding potential is applied between A and B. As this potential is varied, the first derivative of the charge collected at A gives the UP spectrum.

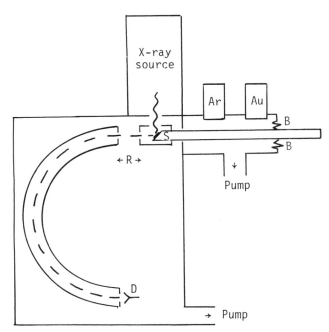

Figure 3 A simple X-ray photoelectron spectrometer. X-rays fall on the sample S and photoelectrons — — — are ejected. They are subject to a retard potential R and those of a specific energy can then traverse the hemispherical analyser to be detected at D. Measurement of the charge collected at D as R is varied produces the XP spectrum. The sample can be withdrawn under vacuum (B = bellows) and subjected to special treatment e.g. Argon ion etching (Ar) for cleaning or depth profiling - gold decoration (Au) for calibration.

experiments. Hence the use of metal vacuum systems which can be regularly baked.

X-ray Emission Spectroscopy[6,7]

The requirements for X-ray emission spectroscopy are much less stringent than for photoelectron spectroscopy. There is no need for a monochromatic source, only the means of creating core vacancies, and no need for such rigorous vacuum conditions as the X-rays will have escape depths ranging (for even the most soft) from tens or hundreds of nanometers up to many microns. X-ray emission is stimulated either by electron bombardment (as in electron microscopes) - the so-called 'direct' method - or by X-ray irradiation, which gives rise to 'X-ray fluorescence'. X-rays are also emitted when matter is bombarded with other particles neutrons, protons, heavy ions etc. For X-ray irradiation a sealed X-ray tube is usually used with the possibility of choosing different metals as anodes (e.g. Sr, Cr, Mo, Rh, W, Au etc.), or, for the better excitation of soft X-rays, an open window gas-discharge source can be used (e.g. CGR Elent-10). The characteristic X-rays that are emitted by sample can either be dispersed by crystals with appropriate 2d spacings ($n\lambda = 2d \sin \theta$) or by gratings and then detected in some form of proportional counter (either gas filled or scintillation). Characteristic X-rays can also be analysed directly using a solid-state energy dispersive detector. Much greater resolution is achieved in the former 'wavelength dispersive' method but the latter is perfectly adequate for the detection and identification of different elements.

In wavelength dispersive spectrometers the highest resolution is achieved by using two crystals or by curved crystals, although very good resolution may be achieved with multiple narrow Soller slits arranged as a collimator, (in this case though it is also necessary to control 'horizontal divergence' which can cause peak tailing to high angles) as shown in fig. 4. In the electronics following the detector it is necessary to have pulse height selection in order that high orders can be rejected, or so that the peaks corresponding to a desired order can be selected. If the spectrum is scanned automatically the ratemeter output can be presented as a chart recording or the region of interest can be steep scanned and the counts per channel stored in and processed by a microcomputer.

Auger Spectroscopy[8]

Auger electrons can most easily be detected, along with photoelectrons, in a photoelectron spectrometer as described above. The Auger electrons may be distinguished from photoelectrons if it is possible to change the X-ray source (dual anodes are now standard) as the energy of the Auger electrons is independent of the irradiation energy. As a monochromatic source is not essential conventional X-ray tubes and electron bombardment can also be used to initiate Auger emission. For this reason Auger spectroscopy is widely used nowadays to characterise the elemental composition of surfaces and to monitor changes during adsorption.

3. PAX SPECTRA

XP Peak Intensities

Peak width and peak intensity are of fundamental importance in all branches of PAX spectroscopy. The first step is the creation of a primary core vacancy. The probability that this will occur is governed

by an integral of the type $\int \Psi_f \underline{P} \Psi_i$ where $\Psi_{f,i}$ are wavefunctions to describe the final and initial states of the atom and $\underline{P}$ is the transition operator. In the simplest, one electron, frozen-orbital approximation Ψ_f and Ψ_i are replaced by the final and initial orbital functions of the electron undergoing the transition (φ_f, φ_i): it is assumed that all other electrons are unaffected by the ejection of one of their colleagues and so their wavefunctions will be unaffected by $\underline{P}$. The critical integral thus reduces to
$$\int \phi_f \cdot \underline{P} \cdot \phi_i \qquad (3)$$

In the electric dipole approximation the transition operator $\underline{P}$ is replaced by $e\underline{r}$ (e, the electron charge, $\underline{r}$ displacement of electron). For photoemission from a core level, ϕ_i is simply the wavefunction for that core orbital and ϕ_f will be a plane wave in the continuum. Thus although the dipole nature of $\underline{P}$ will place restrictions on the types of transition that are allowed (see below for selection rules in X-ray emission) it will always be possible to find a continuum wave function of the appropriate symmetry. It therefore follows that there are no rigorous selection rules in PES which require that certain transitions be forbidden. But the simplicity of (3) does enable qualitative predictions about relative intensities of photoemission peaks to be made. As Price⁹ has graphically demonstrated, the magnitude of (3) will, in effect, be determined by the overlap of ϕ_i and ϕ_f. The wavelength [$\lambda(E)$] of ϕ_f will be determined by its kinetic energy $E(k)$.

$$\lambda(E) = [150.6/E(k) \text{ eV}]^{\frac{1}{2}} \text{ Å}$$

Thus an electron, ejected with a kinetic energy of 10 eV will have an effective wavelength of 3.88 Å whilst for 1000 eV the wavelength is 0.388 Å. If these wavelengths match the nodal properties of the radial part of ϕ_i then ready photoemission can be expected. If however the match is poor then (3) will be small and only weak photoemission peaks will be observed. Thus in XPS where irradiating energies of 1486.6 (Al Kα) are most widely used electrons will be ejected from core orbitals (which will typically have nodes and total effective radial extent in the order of tens of picometers) with energies of hundreds of eV (40-100 pm) whilst from valence band orbitals (nodal pattern etc. of the order of a hundred picometers) higher energy, and therefore shorter wavelength photoelectrons will be generated. This will ensure that much more intense XP spectra are observed from core orbitals than from valence band orbitals. It can also be seen that considerable variation in photopeak intensity will result from different numbers of radial nodes. A specific example of such an effect is the variation observed in the relative intensities of photoelectrons from 2s and 2p orbitals of first row elements in XPS; the probability of photoemission is almost ten times greater from 2s orbitals than from 2p. Conversely in UPS a better match between photoelectron wavelength and the initial orbital is to be had for 2p functions, so that electrons from orbitals with 2p character predominate over those from '2s orbitals'.

If experiments are conducted in which it is possible to vary the irradiating energy then changes in photoemission efficiency can be monitored as a function of photoelectron energy (wavelength) for any given orbital. The minima in such functions are known as Cooper minima.

XP Peak Widths

The natural width (ΔE) of any spectroscopic peak, i.e. the observed width after all experimental broadening factors have been discounted, is directly related to its lifetime (Δt) by the Uncertainty Principle, $\Delta E \cdot \Delta t \approx 3 \times 10^{-16}$ sec. Thus peak widths in XPS are determined by the efficiency of relaxation. Two main channels are open, radiative decay by

X-ray emission, or the non-radiative Auger process. As the probability for the former is related to the cube of the frequency of the radiation it is generally true that X-ray emission predominates for high energy (> 2,000 eV) transitions whilst Auger electron emission is the favoured route at lower energies.

Thus the greater the relaxation energy the shorter the lifetime of the initial state and therefore the broader the XP peak. If initiating energies of only a thousand volts or so are used (Al Kα, MgKα) then the natural widths of most of the observed peaks will be typically of the order of a volt.

There are of course quite specific variations that are observed superimposed upon these generalities, but in keeping with the general principles enunciated above. Thus it is generally observed that when the widths of 2s and 2p peaks are compared for the elements Z < 35 (i.e. those accessible to Al Kα) the 2s are broader than 2p. This is because the 2s - 2p energy difference is usually sufficient to initiate very efficient Auger processes, the 2s holestate lifetime is reduced relative to 2p and so a broader peak results.

XE Peak Widths and Intensities

The probability that a core ionised atom will relax by X-ray emission is determined by functions of both the emission frequency (ν, where hν = E(i) - E(f), E(i,f) energies of the initial and final states) and the integral $\int \Psi_f \underline{P} \Psi_i$, as described above. The transition operator $\underline{P}$ can be expanded as a series of terms such as electric dipole, electric quadrupole, magnetic dipole etc. of which the electric dipole is by far and away the most important (by at least an order of magnitude) unless the wavelength of the emitted radiation is comparable with the size of the emitting atom (i.e. an Ångstrom or less), (ref. 6, p.184-186).

The importance of the electric dipole is that it imposes a rigorous selection rule upon the emission of X-rays which can most simply be expressed (for transitions between otherwise filled shells) as $\Delta \ell = \pm 1$. where ℓ is the angular quantum number (ref. 10, chap. 9). Thus to an s vacancy only transitions from p-type orbitals can occur and to p vacancies, only transitions from s or d orbitals and to a d vacancy only transitions from p or f orbitals. Such transitions are known as 'permitted' or 'diagram' lines. Figure 5 shows these lines for the lighter elements together with the curious idiosyncratic and archaic nomenclature currently used. Non-diagram lines are observed with increasing probability as the atomic number increases and the energy of the X-ray increases. This is because the decreasing wavelength enables magnetic or quadrupole transitions (e.g. 3d $\rightarrow$ 1s, Kβ_5) to occur. Other lines not easily accounted for as due to a simple transition between atomic orbitals are lumped together as 'satellites' for which there are a variety of different causes - some of which will be discussed in more detail below.

The widths of X-ray emission peaks will be determined by the lifetimes of both the initial and final states. The energy associated with the relaxation of the former usually being much greater than the energy of the latter the lifetime of the initial state is usually very much shorter than that of the final state. Thus it may be concluded that, in general, peak widths in XE spectroscopy are determined by the (short) lifetime of the initial state.

For X-rays of more than 10 keV widths of ten up to hundred eV can be observed whilst for softer X-rays of the order of 1 keV widths of just

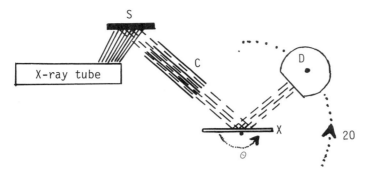

Figure 4 Essential features of a flat crystal X-ray fluorescence spectrometer. The sample S is irradiated with X-rays ≣, and emits its own characteristic X-radiation ≣≣≣≣. After collimation, C, this radiation is reflected by the crystal X to the detector D.

As X is rotated through θ, the detector rotates through 2θ so as to be always in the correction position.

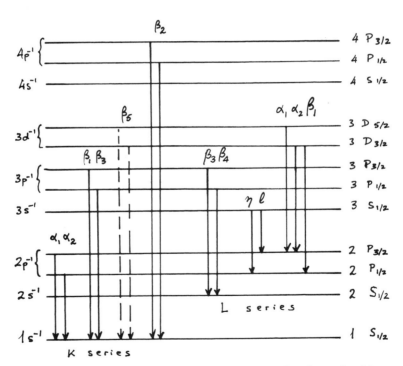

Figure 5 X-ray transitions permitted by the dipole selection rule – "diagram lines". 'Forbidden' transitions are shown as dashed lines.

less than a volt are found. This is of considerable importance to those who would study XE spectra for physical and chemical effects. If these effects cause peak shifts of the order of only a volt then such shifts will only be observable for X-rays with energies of less than 10 kV and only easily observable at energies of less than 1 kV. Whilst the same factors that determine the widths of XP peaks are seen to control XE peak width, the existence of non-radiative decay processes can also cause dramatic changes in XE peak intensities. A specific Auger channel may result in most of a particular type of vacancy being filled before radiative relaxation can occur, thus reducing the intensity of X-rays to that vacancy, an effect to be discussed in more detail in section 4.

Auger Spectra

In the Auger relaxation process the initial core ionised state is transformed into a doubly ionised species, with the second electron being ejected to a continuum state. Viewed as a transition between vacancies the probability of the Auger process is dependent upon an integral of the type,

$$\int\int \Psi(x) . \Psi(y) . \underline{V} . \Psi(i) . \Psi(c)$$

where $\Psi(x)$ and $\Psi(y)$ are the final state, $\Psi(i)$ the initial state and $\Psi(c)$ the continuum, wave functions: $\underline{V}$ is the transition operator. Spin orbit coupling between $\Psi(x)$ and $\Psi(y)$ will usually be important. Such coupling will cause peaks, whose energies can be estimated using eq. (2), to be split, by a few volts for lighter elements but by tens of volts when Z is large. It will also restrict the number and type of possible final states and lead to some possible transitions being forbidden.

A simple example is to be found in KLL spectra (the first letter locates the initial vacancy, the other two letters the final vacancies). In all cases the initial atomic vacancy is $(1s)^{-1}$ which can be combined with any type of continuum state (c) of the correct energy thus

$(1s)^{-1}$	$(cs)^{-1}$	^1S or ^3S	+
$(1s)^{-1}$	$(cp)^{-1}$	^1P or ^3P	−
$(1s)^{-1}$	$(cd)^{-1}$	^1D or ^3D	+

etc.

Possible spectroscopic states and their parities are indicated as well. The parity of a state is determined by the product of the responses of the individual orbitals to inversion through their centres (of symmetry). s and d functions are gerade, even, or positive (+), as they do not change sign when this operation is carried out; by contrast and for the same reason p and f functions are ungerade, odd, or negative (−). For KLL Auger spectra the final configurations, their parities and possible spectroscopic states are,

$(2s)^{-1}$	$(2s)^{-1}$	^1S	+
$(2s)^{-1}$	$(2p)^{-1}$	^1P or ^3P	−
$(2p)^{-1}$	$(2p)^{-1}$	^1S, ^1D or ^3P	+

As the Auger process requires conservation of spin, angular momentum and parity it can be seen that only the transition to the $(2p)^{-2}$, ^3P state is forbidden (by parity in LS coupling). For heavy elements L-S coupling, as outlined here, breaks down, and the spectra become more complex as transitions to individual j states are observed, accompanied by the partial relaxation of the prohibition to $(2p)^{-2}$ ^3P.

If the final states in the Auger process are valence orbitals spectra of extreme complexity result. For their interpretation, even for small molecules, simple 'one-electron' models are, as might be expected, inappropriate. However calculations which take proper account of hole-hole interactions and associated electronic relaxation can give a reasonable account of Auger spectra of molecules.

Configuration Interaction - Shake-up, shake-off - Limitations of the One-electron Model

Complications in interpretation due to many-electron effects are not however confined to Auger spectra. The one-electron model of ground state wavefunctions, whilst useful for a qualitative understanding of XE and XP spectra, shows its limitations when a quantitative comparison with experimental peak energies and intensities is attempted. Furthermore the origins of many 'satellite' peaks can only be understood as arising from 'many-electron' effects (ref. 1. vol. 1, chapter 2). Two equivalent approaches, based on one-electron wavefunctions, have been proposed to consider many electron effects; the shake-up/-off model and configuration interaction.

In the first it is recognised that the ejection of a photoelectron or the relaxation of an electron from an outer to an inner orbital may also cause an 'instantaneous' perturbation of the other electrons in the atom (an effect explicitly omitted in the 'frozen orbital' approach). This perturbation may result in electron excitation (shake-off) or electron ionisation (shake-off) occurring simultaneously with the primary photoemission event. The extra energy required for this excitation reduces the kinetic energy of the ejected electron: this gives rise to 'shake-up' peaks and broad 'shake-off' features on the low kinetic energy side of main XP peaks.

The alternative view is to recognise that any product of one-electron wavefunctions can only be a first approximation to the 'best' molecular wavefunction. Improvements are achieved by allowing ground and excited state functions, of the same symmetry, to interact and requiring that the contributions of the various states minimise the energy of the resulting wavefunction. This process is 'Configuration Interaction' and results in some wavefunctions moving considerably from their anticipated 'one electron' energy and acquiring a complex atomic orbital composition. But not all wavefunctions will be affected equally even in the same atom. In some cases the admixture of excited states may be small so that the 'one-electron' wavefunction is a good approximation to the 'best' wavefunction (that this is often so is why the one electron approximation can ever be claimed to work at all!). In other cases the degree of configuration interaction is so great that no vestige of the 'one electron' picture remains (e.g. Xe $4p_{\frac{1}{2}}$, ref. 3, p.47).

One of the main factors determining the importance of the contributions of excited states to the final wavefunction is their energy: only states of comparable energies interact significantly. By way of example consider a second row atom such as sulphur or chlorine - or more conveniently small molecules derived from such atoms where the ionisation energies of orbitals with 2s character are round about double those with 2p character. Thus states based on configurations such as $(3s)^{-1}$ (^2S) will have almost the same energies as $(3p)^{-2}$ (ns or np)$^{+1}$ (from which configurations, ^2S states can be found). Strong configuration can be anticipated and states corresponding to a 'one-electron' model for $(3s)^{-1}$ will not be found. In fact the (3s) "peak" is observed to be dramatically split into two peaks of unequal intensity, both displaced many volts from the anticipated position[11].

It is therefore always necessary, when considering PAX spectra, to remember that variations from the predictions of one-electron models can often occur and indeed to realise that 'satellite peaks' and 'unexpected' peak shifts, far from being a nuisance are in fact a vital source of information on the relaxation behaviour of electrons in ionised atoms and molecules.

4. CHEMICAL EFFECTS IN ATOMIC PAX SPECTRA

Peak Shifts in XE, XP Spectra

Besides the fundamental variations in peak intensities described in section 3, small but quite significant changes in X-ray, or photoelectron kinetic, energy are found when atoms are in different chemical situations, or have different valencies, spin states or ligand partners. These changes are of great interest and importance to physicists, chemists and analysts. Such changes, involving core orbitals, or transitions between core orbitals will be discussed here; effects concerning valence shell molecular orbitals are the subject of the second lecture.

Core levels shift in energy in response to changes in the occupation of valence shell orbitals - specifically to the effective charge on the atom. This is because the valence orbital - core orbital electron-electron repulsion integral plays a part in determining the binding energy of the core orbital. If this repulsion term is reduced, because electrons are withdrawn from the valence band (because the atom is oxidised to a higher valency or because more electronegative ligands are present) then the core electrons become more tightly bound. Conversely, a reduction in valency or the introduction of electron donating ligands will reduce the core ionisation energies. In an atom of modest size there will be a range of core orbitals - 1s, 2s, 2p, 3s etc., and the overlap with valence orbitals will increase with principal quantum number. Thus the peak shifts discussed above will be least for 1s, a little greater for 2s, 2p, more for 3s, 3p etc. This will also cause shifts in XE spectra (if all core levels altered by the same amount then X-ray energies, being the difference between atomic levels, would not be affected). As it is however, as core orbitals become more tightly bound one must expect that X-rays will show slight shifts to lower energies and vice versa; also these shifts will become larger as the energy becomes larger, i.e. Kβ should show slightly larger shifts than Kα. These arguments only apply of course to core-core transitions - shifts in valence XE spectra are governed by the nature of the valence band molecular orbitals, as well as core orbital changes.

Exchange Effects: Spin-orbit Coupling

When ionisation occurs in an open-shell system the newly created vacancy will interact with the unpaired electrons. In the simplest case where only electron spins are considered, exchange splitting of peaks may be observed and when the complications of orbital angular momentum are also taken into account, a range of possible spectroscopic states results. In XP spectroscopy these states can be observed directly: in XE spectroscopy transitions between the 'split' initial and final states will occur.

A classic example is the case of the 3s and 3p XP spectra of the manganese (II) - $3d^5$, ^6S cation, studied originally by Fadley[6],[12]. The simple exchange splitting of $3s^1$ and the effects of spin orbit coupling for $3p^5$ are shown in Fig. 6. Similar effects have been observed for other transition metal ions with unpaired electrons. Corresponding peak splitting is also observed in XE spectra. Fig. 7 shows $K\beta_{1,3}$ spectra from high spin and low spin complexes of iron: the effect of the presence or absence of unpaired 3d electrons upon the $K\beta'$ feature is dramatic and demonstrates the importance of simple exchange effects[13]. Furthermore for some transition elements the relative intensity of $K\beta'$ to $K\beta_{1,3}$ can be directly related to the number of unpaired 3d electrons (and thus used to determine valency[14]), as shown in fig. 8 for manganese. However, exchange effects cannot explain all that is observed, for as fig. 8 shows for permanganate 3d (and similar spectra are found for CrO_4^{--}, VO_4^{---} etc.) the main $K\beta$ peak is accompanied by a low energy tail spreading over some 10 volts, presumably due to facile excitation of low-lying excited states.

PAX spectroscopy enables effects with ostensibly identical origins to be investigated. Such a direct comparison is shown (fig. 6) for XP and XE spectra for Mn(II) $3p^5$. The states that give rise to $K\beta$ and $K\beta'$ can be clearly seen in the photoelectron spectrum. But, as is shown in fig. 9 the PAX spectra for Ni (II) $3p^8$ do not show the same one-to-one correlation of XE and XP peaks[15], thus demonstrating the power of PAX spectroscopy to confirm, or to deny, proposed electronic mechanisms for peak formation.

Satellite Peaks

Satellite peaks have many different origins: the purpose of this section is to outline chemical effects that are observed in transitions involving core orbitals.

Low (kinetic) energy satellites accompany to a greater or lesser extent, all X-ray photoelectron peaks. They can be of two types, intrinsic-associated with the emitting atom and extrinsic - due to specific energy losses caused by the passage of the photoelectron through the sample. Peaks due to the latter effect can be identified by a comparison with electron energy loss spectra. The intrinsic losses can be thought of as due to 'simultaneous electronic excitation', or to the operation of configuration interaction as described above. Peaks due to such shake-up or shake-off processes can vary in both position and intensity as fig.10 shows for a range of ionic fluorides[16]. Closely related but more dramatic are the changes in such satellites associated with XP spectra of transition metal ions[17]. The presence of low-lying vacancies in the valence d shell appears to greatly facilitate specific shake-up processes. Thus intense 'shake-up' peaks are observed with an energy 6 to 7 eV less than the main photopeaks (fig.11) for Ni $2p_{\frac{1}{2}}$ $2p_{3/2}$ in paramagnetic NiO but not in the low spin dimethylglyoxime complex. High and low-spin states imply small and large splitting of the d shell orbitals respectively so that in NiO empty d levels are more accessible than in the dimethylglyoxime complex. Similar 'shake-up' effects have been found for many other transition metal compounds.

If electronically excited ions are formed as part of the primary excitation process then their relaxation should be amenable to observation. Such relaxation, if radiative, will give rise to an X-ray with a higher

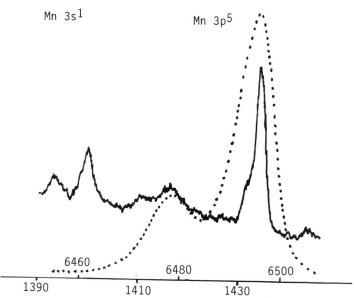

Figure 6   XP (solid line, lower scale for photoelectron kinetic energy) and XE Kβ, Kβ' (dotted line, upper scale for X-ray energy) spectra from manganese (II) difluoride.

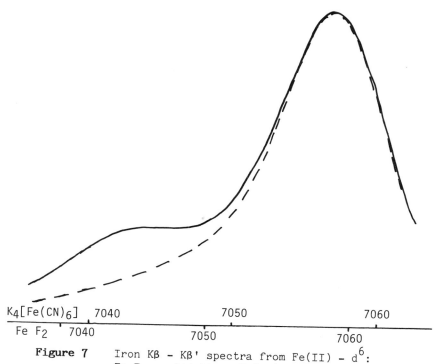

Figure 7   Iron Kβ - Kβ' spectra from Fe(II) - d^6: Fe F_2, solid line; $K_4[Fe(CN)_6]$, dashed line.

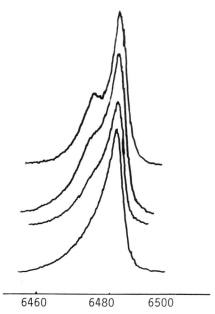

Figure 8

Kβ - Kβ' X-ray emission spectra from manganese compounds with different valencies. From the top, down:-
MnF_2, $Mn(acetate)_3$, MnO_2, $KMnO_4$

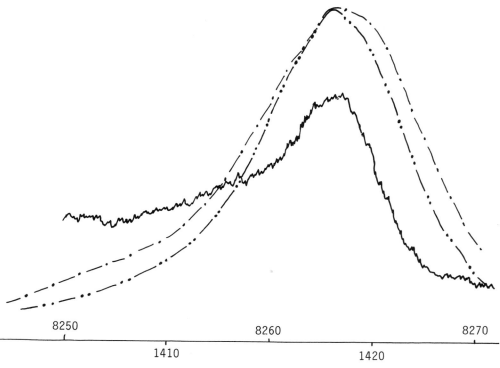

Figure 9 XP (solid line, lower scale for photoelectron kinetic energy), Ni3p spectrum and XE, Kβ spectrum (—•—•—upper energy scale) from nickel (II) hydroxide. Note the tailing to lower kinetic energies for the former and the distinct, if weak, Kβ' feature in the latter at 8252 eV. —••—••—Ni Kβ from the diamagnetic dimethylglyoxime complex: no Kβ'

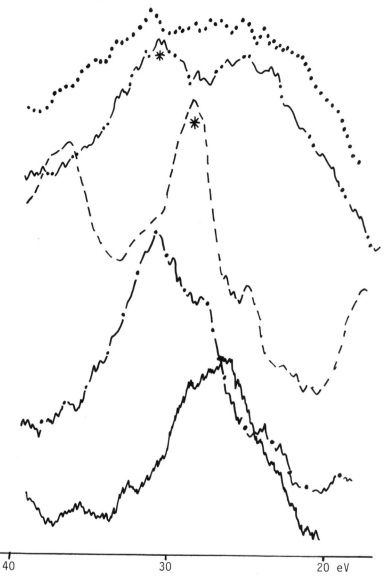

Figure 10 The shake-up, shake-off region on the low kinetic energy side of the F 1s photoelectron peak for alkali-earth fluorides. Energy scale is number of volts from F 1s.

Be F · · · · · · · · ·
Mg F — · · — · · —
Ca F — — — —
Sr F —·—·—·
Ba F ————

Probable shake-up peaks are indicated with an asterisk.

energy than the expected diagram line. An attempt to observe such a process is shown in fig.12 for $Ni(OH)_2$ where the XP spectrum for Ni $2p_{3/2}$ and Ni $L\alpha$ XE spectrum are compared[15].

Other, more dramatic satellites in XE spectra are associated with transitions, not in electronic excited species resulting from 'shake-up' events but in doubly ionised ions produced by 'shake-off'. These satellites vary widely in intensity, relative to the X-ray diagram line, both as a function of primary ionisation (photons < electrons < charged particles) and of local chemistry. The chemical effects are manifest as changes in relative intensities of the component satellite peaks e.g. $K\alpha_3 : K\alpha_4$ for aluminium[18], figure 13 and also as gross changes in the $K\alpha_{3,4} : K\alpha_{1,2}$ intensity ratio, fig. 14.

These various transitions that give rise to satellite peaks are not entirely unrelated. The ion produced by 'shake-off' in XP spectroscopy, if it relaxes by X-ray emission will generate a high energy satellite such as $K\alpha_{3,4}$. And in this particular case the final states include the ^1S and ^1D states of the $KL_{2,3}L_{2,3}$ Auger process. Thus the origins of satellites in XP and XE spectra and also Auger spectra are interconnected. PAX spectroscopy enables these interconnections to be studied. Consider for example the change in the $K\alpha_{3,4} : K\alpha_{1,2}$ intensity ratio in going from NaF to KF shown in fig. 14. It is proposed[3] that this is due to the delocalisation of the 2p fluorine vacancy to the potassium cations, a process not energetically favourable for sodium.[19] If this rationalisation is correct, then comparable effects should be observed in the Auger spectra. And they are, as shown in fig. 15, thus providing support for the original hypothesis.

Coster-Kronig Transitions

Auger processes which involve electron transitions between orbitals of the same principal quantum number and the ejection of a valence shell electron are Coster-Kronig transitions. The involvement of the valence electron means that they are especially susceptible to chemical effects. Because ejected Coster-Kronig electrons have very low kinetic energies they are not usually detected directly, but the effects of the process are manifest in changes to XP peak widths and in XE peak intensities. The possibility of a Coster-Kronig process, of the $L_1 L_{2,3} V$ type, for example, will reduce the lifetime of the L_1 state, increasing the breadth of the XP peak, and because the L_1 vacancy no longer exists, reduce the intensities of all XE peaks involving transitions to $2s^{-1}$, e.g. $L\beta_{3,4}$.

An example of how a C-K channel can be turned on or off by chemical effects is to be found in compounds of iron[20]. High-spin d^5 (ferric) octahedral complexes have electrons in the (t_{2g}) and $(e_g)^*$ orbitals, and the $(e_g)^*$ orbitals have a low enough ionisation energy for the $L_2 L_3 (e_g)^*$ process to occur. Thus the number of L_2 vacancies is reduced and L_3 vacancies is increased from the expected 1:2 ratio. This causes a corresponding increase in $L\alpha : L\beta$ and $L\ell : L\delta$ intensity ratios, fig. 16. But for the low spin $[Fe(CN)_6]^{3-}$ complex where only the (t_{2g}) orbitals are occupied the absence of loosely bound 3d electrons inhibits the C-K process. The $L\alpha : L\beta$ peak intensity ratio is 2:1, which shows that no premature $L_3 \to L_2$ relaxation had occurred.

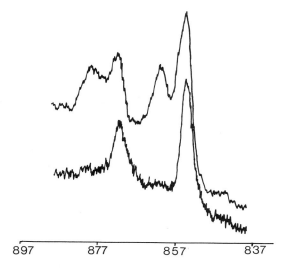

Figure 11

XP spectra, Ni $2p_{\frac{1}{2}}$ - $2p_{3/2}$ region, from paramagnetic Ni O (upper) and diamagnetic nickel dimethylglyoxime complex (lower).

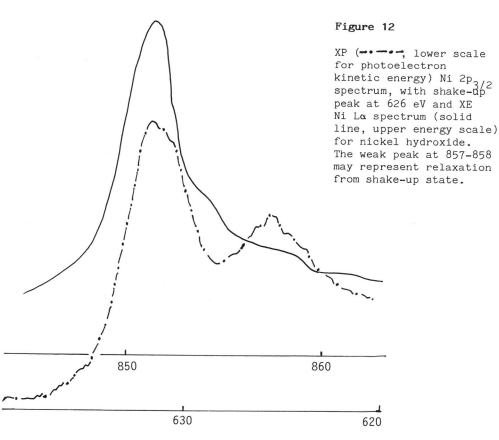

Figure 12

XP (—•—•—, lower scale for photoelectron kinetic energy) Ni $2p_{3/2}$ spectrum, with shake-up peak at 626 eV and XE Ni Lα spectrum (solid line, upper energy scale) for nickel hydroxide. The weak peak at 857-858 may represent relaxation from shake-up state.

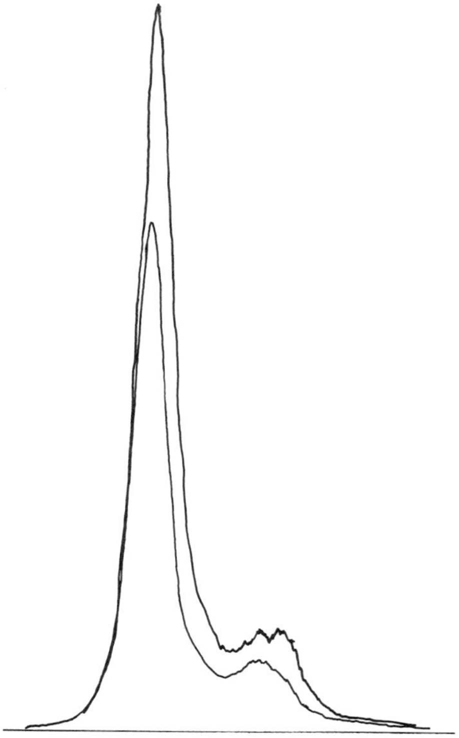

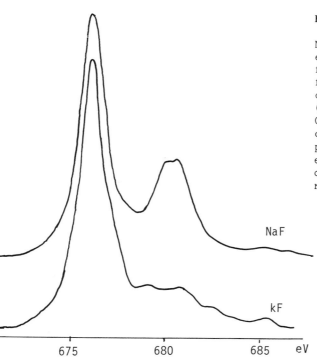

Figure 13

Main $K\alpha_{1,2}$ and high energy satellite, $K\alpha_{3,4}$ peaks from aluminium metal (lower curve) and alumina (upper curve). Note the increase in the intensity of the $K\alpha_4$ peak (relative to $K\alpha_3$) in going from metal to oxide.

Figure 14

Main $K\alpha_{1,2}$ and high energy satellite, $K\alpha_{3,4}$ fluorine X-ray peaks from sodium (upper curve) and potassium (lower curve) fluorides. Changing the counter cation from sodium to potassium has a dramatic effect upon the intensity of the $K\alpha_{3,4}$ satellite, relative to the main peak.

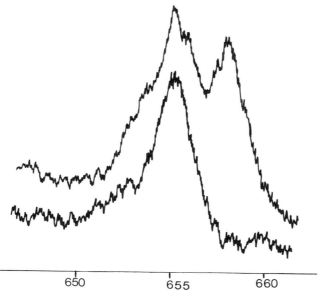

Figure 15 Fluorine Auger ($KL_{2,3}L_{2,3}$) spectra from sodium (lower curve) and potassium (upper curve) fluoride. As for the high energy satellite spectra (fig. 14) dramatic changes in peak profile accompany the change of cation.

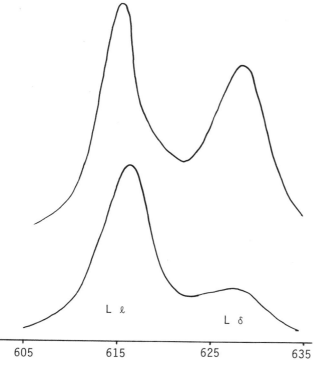

Figure 16

Fe $L\ell$, L_δ XE spectra ($2p_{3/2}$, $2p_{\frac{1}{2}} \leftarrow 3s$) for top - the low spin hexacyano complex ($K_3[Fe(CN)_6]$), and bottom - high spin FeF_3. The presence of less tightly bound d electrons in the latter allows the ($2p_{\frac{1}{2}}$ $2p_{3/2}$ V) Coster-Kronig channel to operate thus converting $2p_{\frac{1}{2}}$ vacancies into $2p_{3/2}$.

5. CONCLUSIONS

The above survey of PAX spectroscopy is not exhaustive but does cover the basic features of the spectra: it also shows how chemical effects manifest themselves and how these effects are related to each other and to the electronic structure of atoms and molecules. PAX spectra from transitions that directly involve valence band orbitals, and how these spectra give direct insight into the nature of molecular orbitals in compounds and complexes, will be the subject of the second lecture.

REFERENCES

1. C.R. Brundle and A.D. Baker (eds.), "Electron Spectroscopy: Theory Techniques and Applications", Vols. 1-5, Academic Press, London, UK, (1977-).

2. K. Siegbahn, C. Nordling, A. Fahlman, R. Nordberg, K. Hamrin, J. Hedman, G. Johansson, T. Bergmark, S.-E. Karlsson, I. Lindgren and B. Lindberg, "ESCA: Atomic Molecular and Solid State Structure Studied by Means of Electron Spectroscopy", Almqvist and Wiksells, Uppsala, Sweden, (1967).

3. K. Siegbahn, C. Nordling, G. Johansson, J. Hedman, P.F. Heden, K. Hamrin, U. Gelius, T. Bergmark, L.O. Werme, R. Manne, and Y. Baer, "ESCA Applied to Free Molecules", North-Holland Pub. Co., Amsterdam, Netherlands, (1969).

4. D.W. Turner, C. Baker, A.D. Baker and C.R. Brundle, "Molecular Photoelectron Spectroscopy", Wiley-Interscience, London, UK, (1970).

5. J.H.D. Eland, "Photoelectron Spectroscopy", Butterworths, London, UK, (1974).

6. L.V. Azaroff, "X-ray Spectroscopy", McGraw-Hill Book Co., New York, USA, (1974).

7. B.K. Agarwal, "X-ray Spectroscopy", Springer-Verlag, Berlin, BRD (1979).

8. M. Thompson, M.D. Baker, A. Christie and J.F. Tyson, "Auger Electron Spectroscopy", J. Wiley & Son, New York, USA, (1985).

9. W.C. Price, A.W. Potts and D.G. Streets, The Dependence of Photoionization Cross-section on the Photoelectron Energy, in: "Electron Spectroscopy", D.A. Shirley, ed., North-Holland Pub. Co., Amsterdam, Netherlands, (1972).

10. E.U. Condon and G.H. Shortley, "The Theory of Atomic Spectra", Cambridge Univ. Press, Cambridge, UK, (1935).

11. W. Domcke, L.S. Cederbaum, J. Schirmer, W. von Niessen and J.P. Maier, "Breakdown of the Molecular Orbital Picture of Ionization for Inner Valence Electrons: Experimental and Theoretical Study of H_2S and PH_3, J. Electron Spec. and Rel. Phenom.", 14:59 (1978).

12. C.S. Fadley, Multiplet Splittings in Photoelectron Spectra, in: "Electron Spectroscopy", D.A. Shirley ed., North-Holland Pub. Co., Amsterdam, Netherlands, (1972).

13. R.A. Slater and D.S. Urch, "The Origin of the Kβ' Satellite Peak in the X-ray-fluorescence Spectra of Iron Compounds: a Correlation with Magnetic Susceptibility", J. Chem. Soc. Chem. Comm., p.564, (1972).

14. D.S. Urch and P.R. Wood, "The Determination of the Valency of Manganese in Minerals by X-ray Fluorescence Spectroscopy", X-ray Spectrometry, 7:9 (1978).

15. P. McClusky and D.S. Urch, "The Non-correlation of X-ray Photoelectron and X-ray Emission Spectra for Transition Metal Compounds", J. de Physique, in press (1988).

16. I. Ikemoto, K. Ishii, S. Kinoshita and H. Kuroda, "Satellite and Loss Structures of Core-electron Peaks in X-ray Photoelectron Spectra of Alkaline-earth Fluorides", J. Electron Spec. & Rel. Phenom., 11:251, (1977).

17. D.C. Frost, A. Ishitani and C.A. McDowell, "X-ray Photoelectron Spectroscopy of Copper Compounds", Molec. Phys., 24:861 (1972).

18. D.W. Fischer and W.L. Baun, "Diagram and Nondiagram Lines in K Spectra of Aluminium and Oxygen from Metallic and Anodised Aluminium", J. Appl. Phys., 36:534, (1965).

19. D.S. Urch, "The Temporary Covalence of Potassium Fluoride (in X-ray and Auger Spectra Processes), J. Chem. Soc., Chem. Comm., p.526, (1982).

20. P.R. Wood and D.S. Urch, "The Lℓ and Lη X-ray Emission Spectra of Some First-row Transition Metals and Compounds, J. Phys. F: Metal Phys., 8:543, (1978).

ELECTRONIC STRUCTURE OF MOLECULES, COMPLEXES AND SOLIDS USING

PAX SPECTROSCOPY

David S. Urch

Chemistry Department, Queen Mary College
Mile End Road, London E1 4NS, UK

1. INTRODUCTION

The first lecture has outlined the basic physical processes that give rise to photoelectron, Auger and X-ray emission spectra. It has also shown how chemical factors, valency, atomic environment etc. can modify both the energies and intensities of the photons and the electrons detected in PAX spectroscopy. In this lecture attention will be focussed on transitions that directly involve the valence electrons and where, understandably, greater 'chemical effects' are observed.

In XP spectra the intensities of peaks corresponding to valence orbitals tend to be the weakest because of the disparity in effective wavelengths between the ejected photoelectron and the orbital from which it came. Much more intense peaks are of course observed in UPS which is therefore more useful in mapping out, on an energy scale, the locations of molecular orbitals. The advantage of XP spectra, in this sense, is that the valence peaks, although weak, are on the same scale as the core orbitals. This is invaluable for the alignment of X-ray emission spectra (see below).

X-ray emission spectra that arise from valence-band to core vacancies show considerable fine structure. This is because the particular type of atomic orbital that could give rise to a specific X-ray may contribute many different molecular orbitals. The structure of the X-ray peak will then reflect the distribution of that particular atomic orbital character amongst the molecular orbitals.[1,2] The great potential of valence-XE spectroscopy is that it is to a great extent atom specific and even orbital specific. This is due to the highly localised nature of the initial core hole and the operation of the dipole selection rule $\Delta \ell = \pm 1$. A valence XE spectrum can therefore be used to investigate the bonding role of a specific type of atomic orbital, and if all possible valence XE spectra for a compound are collected, then the bonding roles of all the atomic orbitals can be studied. When core ionisation energies are known from XPS then all these XE spectra can be aligned on a common energy scale and the atomic orbital structure of each molecular orbital be clearly seen.[3]

In order that these general ideas may be made more precise, recourse must be made to a theoretical model for molecular electronic structure. The predictions of theory, such as peak energies and relative intensities, can then be compared with the observed PAX spectra. The most basic theory of electronic structure is that in which molecular orbitals (MOs), are represented as Linear Combinations of Atomic Orbitals (LCAO), $\Psi_i = \Sigma a_{ri} \Phi_r$, where a_{ri} is a coefficient which describes the contribution of atomic orbital Φ_r to the i th molecular orbital Ψ_i, and in which the Hamiltonian operator is reduced to a series of one-electron operators. Quite gross approximations are made in the name of simplicity but a useful, if rather qualitative, picture emerges in which approximate molecular orbital ionisation energies can be calculated and in which the structure of each molecular orbital is given in terms of constituent atomic orbitals. The use of such M.O's for Ψ_f in equation (1) - lecture I, (the preceding paper) enable the energies and intensities of X-rays from valence bond orbitals to be estimated, as outlined below.

Consider the valence band structure of a molecule AB where A is a 2nd row element and B is from the third row. The molecular orbitals between the two atoms will have the general (LCAO) form,

$$\Psi_f = a(As)_f \cdot \Phi(A2s) + a(Ap)_f \cdot \Phi(A2p) + a(Bs)_f \cdot \Phi(B3s)$$
$$+ a(Bp)_f \cdot \Phi(B3p) + a(Bd)_f \cdot \Phi(B3d)$$

The probability of relaxation, with the emission of X radiation, to an A1s core hole will be proportional to the square of the integral, $\int \Psi_f \underline{P} \Phi(A1s)$. Upon expansion the integral becomes,

$a(As)_f \int \Phi(A2s) \underline{P} \Phi(A1s) + a(Ap)_f \cdot \int \Phi(A2p) \underline{P} \Phi(A1s) +$

$a(Bs)_f \int \Phi(B s) \cdot \underline{P} \Phi(A1s)$ ... etc. where the first term, will be zero by the dipole selection rule, the second term will be non-zero (unless $a(Ap)_f$ is zero) and the remaining terms - 'cross over' terms can usually be neglected since the magnitude of B wavefunctions near the A nucleus is small (but see the lecture by Larkins (and ref. 5) for exceptions when A-B is short and B, like A, is from the second row). The second integral is related to the probability of 2p → 1s relaxation at atom A. When A2p participates in different molecular orbitals (different values of f) with different ionisation energies (E_1, E_2 ... E_f) then the relative intensities of the valence X-ray peaks of energies,

$$(E(A1s) - E_1), (E(A1s) - E_2) \ldots (E(A1s) - E_f)$$

will be proportional to

$$(a(Ap)_1)^2, (a(Ap)_2)^2 \ldots (a(Ap)_f)^2 \qquad (1)$$

respectively (where E(A1s) is the ionisation energy of the 1s orbital of A).

Exactly similar arguments can be advanced for transitions to B core orbitals, (eg B1s, B2p etc.), and so in general, the relative intensities of the peaks found in a valence X-ray spectrum can be taken as a direct measure of the participation of a specific atomic orbital in bond formation. The valence K spectrum of A probes the bonding role of A valence shell p orbitals, and if A lies deeper in the Periodic Table than the second row, the valence $L_{2,3}$ spectrum probes A valence s and d orbitals etc. Similar spectra from B would enable bonding roles played by B valence shell s (valence $L_{2,3}$) and p (valence K) and d (valence $L_{2,3}$) orbitals to be studied. Some examples of the changes wrought in individual valence XE spectra by an atom being in different chemical environments will be discussed in the next section whilst the subsequent section will describe how all XE and XP spectra for a particular compound may be brought together to give a detailed account of the electronic structure of that compound.

2. VALENCE XE SPECTRA

As the pattern and composition of molecular orbitals changes from compound to compound, corresponding changes in valence XE peak profiles and peak energies should be expected. A valence X-ray emission peak will be generated by transitions from all molecular orbitals with a particular type of atomic orbital, the relative intensities of the components of the peak reflecting the contribution of that atomic orbital to each molecular orbital. Examples of this effect are outlined below for compounds of silicon and sulphur; in both cases the valence XE line is $K\beta_{1,3}$.

Chemical Effects in Sulphur $K\beta$ spectra

Figure 1 shows sulphur $K\beta_{1,3}$ spectra from many compounds in which the sulphur valency varies (formally) from -2 to +6 and in which a variety of different ligands is present. The general shift to higher energies with increasing valency is primarily due to the increasing ionisation energy of sulphur 1s (lecture 1) whilst the changes in peak shape and in the fine structure of the peak, are due to the participation of S 3p orbitals in bond formation. It is interesting to note that the $SK\beta'$ peak, which reflects S 3p character in orbitals that are mostly ligand 2s, is separated from the main $K\beta$ peak by an energy difference which varies with the nature of the ligand atom. As can be seen this energy separation is constant enough for the position of $K\beta'$ relative to $K\beta$ to be used for ligand identification. The presence of $K\beta'$ is, then, diagnostic of the presence of specific types of sulphur-ligand bonds (the argument is of course of more general application than just to sulphur compounds). Furthermore the relative intensity of $K\beta'(:K\beta)$ can also be related to the number of ligands of a particular type.

In combination with unsaturated organic groups, the shape of the $SK\beta_{1,3}$ peak can be used to identify S 3p participation in σ and in π bonds. Thus for thiourea it seems reasonable to identify the low intensity peak at low X-ray energy with S 3p character in the σ bond to carbon whilst the main peak represents the remaining 3p electrons in the C-S π bond and the lone pair. Conversely when sulphur forms two σ bonds to two aromatic rings as in dibenzofuran the picture is reversed. The broad main peak is now related to the participation by two S 3p orbitals in the formation of two C-S σ bonds and the presence of S 3p character in a range of σ molecular orbitals. The remaining 3p orbital, by interacting with the π system, allows the two electrons to be extensively delocalised over the aromatic rings. This reduces the charge density at sulphur and hence reduces the intensity of the highest energy X-ray peak. In the aliphatic systems, cystine and cysteine such delocalisation is not possible and the broad $SK\beta$ that is observed can be rationalised as due to two S 3p electrons in σ bonds and two S 3p electrons in a lone pair orbital. A similar broad peak is observed for the S_2^{--} anion found in pyrites. Here the breadth is due to S 3p character in σ, π and π* orbitals whose ionisation energies are spread over many electron volts. When no S-S bonds are present, as in S^{--}, the $SK\beta$ peak is without structure, as would be expected; but it is perhaps surprising that the 'sulphide' $K\beta$ peak is not as narrow as that for sulphate. That it is not probably indicates polarisation of sulphide by the cation, i.e. some degree of covalency being present in the bond.

Chemical Effects in Silicon $K\beta_{1,3}$ Spectra

That the dramatic changes in peak shape and profile are not confined to valence XE spectra from sulphur is shown in fig. 2 where similar effects can be observed for silicon $K\beta$. Again the separation of $K\beta'$

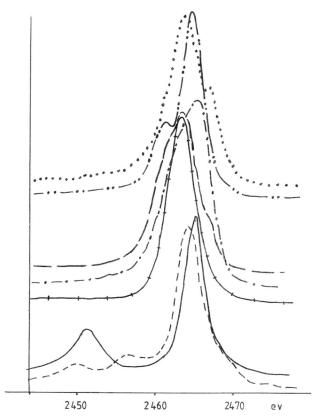

Figure 1 SKβ XE spectra from various sulphur containing compounds. From the top, down: dibenzofuran, dotted line; thiourea, 2 dots-dash line; pyrite, long dashes line; cysteine, dot-dash line; lithium sulphide, solid line with bars; sodium sulphate, solid line; dimethyl sulphoxide, dashed line.

from the main Kβ peak can be correlated with the ligand atom (and if two distinct ligand atoms are present,[7] as in topaz, then two peaks, each at its characteristic distance, are seen). When gross changes are made, as in going from carbide to nitride to oxide, correspondingly large changes are observed in the SiKβ peak reflecting the quite different bonding situations in these compounds. But very significant changes are also observed depending upon whether the silicon is four or six coordinated and even whether the groups attached to the ligand atoms are unsaturated or not.[8] Thus the acetylacetonate complex and the catechol complex have a very intense 'high energy' peak at ~1836 eV, presumably due to interaction between Si3p and the π-system. Other compounds and complexes show emission in this region but it is much less intense. Also this weaker emission seems to be related to Si - O - X bond angles[9] as well as the chemical nature of X. Even for minerals where the silicon coordination is exclusively fourfold by oxygen variations are found in SiKβ which can be correlated with mineral type and even polarising power of the cation.[10]

3. PAX SPECTROSCOPY

Introduction. If the full potential of PAX spectroscopy for investigating electronic structure is to be exploited then it is necessary to combine all the valence XE spectra together with the XP valence band spectrum from the sample onto the same energy scale.[3] Then all the peaks reflecting the presence of different atomic orbitals in any one molecular orbital will be found at the same point. As each X-ray is the result of a transition from an initial (core) state to a final (valence band) state the energy of the core state must be measured by XPS so that the ionisation energy of the final state(s) can be calculated. As these final states are molecular orbitals all the XE spectra can now be aligned so that the peaks in the fine structure of each spectrum correspond to the appropriate molecular orbital. As explained above, eq.(1), the relative intensities of the component peaks in any one spectrum can be correlated with the squares of the coefficients for that atomic orbital in the different molecular orbitals of the compound being studied. If, as is possible in some cases, one peak can be identified with reasonable confidence with an orbital of known electronic composition (eg. a lone pair orbital, with two electrons localised on one atom) then the relative areas of peaks in this spectrum can be used to quantify the distribution of charge associated with a specific atomic orbital. And whilst it is always possible to correlate ratios of areas of peaks with the squares of LCAO coefficients in any one spectrum it is of course *not* possible to compare one spectrum with another unless both the transition integrals $\int \varphi$ (valence AO)$\underline{P}\varphi$ (core AO) are known. A further complication arises for non K spectra since two possible types of valence AO can initiate transitions to the core vacancy eg. transitions from s and d orbitals are possible to a p vacancy. In this case it is *not* possible to use relative XE peak areas to compare s and d contributions to m.os unless the relevant integrals can be estimated - *not* an easy task! Despite these limitations PAX spectra can be used to build up a semi-quantitative picture of electronic structure. Furthermore where theoretical calculations are available their predictions, both of molecular orbital energies and of atomic orbital composition of molecular orbitals, can be verified. Examples are discussed below.

Sulphate anion SO_4^{--}

The sulphate anion has tetrahedral (T_d) symmetry. The constituent valence atomic orbitals (O 2s, 2p: S 3s, 3p, 3d) can therefore be

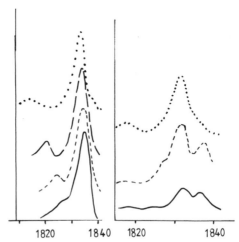

Figure 2 SiKβ XE spectra from various compounds.
Left hand diagram, from the top down:
Na_2SiF_6, dotted line; SiO_2, long-dashes line; Si_3N_4, short-dashes line; $Si C$ solid line.
Right hand diagram, from the top down:

$[NH_3(CH_2)_6NH_3]^{++}$ $[C_2H_5OSi(OCH_2CH_2O)_2]_2^-$, dotted line;

$[Si(CH_3\ CO\ CH\ CO\ CH_3)_3]^+$ $[HCl_2]^-$, dashed line;

$[NH(C_2H_5)_3]_2^+$ $[Si(C_6H_4O_2)_3]^{2-}$, solid line

classified into many different irreducible representations (Table 1) which greatly facilitates the construction of a qualitative molecular orbital energy level diagram, fig. 3.

Table 1

Classification of Sulphur and Oxygen Valence Shell Orbitals in tetrahedral point group symmetry.

S 3s	a_1	O 2s	$a_1 + t_2$
S 3p	t_2	O 2p (O - S)	$a_1 + t_2$
S 3d	$e + t_2$	O 2p (O $\perp$ S)	$e + t_1 + t_2$

(O - S) O 2p orbitals orientated along O - S directions
(O$\perp$S) O 2p orbitals orientated perpendicular to O - S directions

In drawing the diagram it has been assumed that because the overlap between t_2 orbitals derived from oxygen 2p orbitals perpendicular to S-O bonds and S 3d t_2 orbitals will be good (π-bonding) and the overlap between oxygen 2p t_2 orbitals pointing towards sulphur and sulphur 3p orbitals (σ-bonding) will also be good, that it is possible to consider these σ and π interactions separately. In practice, however, there is some slight $\sigma - \pi$ mixing which is discussed below. Figure 3 also shows all the electronic transitions which should give rise to peaks in the XE spectra. Corresponding PAX spectra for sulphate are collected in figure 4 (c.f. ref. 11). In order to include the SKβ spectrum, for which it is not possible to measure the S 1s ionisation energy by conventional XPS, the sulphur 1s ionisation energy must be calculated from the XP determined S $2p_{3/2}$, $2p_{1/2}$ ionisation energies and the energy of S K$\alpha_{1,2}$. Then the energy of the MO that corresponds to a particular peak can be calculated from the general expression MO ionisation energy = (core level ionisation energy) - (X-ray energy).

The most intense peak in the valence XP spectrum, at 27-24 eV can be identified by its energy and relative intensity as being due to orbitals mostly O 2s in character, corresponding to $4a_1$ and $3t_2$.

The location of S Kβ' at 24.5 eV enables the position of $3t_2$ to be determined. Similar peak coincidences at 14.5 eV show that this XP peak arises from $5a_1$, a σ orbital with S 3s and O 2p character. The main S Kβ at 11 eV identifies this peak as due to the σ orbital which has $4t_2$ mixed S 3p, O 2p character. The intense O Kα peak spans 5 to 8.5 eV clearly showing that orbitals in this ionisation range are mostly O 2p, i.e. 1e, $5t_2$ and $1t_1$. This peak has a low energy shoulder (~ 6.5 eV) which aligns with the high-energy X-ray peak in the S $L_{2,3}$ M spectrum. As there are no other a_1 orbitals possible in the tetrahedral unit this peak must be due to S 3d character. Thus molecular orbitals between 6.5 and 8.5 eV will have S 3d admixed with O 2p. This is possible for 1e and $5t_2$ orbitals as a result of 'back-bonding' from oxygen to sulphur. Such bonding helps to stabilise the anion both by the formation of π-bonds and by charge delocalisation. The small 'foot' in the S Kβ spectrum at 6.5 eV enables the $5t_2$ orbitals to be located. Its presence indicates that the separation into "σt_2" and "πt_2" orbitals proposed above on overlap grounds is only an approximation and indeed its intensity (which is small) can be used to measure the degree of $\sigma-\pi$ mixing present in sulphate.

The detailed analysis of the SO_4^{--} PAX spectra and comparison with ab-initio calculation of atomic contributions to various MO's by Kosuch,

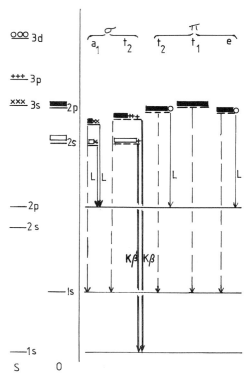

Figure 3 Molecular orbital energy level diagram for the sulphate. Atomic orbitals are on the left and molecular orbitals, classified according to the appropriate irreducible representations for T_d symmetry, are on the right. Approximate molecular orbital structure, in terms of constituent atomic orbitals is shown by the appropriate symbols. Possible X-ray transitions are shown, as dashed lines for O Kα, as solid lines marked L for SL$_{2,3}$M and as double lines for SK$\beta_{1,3}$ and SKβ'.

Wiech and Faessler[11] shows good agreement between theory and experiment. It is necessary to appreciate that peak intensities must be related to area (and not height) as component peak widths (in O Kα for example) vary. This is probably due to different degrees of vibrational structure being present in bonding and non-bonding orbitals.

<u>Sulphur Hexafluoride</u>. PAX spectra for SF_6 are shown in figure 5. The $S\ L_{2,3}M$ spectrum has been divided into L_2 and L_3 components and the latter aligned to the other spectra using the L_3 (S $2p_{3/2}$) ionisation energy. The high symmetry of SF_6 enables the atomic orbitals from sulphur and fluorine and the molecular orbitals that result from their interaction to be allocated to many different irreducible representations, as shown in Table 2.

As in the sulphate anion the S L spectrum again provides direct evidence for the participation of S 3d orbitals in bond formation. S 3s character can only be anticipated in a_{1g} orbitals and the peak at 27 eV can be identified with $2a_{1g}$. The other peaks at 19.5 eV and 18.3 eV must therefore be due to S 3d character in $2e_g$ and $1t_{2g}$ orbitals that are otherwise mostly F 2p. The twelve fluorine lone pair orbitals (two from each fluorine atom perpendicular to the fluorine-sulphur bond) will be split by mutual repulsion into four triply degenerate sets, $1t_{2g} - 3t_{1u} - 1t_{2u} - 1t_{1g}$ (in order of decreasing binding energy). But the t_{1u} set ($3t_{1u}$) belongs to the same irreducible representation as the σ bonds $2t_{1u}$ so that σ-π interaction is possible and would lead to S 3p character being present in $3t_{1u}$ as well as the stabilisation of $2t_{1u}$ and the destabilisation of $3t_{1u}$. Thus it is observed that $3t_{1u}$ and $1t_{2u}$ have just about the same ionisation energy, with the peak at 17.0 eV in the XP spectrum having double the intensity of the peaks that may be ascribed to $1t_{2g}$ at 18.5 eV and $1t_{1g}$ at 15.7 eV. That the peak at 17.0 eV has t_{1u} character is confirmed by the minor peak in the S Kβ XE spectrum at 2472.3 eV.

Hexafluorophosphate PF_6^-
‾‾‾‾‾‾‾‾‾‾‾‾‾‾‾‾‾‾‾‾‾‾‾‾‾‾

The same MO model (table 2, replace S by P) can be used for this anion as for isoelectronic SF_6. The PAX spectra (XPS, F Kα, P Kβ and $P\ L_{2,3}M$) are aligned on a common energy scale in Fig. 6. The relative positions of P Kβ and $P\ L_{2,3}M$ peaks show clearly the partitioning of P 3s P 3p and P 3d character into distinct orbitals. The high energy peak in the $L_{2,3}M$ spectrum can only be easily explained as due to P 3d participation in bonding, stabilising $2e_g$ and $1t_{2g}$ (σ & π respectively) orbitals on fluorine. The weak shoulder at about 12.5 eV in the P Kβ spectrum locates $3t_{1u}$ and presumably arises from σ($2t_{1u}$)-π($3t_{1u}$) interaction.

What is not clear from this consideration of bonding based on hexafluoride PAX spectra is the reason for the low relative intensity of F Kα in $2t_{1u}$ and especially $2a_{1g}$, implying as it does, rather low fluorine contributions to these bonding orbitals. It may be due to the use of an over-simplified model in which relaxation effects are excluded (see lectures by Prof. Larkins), or to sample decomposition (observed, but only monitored[12] for $P\ L_{2,3}M$) or it may reflect the reality of the S-F and P-F bonds, in which ionic structures *are* important and in which the 3s orbital especially plays little active part in bonding.

Periclase (MgO) and brucite ($Mg(OH)_2$)
‾‾‾‾‾‾‾‾‾‾‾‾‾‾‾‾‾‾‾‾‾‾‾‾‾‾‾‾‾‾‾‾‾‾‾‾‾

Periclase has a simple rock salt structure and might be expected to give an equally simple set of PAX spectra. What is observed is

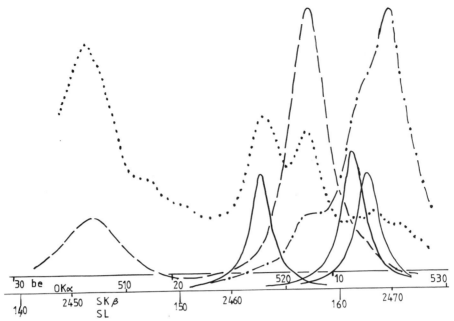

Figure 4 PAX spectra for the sulphate anion XPS, dotted line; OKα dot-dash line; SKβ, dashed line; S L$_3$M, component peaks of this spectrum shown as solid lines.

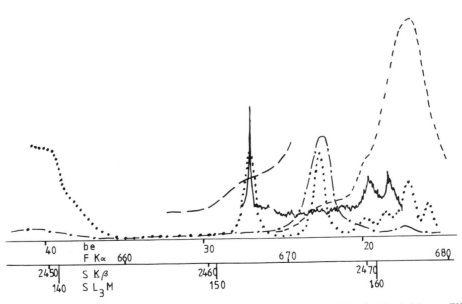

Figure 5 PAX spectra for sulphur hexafluoride. XPS, dotted line; FKα dashed line (664-670 eV, shown as long dashes, intensity X10); SKβ, dot-dash line; S L$_3$M, solid line (features in the original experimental trace due to S L$_2$M have been omitted)

TABLE 2

Classification of Atomic and Molecular Orbitals for SF_6: Octahedral (O_h) Symmetry

Atomic Orbitals (AO)	Irreducible Representations (I.R.)	Valence Shell Molecular Orbitals (I.R.) increasing binding energies on going down	Major AO Components
S 3s	a_{1g}	$1t_{2u}$	F 2p
S 3p	t_{1u}	$1t_{1g}$	F 2p
		$3t_{1u}$	F 2p–(S 3p)*
S 3d	$e_g + t_{2g}$	$1t_{2g}$	π F 2p – S 3d
		$2e_g$	σ F 2p – S 3d
F 2s	$a_{1g} + t_{1u} + e_g$	$2t_{1u}$	σ F 2p – S 3p
		$2a_{1g}$	σ F 2p – S 3s
F 2p (F – S)	$a_{1g} + t_{1u} + e_g$	$1e_g$	F 2s
		$1t_{1u}$	F 2s
F 2p (F ⊥ S)	$t_{1g} + t_{1u} + t_{2g} + t_{2u}$	$1a_{1g}$	F 2s

(F – S) F 2p orbitals orientated towards sulphur
(F ⊥ S) F 2p orbitals perpendicular to F – S direction

* σ – π mixing due to interaction between $2t_{1u}$ and $3t_{1u}$ allows S 3p character to enter $3t_{1u}$

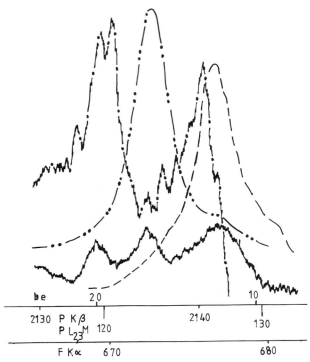

Figure 6 PAX spectra for the PF_6^- anion. XPS, solid line (at bottom of figure); F Kα, dashed line; P Kβ two dots-dash line; P $L_{2,3}$M solid line with single dots. Appropriate energy scales are given at the bottom of the diagram. Molecular orbital binding energies are given by the be scale.

shown in fig.7. The splitting of the MgKβ peak and the shoulder at 522 eV in the O Kα spectrum show that the bonding between magnesium and oxygen is more complex than a simple ionic model would predict. The very presence of a Mg Kβ spectrum shows that there must be some degree of covalent bonding between the magnesium and the oxygen.

A simple model for the electronic structure may be constructed from $Mg_4 O_4$ in which alternate magnesiums and oxygens are placed at the corners of a cube.[13] As the Mg 3s and 3p orbitals have not too dis-similar ionisation energies sp hybrids can be constructed, and one from each Mg points to the centre of the cube. Under the T_d symmetry of Mg_4 they transform as a_1 and t_2. For oxygen the 2s orbital is relatively tightly bound and so can be ignored at this stage. If four O 2p orbitals, one from each atom of the O_4 unit, are also orientated to the centre of the cube they too transform as a_1 and t_2. A detailed consideration of the interaction of the Mg_4 and O_4 a_1 and t_2 orbitals provides an explanation for the observed magnesium and oxygen PAX spectra. The model may be extended (by the addition of other cubes with corners common to the original) to $Mg_{32} O_{32}$ and thus indicate how the localised MOs of $Mg_4 O_4$ will transform into bands in solid periclase. The PAX spectra of brucite $Mg(OH)_2$ fig. 8, turn out to be determined by the electronic structure of the hydroxyl group.[14] Each oxygen has two 2p lone pair orbitals (e) and one σ bond to hydrogen (a_1). Thus O Kα has a sharp peak at 525.4 eV (12.6 eV) and a shoulder at 521.0 eV. It is interesting that the XP spectrum, which would be expected to be very similar to O Kα, has a much greater relative intensity in the 'O-H bond' peak at 17.0 eV. This is probably due to a few percent of O 2s character. The Mg Kβ spectrum, however, can be understood as simply due to the donation of some charge from the hydroxyl e and a_1 **orbitals,** to the magnesium cation.

Thiourea

A correlation diagram for the PAX spectra (XPS, S Kβ, S $L_{2,3}$ M N Kα and C Kα) of thiourea is shown in fig. 9. The resolution of the N Kα is not good enough for unambiguous deconvolution but structure can be seen which aligns with main features in other spectra. In particular the peak at 396 eV would seem to make a reasonably large contribution to the orbitals at 3.5 eV which also have considerable S 3p character. The alignment of C Kα and S Kβ (satellite) at 12.5 eV suggests a σ bond but then it is difficult to reconcile the number of peaks and their positions with S 3p contributions to C-Sσ, C-Sπ and S lone pair orbitals. Whilst the deconvolution proposed for 3p cannot be unique the relative intensities suggest that most (~ 70%) of the S 3p character is to be found in the least tightly bound orbitals. If the S 3p lone pair is located at 2465 eV (8 eV) then C-Sπ must be at 2.464 eV (9 eV) and have about 80% S 3p character, which seems unreasonable. A less awkward rationalization can be found by admitting the nitrogen 2p lone pair orbitals into the π bonding scheme which then becomes approximately isoelectronic with the π-system of carbonate and nitrate. In this case the simplest Hückel type MO calculations suggest that C 2p should only be found in the bonding π-orbital (50%) with one third of an electron located on S and on each of the nitrogens.

The four electrons in the degenerate 'non-bonding' orbitals would then be located one and one-thirds each on the three ligand atoms. This model simply and easily accounts for the large electron count in the 'lone pair' peak. A separate π peak is not observed in the carbon Kα spectrum because the extended π bonding gives rise to a more tightly bound π orbital, comparable in ionisation energy with the σ bond. The large amount of carbon character predicted for this orbital presumably augments the carbon peak at 280-281 eV (12.5 eV). The σ bonds form a more complicated

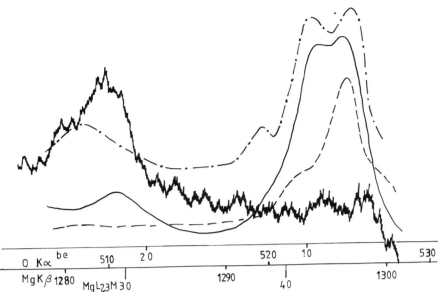

Figure 7 PAX spectra for MgO (periclase).
XPS, solid line (with noise): O Kα, dashed line: Mg Kβ, solid line (smooth): Mg $L_{2,3}$M dot-dash line; (peak at 39 eV probably Mg $L_1 L_{2,3}$)

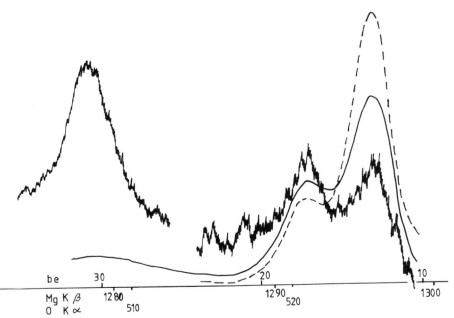

Figure 8 PAX spectra for $Mg(OH)_2$ (brucite)
XPS, solid line (with noise), note scale change (x3) at 25 eV: O Kα, dashed line: Mg Kβ, solid line (smooth).

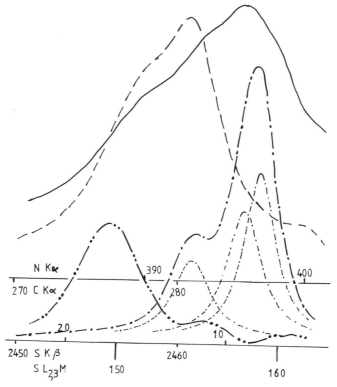

Figure 9 X-ray emission spectra for thiourea. N Kα, solid line: C Kα, dashed line: S L$_{2,3}$M two dots-dash line: S Kβ dot-long dash line (component peaks shown by dot-dash lines).

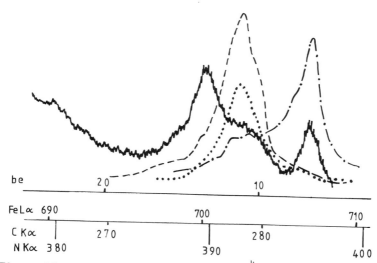

Figure 10 PAX spectra for [Fe(CN)$_6$]$^{4-}$ XPS, solid line: Fe Lα, dot-dash line: C Kα, dashed line: N Kα, dotted line

set of orbitals whose interactions cannot easily be determined from the spectra. The S 3s orbital would however seem to participate only slightly in bond formation. The main peak at 17.5 eV overlaps with a small C 2p component whilst the main C-Sσ bond at 16 eV would seem to have a small complimentary contribution from S 3s.

Transition Metal Complexes

Ferrocyanide [Fe(CN)$_6$]$^{4-}$. PAX spectra for the potassium salt of this anion are shown in fig.10 (XP, Fe Kβ, Lα, CKα, NKα).[15] The conventional ligand field model for the bonding suggests that ligand orbitals (C sp 'lone pairs' from cyanide) should interact with Fe 4s (a_{1g}), Fe 4p (t_{1u}) and Fe 3d ($d_{x^2-y^2}$, d_{z^2} : e_g) to form reasonably tightly bound σ bonds. Such orbitals can be found in the XP spectrum in the range 6 - 16 eV and evidence for iron and carbon character is seen in the XE spectra. The main peak in the valence XP spectrum aligns with the Fe Lα maximum, thus identifying it with the filled Fe 3d t_{2g} orbital. The electronic structure of this anion is also of interest because of the possibility of π- 'back-bonding' from the Fe 3d (t_{2g}) orbitals to empty π* orbitals of the same symmetry. There will be twelve π* orbitals on the six cyanide ligand groups which transform as t_{1g}, t_{2g}, t_{1u}, and t_{2u}. If it were to occur then some evidence should be found in the NKα (and to a lesser extent CKα) with a peak which aligns with the Fe Lα maximum. That no such feature is observed indicates that 'back-bonding' is not present to any great extent in the ferrocyanide anion.

Ruthenium-ethene bonding Another example of organo-metallic bonding is present in the complex [(C$_2$H$_4$)$_2$ RuCl]$_2$ where the centre of each ethene molecule lies in the plane which contains the two ruthenium atoms and the two bridging chlorine atoms: the four ethenes are perpendicular to this plane. A qualitative model of the Ru-ethene bond suggests that charge is donated from the π-bonds of ethene to empty Ru 4d orbitals (forming σ bonds) and that charge is also 'back-bonded' from filled 4d orbitals to the empty π* orbitals of ethene (π- bonds). The PAX spectra, before and after complex formation, which are shown in fig. 11 support this model;[16] especially large changes are observed in the high energy part of C Kα indicative of profound changes in the ethene π orbitals. Deconvolution of the C Kα for ethene is guided by the UP spectrum which indicates the binding energies of the σ and π orbitals. When the complex is formed it is possible to deconvolve the new C Kα spectrum in terms of an unchanged σ portion, a π peak of slightly reduced intensity which has been displaced to a lower X-ray energy (i.e. the corresponding MO has become more tightly bound), from 280 eV to 278.8 eV and a new peak at 281.2 eV. This new peak aligns with the main peak in the XP valence band spectrum which by comparison with the Ru Lβ$_{2,15}$ peak can be demonstrated to be due to orbitals that are mostly Ru 4d in character.

Thus complex formation leads to a reduction in the C 2p character in, and to increase in ionisation energy of, the ethene π bond together with the formation of a new, less tightly bound, bond with both C 2p and Ru 4d character. This new bond provides direct experimental evidence for 'back bonding' and shows how PAX spectroscopy could be used to quantify this concept. The changes observed in PAX spectra are wholly in accord with the predictions of the synergic model outlined above and, thus, confirm its validity.

Metal-Metal Multiple Bonding [Mo$_2$Cl$_8$]$^{4-}$. One of the most dramatic developments of the past few decades in inorganic chemistry has been the preparation and characterisation of innumerable compounds containing directly linked metal atoms. Both single and multiple metal-metal bonds

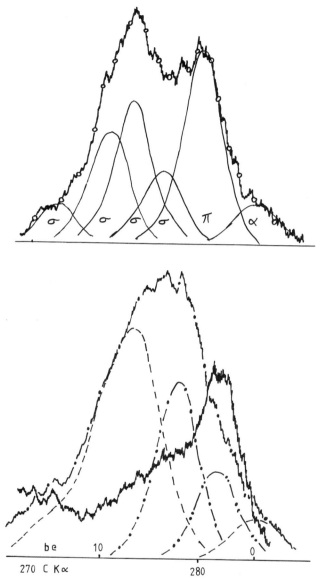

Figure 11 Upper diagram shows C Kα X-ray emission spectrum for ethene (77K). Component peaks, based on UPS data, are shown by solid lines. σ, C-C and C-H bonds; π, pi-bond; α, Kα$_{3,4}$ satellite peak.
Lower diagram shows the C Kα X-ray emission spectrum from the complex [Cl Ru (C$_2$H$_4$)$_2$]$_2$ as a solid line with dots and noise.
Component peaks are indicated as; σ and α, dashed lines: displaced π, dot-dash line: new π, two dots-dash line. The XP spectrum (which is dominated by Rh 4d character) is shown as a solid line (with noise).

have been found. The stereochemistry of the octochlorodimolybdenum (II) anion is particularly interesting: the two square planar Mo Cl_4 units are parallel and eclipsed. To rationalise this observation a quadruple bond between the 4d orbitals of the two molybdenum atoms has been proposed[17] involving one σ, two π and one δ bond.

To test this hypothesis the PAX spectra of $K_4[Mo_2 Cl_8]$ were measured, (fig. 12). The Mo $Lβ_{2,15}$ [4d → 2p] spectrum, in effect, maps out on an energy scale the molecular orbitals with Mo 4d character. It can be resolved into four components, of which the lowest (X-ray) energy can be seen, by virtue of its alignment with the low energy side of the Cl Kβ peak, to arise from Mo 4d character in molybdenum-chlorine bonds. The other three components however arise from orbitals less tightly bound than chlorine orbitals, they must therefore be exclusively concerned with Mo-Mo bonding. The three peaks have relative intensities approximately 1:2:1 which may be correlated with the σ, two π, and δ bonds. This XE spectrum thus provides direct confirmation of the proposed quadruple bond.[18]

Sulphides (i) Pyrite Fe S_2. The PAX spectra shown in fig. 13 enable the various parts of the XP valence band spectrum to be identified as due to orbitals of broadly different types. Thus the least tightly bound peak at 1.5 eV is due to orbitals that are mainly Fe 3d (alignment with main peak in Fe Lα), the next peak 4 - 8 eV can be identified with orbitals that are mostly S 3p whilst the most tightly bound of the valence peaks, at 13.5 - 17.0 eV is, due to S 3s orbitals (S $L_{2,3}$ M).

The structure of the S Kβ XE peak reflects the molecular orbital structure of the disulphide anion with filled σ, two π and two π* orbitals.[19] Because the symmetry of the 3s σ orbital is the same as the 3p σ orbital interaction between them is possible leading to the stabilization of 3s σ, and the destabilization of 3p σ, 3p character should therefore be found in "3s σ" and 3s in "3p σ". The destabilization of 3p σ leads to the ordering of the orbitals in S_2^{--} being changed to π, σ, π*. Only if this ordering is accepted can the S Kβ peak profile be rationalised.

Other features in S Kβ and Fe Lα are associated with covalent interactions between Fe^{++} and S_2^{--}. Thus some S 3p is found in the 'Fe 3d' orbitals at 1.5 eV (S Kβ peak at 2466.4 eV) and Fe 3d character can be seen in the S π* orbitals at 4.2 eV as a peak in the Fe Lα spectrum at 703.5 eV.

Sulphides (ii) Molybdenum Disulphide Mo S_2. The structure of Mo S_2 is quite different from that of Fe S_2. Layers of molybdenum atoms are sandwiched between layers of sulphur atoms so that each Mo is at the centre of a triangular prism of six sulphurs. PAX spectra (fig. 14) can again be used to identify the predominant atomic orbital character present in the various molecular orbital bands: 4-12 eV Mo 4d, 8-12 eV Mo 5p, 5-12 eV S 3p, 16-19 eV S 3s. However as can be seen from the figure there is considerable overlap of XE spectra with respect to orbital energy, indicative[20] of covalent bonding between S 3p and molybdenum 4d and 5p orbitals.

4. FURTHER DEVELOPMENTS

Anisotropic X-ray Emission All the XE spectra discussed above have been from powdered samples or have been taken from samples where isotropic emission has been assumed. But X-ray emission is by no means an isotropic

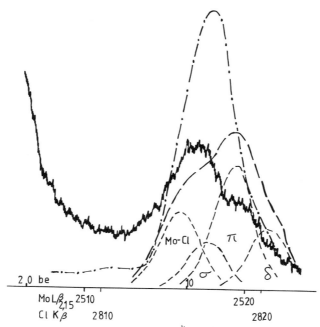

Figure 12 PAX spectra from $[Mo_2Cl_8]^{4-}$ (potassium salt). XPS, solid line (peak at be>20eV is K 3p); Cl Kβ dot-dash line; Mo L$\beta_{2,15}$ dashed line, component peaks indicated by short-dash lines.

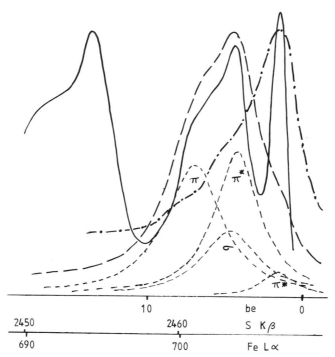

Figure 13 PAX spectra from pyrite, FeS_2. XPS, solid line; Fe Lα, dot-dash line; SKβ, dashed line with component peaks indicated by short-dash lines.

process as experiments from single crystals have clearly shown.[21] The simplest examples are to be found in layered compounds such as graphite,[22] boron nitride[23] and molybdenum disulphide.[24] The sort of variation of X-ray spectra as a function of angle of emission that can be expected is shown in fig. 15 for graphite. As the electric vector is perpendicular to the direction of propagation, X-rays that result from the relaxation of electrons in the π orbitals of graphite to C1s vacancies will be emitted exclusively in the xy plane.

Variation of "take off" angle from this plane to a direction perpendicular to the graphite planes will therefore progressively reduce the contribution to the C Kα spectrum that can be ascribed to the transitions from π orbitals. With suitable normalisation it is therefore possible to separate the spectrum into σ- and π- components and to study them separately.[22]

A similar approach is possible for MoS_2 but here the more complex angular properties of d orbitals give less clear-cut results than for p orbitals. Even so it is possible to identify a specific section of the $Mo\ L\beta_{2,15}$ spectrum as being due to transitions from orbitals with Mo $4d_{z^2}$ character.[24]

The power of this approach has also been demonstrated in its application to single crystals of naphthalene and calcite, where σ and π contributions to the electronic structure have been successfully resolved.[25]

Surface Studies. X-rays with wavelengths of a few angstroms or less are characterised by their ability to transverse many millimetres of solid matter, but as wavelengths grow longer and the photon energies become less, so the 'escape depths' become shorter. When the X-ray wavelengths are greater than 10 Å (E < 1 keV) escape depths in most materials are of the order of microns. Thus as wavelengths increase the characteristic X-rays become more and more surface sensitive. And this sensitivity can, of course, be exaggerated by orientating the sample so that only those emitted at a glancing angle to the surface are detected, an approach exploited by Andermann and his group.[26]

An alternative approach to enhancing the surface sensitivity of soft X-ray spectroscopy is to control the penetration depth of excitation.[27] This may most conveniently be done by using electron rather than photon excitation since at energies of a thousand volts or so the mean free paths of electrons in most materials are about three orders of magnitude less than for photons of the same energy. So, if electrons with energies in the range 500 - 10,000 eV are used to initiate X-ray emission then those X-rays will come from the surface layers of the sample only tens or hundreds of nanometers thick. LEE IXS (Low-Energy Electron Induced X-ray Spectroscopy)[27] thus compliments XPS for the comprehensive investigation of surface layers. LEE IXS provides a non-destructive method of depth profiling the elemental composition and chemical state of elements in surface and sub-surface layers. The existing methods of Auger and photoelectron studies can now be complimented by this X-ray spectroscopic technique: another application of PAX spectroscopy.

Conclusions

Photoelectron spectroscopy enables the ionisation energies of molecular orbitals to be measured directly and X-ray emission spectroscopy their composition and structure, in terms of constituent atomic orbitals, to be determined. Core orbital ionization energies measured by XPS can be

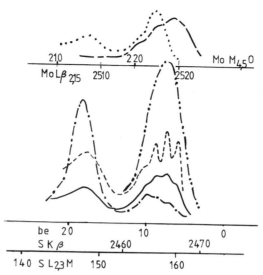

Figure 14 PAX spectra from Mo S_2: Upper section; Mo $L\beta_{2,15}$ dashed line: Mo $M_{4,5}O$, dotted line. Lower section: XPS, solid line: UPS, dashed line: S $L_{2,3}M$ dot-dash line: $SK\beta$ two-dots dash line.

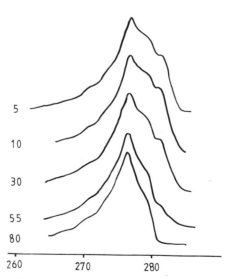

Figure 15 Carbon $K\alpha$ X-ray emission spectra from graphite. Take-off angles, from the crystal surface, are given on the left of the diagram.

used to indicate the effective charge of an atom and are essential for the alignment of different XE spectra onto a common energy scale. Auger spectra can be used to study transitions to doubly ionized states. PAX spectroscopy thus provides a direct and comprehensive method for the experimental determination of the electronic structure of molecules, complexes and solids.

Acknowledgements. It is a pleasure to acknowledge the help and stimulation I have received from research students past and present, C. Nicholls, S. Webber, J. Jones, E. Esmail, M. Al-Kadier, P. Wood, D. Haycock, P. Norman, J. Arber, K. Gillies, R. Horn, J. Purton, S. Luck and P. McClusky and fruitful collaboration with Professors Wiech, Meisel, Mazalov, Andermann and Henke. I am also deeply grateful to The Royal Society, The Science and Engineering Research Council, the Central Research Fund of London University, to the Scientific Section of NATO and to the European Commission for funds to support much of the research which has been reviewed in these two lectures.

References

1. (a) D.S. Urch, The Origin and Intensities of Low Energy Satellite lines in X-ray Emission Spectra: a Molecular Orbital Interpretation, J. Phys. C-Solid State Phys., 3:1275 (1970).

    (b) R. Manne, Molecular Orbital Interpretation of X-ray Emission Spectra: Simple Hydrocarbons and Carbon Oxides, J. Chem. Phys., 52:5733 (1970).

2. (a) C.G. Dodd and G.L. Glen, Chemical Bonding Studies of Silicates and Oxides by X-ray Emission Spectroscopy, J. Appl. Phys., 39:5377 (1968).

    (b) A. Meisel, G. Leonhardt and R. Szargan, "Röntgenspektren und Chemische Bindung", pub. Akademische Verlagsgesellschaft, Leipzig DDR, (1976).

3. D.S. Urch, X-ray Emission Spectroscopy in: "Electron Spectroscopy, Theory, Techniques and Applications", Eds. C.R. Brundle and A.D. Baker, Academic Press, London, UK, 3:1, (1979).

4. D.S. Urch, X-ray Spectroscopy and Chemical Bonding in Minerals, in: "Chemical Bonding and Spectroscopy in Mineral Chemistry", Eds. F.J. Berry and D.J. Vaughan, pub. Chapman and Hall, London, p.31, (1985).

5. F.P. Larkins and T.W. Rowlands, Importance of Interatomic Contributions to Molecular X-ray Emission Processes, J. Phys. B: Alt. Mol. Phys., 21: in press, (1988).

6. R. Horn and D.S. Urch, Chemical Effects in the X-ray Emission Spectra of Sulphur, Spectrochimica Acta, 42B:1177, (1987).

7. E.I. Esmail, C.J. Nicholls and D.S. Urch, The Detection of Light-Elements by X-ray Emission Spectroscopy with Use of Low-Energy Satellite Peaks, The Analyst, 98:725, (1973).

8. A.J.C.L. Hogarth and D.S. Urch, A Study of the Bonding in Some Five- and Six-Coordinate Compounds of Silicon Based on Si K$\beta_{1,3}$ X-ray Emission Spectra, J. Chem. Soc. Dalton Trans., p.794, (1976).

9. A. Alter and G. Wiech, X-ray Spectroscopic Studies on the Electronic Structure of Binary Glasses, Jap. J. Appl. Phys., 17-2:288, (1978).

10. J. Purton and D.S. Urch, Silicon Kβ X-ray Spectra and Crystal Structure, Mineralogical Mag., to appear, (1988).

11. N. Kosuch, G. Wiech and A. Faessler, Oxygen K-Spectra and Electronic Structure of the Oxyanions SO_3^{--}, SeO_3^{--}, TeO_3^{--} and SO_4^{--}, SeO_4^{--}, TeO_4^{--}, J. Electron. Spec. & Rel. Phenom., 20:11, (1980).

12. G. Andermann, B. Henke, D.S. Urch and G. Wiech, Soft X-ray Spectroscopy: Excitation and Dispersion, Japanese J. Appl. Phys., 17-2:428, (1978).

13. C.J. Nicholls and D.S. Urch, X-ray Emission and Photoelectron Spectra from Magnesium Oxide: a Discussion of the Bonding Based on the Unit Mg_4O_4, J. Chem. Soc. Dalton Trans., p.2143, (1975).

14. D.E. Haycock, M. Kasrai, C.J. Nicholls and D.S. Urch, The Electronic Structure of Magnesium Hydroxide (Brucite) using X-ray Emission, X-ray Photoelectron and Auger Spectroscopy, J. Chem. Soc. Dalton Trans., p.1791, (1979).

15. L. Bergknut, G. Andermann, D. Haycock, M. Kasrai and D.S. Urch, Electronic Structure of Cyanide and Hexacyanoiron (II) Anions Using X-ray Emission and X-ray Photoelectron Spectroscopies, J. Chem. Soc. Faraday Trans. II, 77:1879, (1981).

16. J.M. Arber, D.S. Urch, D.M.P. Mingos and S.J. Royse, The Metal-Alkene Bond in Di-μ-chloro-tetra (Ethene) Dishodium (I): Direct Evidence for the Dewar-Chatt-Duncanson Model Using X-ray Emission and X-ray Photoelectron Spectroscopies, J. Chem. Soc., Chem. Comms., p.1241, (1982).

17. D.E. Haycock, D.S. Urch, C.D. Garner, I.H. Hillier and G.R. Mitcheson, Direct Evidence for the Quadruple Metal-metal Bond in the Octachlorodimolybdenum (II) Ion, $[Mo_2 Cl_8]^{4-}$, using X-ray Emission Spectroscopy, J. Chem. Soc. Chem. Comm., p.262, (1978).

18. F.A. Cotton, Quadruple Bonds and Other Multiple Metal to Metal Bonds, Chem. Soc. Rev., 4:27, (1975).

19. G. Wiech, W. Köppen and D.S. Urch, X-ray Emission Spectra and Electronic Structure of the Disulphide Anion, Inorg. Chimica Acta, 6:376, (1972).

20. D.E. Haycock, D.S. Urch and G. Wiech, Electronic Structure of Molybdenum Disulphide, J. Chem. Soc. Faraday Trans. II, 75:1692, (1979).

21. G. Wiech, Anisotropic Emission of X-radiation in: "Inner Shell and X-ray Physics of Atoms and Solids", D.J. Fabian, H. Kleinpoppen and L.M. Watson eds., Plenum Press, New York, USA, p.815, (1981).

22. G. Wiech and W. Zahorowski, Studies on the Anisotropic X-ray Emission of Graphite in: "Inner Shell and X-ray Physics of Atoms and Solids", D.J. Fabian, H. Kleinpoppen and L.M. Watson eds., Plenum Press, New York, USA, p.833, (1981).

23. E. Tegeler, N. Kosuch, G. Wiech and A. Faessler, Anisotropic Emission of the X-ray K Emission Band of Nitrogen in Hexagonal Boron Nitride, Phys. Stat. Sol., B84:561, (1977).

24. (a) D.E. Haycock, M. Kasrai and D.S. Urch, Electronic Structure of Molybdenum Disulphide: Angular Resolved X-ray Spectroscopy, Japanese J. Appl. Phys., 17-2: 138, (1978).

    (b) W. Ormerod and D.S. Urch, Angular Resolved X-ray Emission from Molybdenum Disulphide in: "Inner Shell and X-ray Physics of Atoms and Solids", D.J. Fabian, H. Kleinpoppen and L.M. Watson eds., Plenum Press, New York, USA, p.829, (1981).

25. E. Tegeler, N. Kosuch, G. Wiech and A. Faessler, Molecular Orbital Analysis of the Carbonate Anion by Studies of the Anisotropic X-ray Emission of its Components, J. Electron. Spec. and Related Phenom., 18:23, (1980), and Studies of K-emission Spectra of Carbon in Aromatic Hydrocarbons, Jap. J. Appl. Phys., 17-2:97, (1978).

26. W. Worthy (reporting the work of Prof. Andermann), X-ray Technique may Provide New Way to Study Surfaces, Films, Chem and Eng. News, p.28, (April 8, 1985).

27. M. Romand, R. Bador, M. Charbonnier and F. Gaillard, Surface and Near-surface Chemical Characterization by Low-energy Electron-induced X-ray Spectrometry (LEEIXS), A Review, X-ray Spectrometry, 16:7, (1987).

IMPORTANCE OF INTRA- AND INTER-ATOMIC CONTRIBUTIONS

TO MOLECULAR X-RAY EMISSION PROCESSES

Frank P. Larkins

Department of Chemistry
University of Tasmania
Hobart, Tasmania, Australia 7001

ABSTRACT

The factors which influence the ab initio calculation of X-ray emission transition probabilities for molecules are reviewed. It is demonstrated that in order to calculate absolute transition probabilities inter alia both multi-centre and electronic relaxation effects must be included, especially for first row polyatomic molecules. The influence of choice of basis set, of length or velocity dipole operator forms and the use of the adiabatic approximation are discussed.

INTRODUCTION

The past two decades have seen a steady development in both the experimental measurement and the theoretical analysis of the X-ray emission spectra (XES) of isolated molecules. In the 1960's experimental techniques were developed to yield gas phase spectra virtually free of intermolecular interactions.[1,2] Throughout the 1970's and 1980's there has been a steady improvement in energy resolution and detection efficiency,[3,4] however, most of the spectra reported represent radiative de-excitation following core ionisation by probes at energies well above core hole ionisation threshold. Consequently, in many cases there is a significant satellite contribution. The satellite lines often have transition energies in the same range as the diagram lines and may have considerable intensity. The importance of initial-state, final-state, multiply-excited and multiply-ionised correlation satellites for atoms has been discussed in detail in the context of the argon atom.[5] The same features are to be expected for molecular systems under appropriate experimental conditions, making the interpretation of such spectra very difficult. Recently, the availability of monochromatic synchrotron radiation and suitable crystal spectrometers has provided the technology to unravel single vacancy spectra from associated satellite spectra through near threshold selective excitation.[6,7] It is now possible to obtain almost pure diagram line spectra against which theoretical models can be more rigorously tested. Furthermore, since excited resonant state spectroscopy is possible participator and spectator spectra can be used as probes for the electronic structure of molecular excited states and as a further test of theoretical developments.

The earliest theoretical interpretation of XES was based upon a one-centre localized transition model.[8] The one-centre model has been widely used. It incorporates the assumption that the interatomic X-ray transition moment contributions, sometimes known as cross-over transitions, are some orders of magnitude less than the single centre intra-atomic transition moment contributions.[9] Recent theoretical work[10-12] has shown that this generalised assumption is not valid, especially for first row polyatomic molecules involving valence electrons. Several researchers had previously undertaken full frozen orbital type ab initio transition moment calculations,[13-16]. For the molecules considered the relative intensity predictions were assessed to be in fair agreement with the one-centre model and with experiment. These analyses represent an over-simplification of a complex problem. Furthermore, the more fundamental question of the relationship between the absolute transition rates from various models had not been addressed until recently.[10-12]

In this paper we examine the factors which influence the ab initio calculation of molecular X-ray emission transition probabilities with particular reference to the role of intra-atomic (one-centre), interatomic (multi-centre) and electronic relaxation effects. They include: i) basis set dependence; ii) the use of the length or velocity dipole operator forms; iii) single-centre or multi-centre approaches; iv) frozen orbital or relaxed orbital Hartree-Fock representations and v) the adiabatic approximation relating to vibrational effects.

The calculation of transition energies, while complex, is more straightforward involving the difference between the total energies of the initial and final hole states. At a minimum relaxed Hartree-Fock calculations are necessary. For a more refined approach electron correlation effects should be included and changes in molecular geometry may need to be taken into account.

THEORETICAL CONSIDERATIONS

The normal X-ray emission process following selective ionisation of an electron from core orbital W may be represented as

$$M^+[W] \rightarrow M^+[X] + h\nu \tag{1}$$

The probability for the transition in a molecular system from an initial state i with a hole in orbital W, represented by the wavefunction $\psi^i[W]$, to a final state f with a hole in orbital X, represented by the wavefunction $\psi^f[X]$, with the associated emission of an X-ray photon of energy E_{if}, is given by the expression

$$\Gamma_{X \rightarrow W} = \frac{4}{3} (E_{if})^3 \alpha^3 |M_{if}|^2 . \tag{2}$$

Within the dipole approximation for orthonormal wavefunctions the transition moment $\underset{\sim}{M}_{if}$ is given by

$$\underset{\sim}{M}_{if} = \langle \psi^i[W] | \sum_j \underset{\sim}{d}_j | \psi^f[X] \rangle \tag{3}$$

where $\underset{\sim}{d}_j$ is the one-electron dipole operator. For wavefunctions of a single Hamiltonian, the equivalence of the length and velocity forms of the dipole operator is such that

$$\langle\psi^i[W]|\underline{d}_j|\psi^f[X]\rangle = \langle\psi^i[W]|\underline{r}_j|\psi^f[X]\rangle$$
$$= E_{if}^{-1}\langle\psi^i[W]|\underline{\nabla}_j|\psi^f[X]\rangle \qquad (4)$$

For approximate wavefunctions derived with different Hamiltonians this equivalence is not usually obtained.

In the Born-Oppenheimer approximation the state function ψ is separated into nuclear and electronic parts

$$\psi = \psi_N \psi_e \,. \qquad (5)$$

The dipole operator can also be expressed as a sum of nuclear and electronic components,

$$\underline{\mu} = \underline{\mu}_N + \underline{\mu}_e = \sum_J^M Z_J \underline{d}_J - \sum_j^n \underline{d}_j \,, \qquad (6)$$

where there are M nuclei and n electrons in the system. The negative sign arises from the charge of the electron. Substituting (6) and (7) into (2), we obtain

$$\underline{M}_{if} = \langle\psi_N^f|\psi_N^i\rangle\langle\psi_e^f|\underline{\mu}_e|\psi_e^i\rangle + \langle\psi_e^f|\psi_e^i\rangle\langle\psi_N^f|\underline{\mu}_N|\psi_N^i\rangle \,. \qquad (7)$$

In the fixed nuclei approximation with no geometric relaxation, the initial and final nuclear wavefunctions are assumed to be identical and are represented as delta functions, or point charges. If there is no vibrational motion, we have

$$\langle\psi_N^f|\psi_N^i\rangle = 1 \,. \qquad (8)$$

The nuclear dipole term becomes

$$\langle\psi_N^f|\underline{\mu}_N|\psi_N^i\rangle = \sum_{J=1}^M Z_J \underline{R}_J \quad \text{(length form)}, \qquad (9)$$
$$= 0 \quad \text{(velocity form)},$$

where $\underline{R}_J$ is the position vector of nucleus J.

If the wavefunctions are orthogonal then the second term in (7) vanishes since the overlap $\langle\psi_e^f|\psi_e^i\rangle = 0$ and only the electronic dipole term remains,

$$\underline{M}_{if} = \langle\psi_e^f|\underline{\mu}_e|\psi_e^i\rangle \,. \qquad (10)$$

This is the basis of a <u>frozen orbital</u> model. $\underline{M}_{if}$ reduces to a one-electron dipole matrix element for a single set of orthonormal molecular orbitals, ϕ_i,

$$\underline{M}_{if} = \langle\phi_X|\underline{\mu}_e|\phi_W\rangle \qquad (11)$$

When the one-electron molecular orbitals are expressed as linear combinations of atomic functions, χ_α, over all the atoms in the molecule

$$\phi_i = \sum_\alpha c_{\alpha i} \chi_\alpha \tag{12}$$

then equation (11) becomes

$$\underset{\sim}{M}_{if} = \sum_{\alpha,\beta} c_{\alpha X} c_{\beta W} \langle \chi_\alpha | \underset{\sim}{\mu}_e | \chi_\beta \rangle \tag{13}$$

This is the <u>multi-centre approximation</u> which includes inter-atomic as well as intra-atomic moment contributions. In the <u>one-centre approximation</u> only those terms in the summation involving basis functions on the same atomic centre as the core hole (say A) are retained, i.e. the summation includes only intra-atomic moment elements. Hence

$$\underset{\sim}{M}_{if} = \sum_{\alpha,\beta} c^A_{\alpha X} c^A_{\beta W} \langle \chi^A_\alpha | \underset{\sim}{\mu}_e | \chi^A_\beta \rangle \tag{14}$$

With Slater-type orbitals further simplification is possible.[10]

If hole state <u>electronic relaxation effects</u> are included such that ψ^f_e and ψ^i_e are non-orthogonal then equation (7) with or without the correction for nuclear motion, equations (8) and (9), must be used.

We will now consider the validity of these various approaches, but first it must be recognised that the accuracy of any ab initio calculation is related inter alia to the quality of the basis set chosen.

BASIS SET DEPENDENCE OF TRANSITION RATES

The basis set dependence of transition energies and transition rates have been considered in detail for the CO molecule.[12] Five different Gaussian basis sets were used as previously described. A summary of the results for the total C-K and O-K rates using a relaxed Hartree-Fock model and the length form of the dipole operator along with the individual O-K rates for the [1π] and [4σ] final state holes are presented in Table 1. They are indicative of the results obtained for various first row polyatomic molecules.

The total rates decrease with the contraction of the basis set and with the inclusion of polarisation functions; however, the changes are small, being <4 per cent for C-K and <2 per cent for O-K as deduced from column 2 to column 6 in Table 1. Conclusions should not be drawn simply from the total rates since the individual transition rates do show more variability. For example, the O-K [4σ] rate increases by 13 per cent and the [1π] rate decreases by 3 per cent across the table. Overall, it would seem that the larger the rate, the smaller the percentage change in absolute rate as the basis set flexibility is increased. Inevitably, the selection of a basis set is dependent upon the size of the polyatomic molecules being considered and the amount of computer time available. Our experience is that for first row atoms a (9s5p) basis contracted to a [5s3p] basis augmented by at least one d function is reasonable, while for second row atoms a (12s9p) basis contracted to a [6s4p] basis augmented by diffuse s, p and d functions, is adequate for addressing fundamental questions related to molecular X-ray emission processes.

Table 1. Basis Set Dependence of Selected Absolute Transition Probabilities (x 10^{-6} au) for the O-K and C-K Spectra of the CO Molecule.[a]

Transition	(9s 5p)	(9s 5p)	[5s 3p]	[5s 3p 1d]	[6s 4p 2d]
C-K Total	7.65	7.65	7.49	7.37	7.35
O-K Total	44.42	44.43	43.76	43.56	43.56
O-K [4σ]	34.69	34.70	34.12	33.79	33.75
O-K [1π]	5.93	5.92	5.88	6.23	6.71

[a] See Ref. 12 Table 2 for more details. Length form of dipole operator used.

LENGTH AND VELOCITY DIPOLE OPERATORS

When approximate wavefunctions are used to represent the molecular single hole states the moment values $\underset{\sim}{M}_{if}$ calculated with the length and velocity forms of the dipole operator (equation 4) are not equal. A necessary, but not sufficient, condition to assess the quality of the wavefunctions used is the level of agreement between the transition rates calculated with the two forms of the operator. The variability is illustrated by reference to the N-K spectrum of the HCN molecule[12] using a [5s3p1d] Gaussian basis at the relaxed Hartree-Fock level (Table 2). In general, for the most important transition experience has shown that the absolute rate agrees within ±20 per cent. For the small transitions greater variations are possible. Furthermore, the variability persists independent of whether a one-centre, frozen orbital or relaxed orbital model is chosen. For example, with the above mentioned three models the N-K total rates for the HCN molecule vary by -6, -12 and -15 per cent respectively from the length to the velocity form. The relative rates are usually less sensitive to the form of the dipole operator.

When performing calculations of transition rates it is highly desirable to consider both the length and velocity forms of the dipole operator.

INTRA-ATOMIC AND INTER-ATOMIC CONTRIBUTIONS

In this context it is useful to partition the problem. Firstly, for a single set of orthonormal molecular orbitals, a frozen orbital set, the contribution of one-centre and multi-centre terms can be investigated in essence using equations (14) and (13) respectively. Secondly, the effects of electronic relaxation due to the presence of the single hole in different molecular orbitals for the initial and final states resulting in a change in electron screening and hence polarisation of the electrons in the molecule must be considered. This latter effect is a manifestation of the fact that electrons are correlated. Equations (7)-(9) are used for the evaluation in this case.

Multi-Centre and One-Centre Models

It is frequently stated that the X-ray emission process is dominated by contributions from the atomic centre with the core-hole. The results presented in Table 3 serve to demonstrate that such is not the case for

Table 2. Absolute and Relative Transition Probabilities Calculated for the N-K Spectrum of the HCN Molecule Using the Length and Velocity Forms of the Dipole Operator at the Relaxed Hartree-Fock Level.[a]

Final Hole State	Absolute (x 10^{-6} au)			Relative	
	A_L	A_V	$\Delta\%$[b]	I_L	I_V
N-K Spectrum					
[3σ]	1.28	0.68	-47	0.09	0.06
[4σ]	0.23	0.28	19	0.02	0.02
[5σ]	6.26	5.87	-6	0.44	0.50
[1π]	14.17	11.82	-17	1.00	1.00
Total	21.96	18.65	-15		

a See Ref. 12 Tables 5 and 6.

b $\Delta(\%)$ = ((velocity - length) / length) x 100

first row polyatomic molecules. The total absolute rates change by as much as 70 per cent when ground state frozen orbital wavefunctions are used. Furthermore, there is not a consistent pattern across the individual orbital transitions as shown for the CO molecule in Table 4. Changes are variable and very substantial.

The importance of multicentre terms depends inter alia on the bond length between the atom with the core hole and its neighbours. A simple model has been used to illustrate the importance of multicentre effects in XES processes.[10] The ratio of the moment integrals $\langle x^Z_{np_z} | r | x^C_{1s} \rangle / \langle x^C_{2p_z} | r | x^C_{1s} \rangle$ as a function of internuclear distance has been calculated using Slater-type functions for various first (n=2) and second row (n=3) atoms bonded to a carbon atom. The findings are summarized in Fig. 1. For example, the C-O bond distance in the CO molecule is 2.132 au and the moment ratio is ~0.23, whereas, the C-Cl bond distance in CH_3Cl is 3.366 au and the moment ratio is ~0.08. Ab initio calculations, Table 3, confirm that multi-centre contributions are not important for the Cl K_β spectrum from the CH_3Cl molecule.

Table 3. Importance of Multi-Centre Contributions to the Absolute Total X-ray Transition Rates of Some Selected Molecules: Percentage Change in Values.[a]

CO[b]		HCN		CO_2	CH_3Cl[c]
C	O	C	N	C	Cl K_β
59	15	49	21	71	-1

a Percentage change = ((multicentre - one centre)/one centre) x 100.

b Details, see Ref. 12, Tables 3, 5 and 7, length form used.

c Phillips and Larkins, unpublished.

Table 4. Importance of Multi-Centre Contributions to the Individual Final Hole State X-ray Transition Rates for the CO Molecule. Percentage Change In Values.[a]

Final State Hole	O-K[b]	C-K
[3σ]	65	223
[4σ]	17	147
[1π]	12	51
[5σ]	7	39
Total	15	59

[a] See Footnote Table 3.
[b] Details, see Ref. 12, Table 3.

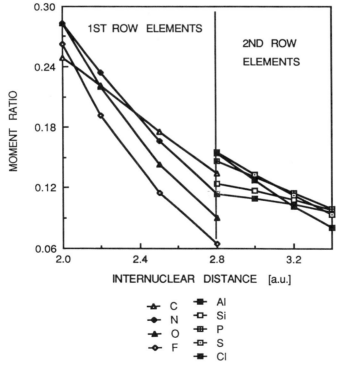

Fig. 1. Ratio of the moment integrals as a function of internuclear distance for various first and second row atoms bonded to a carbon atom.

Table 5. Importance of Electronic Relaxation Effects on the Absolute Total X-ray Transition Rates of Some Selected Molecules: Percentage Change in Values.[a]

CO[b]		HCN		CO_2	CH_3Cl[c]
C	O	C	N	C	Cl K_β
61	40	45	68	73	63

[a] Percentage Change = ((Relaxed-Frozen)/Frozen) x 100.
[b] Details, see Ref. 12 Tables 3, 5 and 7. Length form used.
[c] Phillips and Larkins, unpublished.

Electronic Relaxation Effects

A second contribution to inter-atomic effects is the polarisation of the molecule due to the presence of the electron hole in different orbitals for the initial and final hole states as discussed above. The electronic relaxation effect includes a single-centre contribution as well as the multi-centre contribution. The effect may be modelled theoretically by using a relaxed orbital model to separately optimise wavefunctions for the initial and final hole states. Non-orthogonal wavefunctions ψ_e^f and ψ_e^i result, therefore equations (7)-(9) must be evaluated to determine the transition moment. A comparison of the percentage change in absolute total transition rates for selected molecules from using a frozen orbital to a relaxed orbital model are given in Table 5. It is evident that the re-arrangement of electron density due to the presence of the core or valence hole has a major effect on the absolute total transition rates with changes typically >40 per cent. It is interesting to note that the Cl K_β total rate for the CH_3Cl molecule increases by 63 per cent even though multi-centre effects, as defined earlier, were small. This case provides evidence that one-centre and multi-centre electronic relaxation effects are important.

Table 6. Importance of Electronic Relaxation Effects on the Absolute X-ray Transition Rates for Individual Final Hole States of the CO Molecule. Percentage Change in values.[a]

Final State Hole	O-K[b]	C-K[b]
[3σ]	14	-23
[4σ]	-19	-95
[1π]	71	101
[5σ]	-9	64
Total	40	61

[a] See Footnote Table 5.
[b] Details, see Ref. 12, Table 3.

The extent of the change in absolute rates is not uniform across the hole states for a particular molecule when the orbital relaxation effects are included. Data are presented in Table 6 for the CO molecule. Some rates increase by as much as 100 per cent, while others decrease by a similar amount. The results above provide clear evidence that multi-centre and electronic relaxation contributions should be included when calculating molecular X-ray transition rates.

RELATIVE INTENSITIES

To date, experimental molecular X-ray spectra have not measured absolute transition rates. Consequently, it has been sufficient to obtain relative intensities to compare with experiment. Furthermore, because satellite structure is often present in the same energy region as diagram lines there have been limited opportunities for detailed comparisons. It can be confidently predicted that this situation will change rapidly in the future for reasons discussed earlier. If one considers only relative intensities for major transitions in some selected molecules (Table 7) then the variability between the three models is not as great as when absolute values are considered. Hence, relative intensities do not provide as sensitive a discriminator between the various models as do absolute intensities. It is because of the almost exclusive use until recently of relative intensities in the literature that the deficiencies of various models addressed in this paper have not been widely examined.

VIBRATIONAL EFFECTS

There is clear evidence in many of the molecular X-ray emission spectra for the presence of vibrational effects corresponding to the change in internuclear distance during the ionisation and radiative processes. In order to accurately calculate the spectral band profile such effects must be accounted for within a chosen model.

It is usual to assume that the adiabatic approximation is valid which implies that the electronic transition moment, $\langle \psi_e^f | \mu_e | \psi_e^i \rangle$, is independent of internuclear distance within normal vibrational limits. For orthogonal electronic wavefunctions, from equation (7)

Table 7. Relative Intensity for the Two Most Intense X-ray Transitions for Selected Molecules Calculated with Various Models.[a]

Molecule[b]	Transition Ratio	One Centre	Frozen Orbital	Relaxed Orbital
*CO	$5\sigma:1\pi$	1.20	1.11	0.91
CO*	$4\sigma:1\pi$	0.41	0.42	0.20
HC*N	$4\sigma:1\pi$	0.29	0.38	0.24
HCN*	$5\sigma:1\pi$	0.68	0.69	0.44
*CO$_2$	$3\sigma_u:1\pi_u$	0.40	0.51	0.41
CH$_3$Cl*	$7a_1:3e$	0.36	0.35	0.39

[a] Details, Ref. 12 Tables 4, 6, 8; Length form.
[b] * denotes centre with core hole.

$$|M_{if}|^2 = |\langle\psi_N^f|\psi_N^i\rangle|^2 |\langle\psi_e^f|\mu|\psi_e^i\rangle|^2 \tag{15}$$

The squares of the nuclear wavefunction overlaps are the Franck Condon factors. These factors are then evaluated to determine the band shape. The adiabatic approximation has been used when either direct lifetime broadening effects, as outlined above, or more recently when lifetime-vibrational interference effects, which are especially relevant to resonant excitation X-ray spectra,[17,18] are considered.

The variation of the electronic transition moment $\langle\psi_e^f|\mu|\psi_e^i\rangle$ with internuclear distance has been calculated ab initio for the CO molecule corresponding to an initial core hole on the oxygen, [1σ] and the carbon [2σ] atom.[11] The results are shown in Fig. 2a and 2b respectively, calculated at the frozen orbital multicentre level using neutral molecule orbitals with both the length and velocity dipole forms. The [1σ]-[1π] and [2σ]-[5σ] transition moments in Fig. 2a and 2b respectively show little dependence on bond length variation. The other transitions show a stronger dependence on bond length. Most decrease in magnitude with increasing bond length reflecting the reduced importance of two-centre contributions as the bond length increases. The notable exception is the O-K [1σ]-[5σ] transition which increases with increasing bond length in the range investigated because of the destructive interference of one- and two-centre terms. Since all transitions do not show the same bond length dependence, the application of the adiabatic approximation coupled with a Franck-Condon vibrational analysis must be used with caution.

AN EXAMPLE : THE N_2O MOLECULE

An interesting molecule for study is N_2O, since it has two nitrogen atoms in different chemical environments. The X-ray emission spectra for the central nitrogen, N_C and the terminal nitrogen, N_T, have recently been calculated by ab initio molecular orbital methods using a relaxed Hartree-Fock model.[19] They are shown in Fig. 3a and 3b respectively. A comparison of the theoretical and experimental[4] total nitrogen spectra is shown in Fig. 3c. Experimental energies have been used, although theoretical values are within 2-3 eV. In addition, experimental FWHM values along with theoretical intensities and Lorentzian line shapes are used in the synthesis of the spectrum. The agreement is most encouraging, especially since there are probably some satellite contributions present in the experimental spectrum. A more complete study should include further electron correlation effects and a first principles vibrational analysis to determine the individual band shapes.

CONCLUSION

The calculation from first principles of X-ray emission spectra involves a consideration of several factors. Among the most important are multi-centre and electronic relaxation contributions to the absolute transition probabilities. For most molecules the one-centre approximation is not adequate. Recent refinements in experimental measurements and theoretical techniques promise to continue to provide a valuable insight into the electronic properties of ionised molecules.

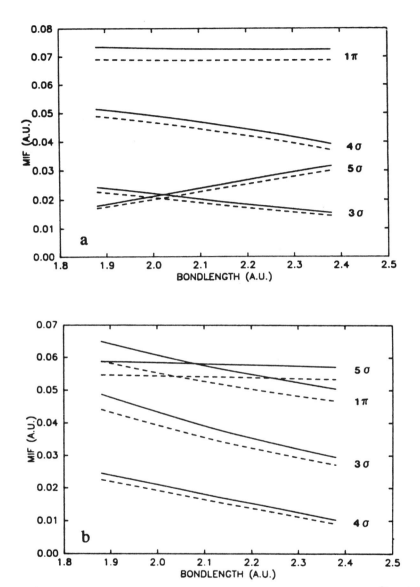

Fig. 2. Variation of the electronic transition moment $\langle \psi_e^f | \underline{\mu}_e | \psi_e^i \rangle$ with internuclear distance for the CO molecule. L: length form. V: velocity form. a) O-K [1σ] initial hole state; b) C-K [2σ] initial hole state. (After ref. 11).

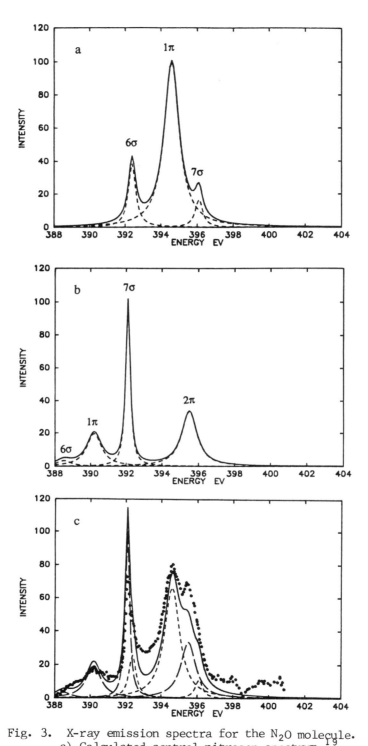

Fig. 3. X-ray emission spectra for the N_2O molecule.
a) Calculated central nitrogen spectrum.[19]
b) Calculated terminal nitrogen spectrum.[19]
c) Comparison of theoretical and experimental[4] total nitrogen spectra.

ACKNOWLEDGEMENT

The contributions of Mr T.W. Rowlands and Dr R.A. Phillips to this work are gratefully acknowledged. The work was undertaken with the financial support of the Australian Research Grants Scheme.

REFERENCES

1. R.E. LaVilla and R.D. Deslattes, J. Chem. Phys., 45:3446 (1966).
2. R.A. Mattson and R.C. Ehlert, J. Chem. Phys., 48:5465 (1968).
3. J. Nordgren, H. Ågren, L.O. Werme, C. Nordling and K. Siegbahn, J. Phys. B: Atom. Molec. Phys., 9:295 (1976) and references therein.
4. L. Pettersson, M. Bäckström, R. Brammer, N. Wassdahl, J.E. Rubensson and J. Nordgren, J. Phys. B: Atom. Molec. Phys., 17:L279 (1984).
5. K.G. Dyall and F.P. Larkins, J. Phys. B: Atom. Molec. Phys., 15:1811 (1982).
6. R.C.C. Perera, J. Barth, R.E. LaVilla, R.D. Deslattes and A. Henins, Phys. Rev. A, 32:1489 (1985).
7. R.D. Deslattes, Aus. J. Phys., 39:845 (1986).
8. R. Manne, J. Chem. Phys., 52:5733 (1970).
9. D.S. Urch, J. Phys. C: Solid State Phys., 3:1271 (1970).
10. F.P. Larkins and T.W. Rowlands, J. Phys. B: At. Mol. Phys., 19:591 (1986).
11. T.W. Rowlands and F.P. Larkins, Theor. Chim. Acta, 69:525 (1986).
12. R.A. Phillips and F.P. Larkins, Aus. J. Phys., 39:717 (1986).
13. A. Støgard, Chem. Phys. Lett., 36:357 (1975).
14. H. Ågren and R. Arneberg, Phys. Sci., 28:80 (1983).
15. H. Ågren, R. Arneberg, J. Müller and R. Manne, Chem. Phys., 83:53 (1984).
16. H. Ågren and J. Nordgren, Theor. Chim. Acta, 58:111 (1981).
17. A. Flores-Riveros, N. Correia, H. Ågren, L. Pettersson, M. Bäckström and J. Nordgren, J. Chem. Phys., 83:2053 (1985).
18. T.X. Carroll and T.D. Thomas, J. Chem. Phys., 86:5221 (1987). (See this paper for correction to reference 17).
19. F.P. Larkins and R.A. Phillips, J. de Phys., X87 Conference, to be published.

THE ELECTRONIC DECAY OF CORE HOLE EXCITED STATES IN FREE AND CHEMISORBED MOLECULES

W. Eberhardt

Exxon Research and Engineering Company
Route 22 East, Annandale, NJ 08801

Lecture given at the NATO Advanced Science Institute on "X-ray Spectroscopy in Atomic and Solid State Physics", Vimeiro, Portugal, August 30 - September 12, 1987.

ABSTRACT

The electronic decay spectra of neutral and ionic core hole excited states in free and chemisorbed molecules are discussed. A detailed analysis of these spectra reveals the static and dynamic charge distributions and rearrangements around the atomic center where the core hole was created.

INTRODUCTION

The excitation of a core electron in a molecule or a solid creates a highly excited, short lived state, which decays via various secondary processes. In light elements the predominant channel is via a radiationless two electron transition, whereas for heavy elements the emission of a fluorescence photon generated in a one electron transition becomes more and more probable. The secondary emission spectrum is a unique fingerprint of the atom and of the chemical environment, where the core hole was created. Therefore these secondary spectroscopies are widely applied in materials characterization techniques. For example, scanning Auger spectroscopy yields the distribution of elements on or near the surface of a material and the x-ray fluorescence microprobe is used to determine bulk element compositions and impurity distributions in solids.

Here I want to demonstrate on a few examples that a detailed understanding of these relaxation processes yields information reaching far beyond the pure identification of the element the process occurs in. The intensity distribution of the lines observed in these spectra reveals static and dynamic change distributions of the valence electrons around the atomic center where the core hole was created. Additionally the electrons involved in these screening processes of the hole are highlighted and in high resolution spectra we even observe the effects of the nuclear motion on the electronic structure and the results of the geometry changes occurring in these systems upon the excitation of a core electron.

In a way the situation in these experiments is similar to a pump-probe experiment with an ultrafast time resolution. We create a state, the core hole excited state, and watch its fate, the decay, within a very short time after the initial excitation. Typical core hole lifetimes in light elements like C, N, or O are about 10^{-14} sec. This time is long enough for the electronic relaxation to occur to the situation in the excited state and just about long enough for some geometric relaxation process involving a change of the mean nuclear coordinates. The decay spectrum then gives a snapshot of the situation after these relaxation processes have occurred. Obviously we cannot vary the time in our "time resolved" snapshot of the evolution of a system, but studying similar process around different nuclei within the same molecular system will also give us the timing information at the different discrete lifetimes of the various core excited states.

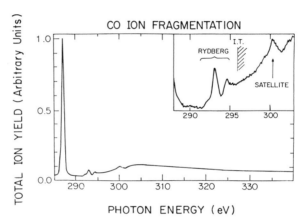

Fig. 1. Total electron yield spectrum of CO as a function of photon energy near the onset of the C 1s transitions. The total electron yield closely resembles the absorption spectrum of the molecule.

BASIC PRINCIPLE

In Fig. 1 the total ion yield of CO is shown as a function of photon energy near the threshold of the C 1s core electron excitation. The total ion yield is proportional to the absorption of the sample. We see several sharp structures below the ionization threshold of a C 1s electron at 296 eV. The most prominent one at 287 eV corresponds to the excitation of the C 1s electron into the lowest unoccupied molecular orbital, the 2π orbital. The weaker peaks belong to a series of Rydberg excitations converging to the ionization threshold. The peak marked "satellite" is attributed to a two electron transition into a $(1s)^{-1} (1\pi)^{-1} (2\pi)^2$ final state configuration.

From this absorption spectrum we recognize that several possible core hole configurations can be optically created. We can excite the C 1s electron into an unoccupied molecular orbital, a Rydberg orbital or directly into the continuum. All these configurations have different excitation energies and electron occupations such that the decay spectra can be expected to be different for these states. As we will see later even the C 1s ionization will not generally result in a unique electronic state, but as apparent from the satellites in the photoemission spectrum, several 1s ionized states may be created according to their excitation probabilities if the energy of the absorbed photon exceeds the ionization potential of a C 1s electron.

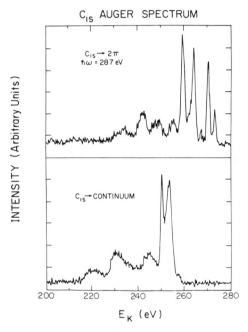

Fig. 2. Electron emission spectra of CO generated in the decay of a neutral C 1s → 2π excitation (DES) and in the decay of a C 1s ionized core hole state. Both spectra are plotted on a kinetic energy scale.

The difference in the electronic decay spectra for the two most prominent core hole states, in CO, C 1s ionization and the decay of the C 1s → 2π excitation, is illustrated in Fig. 2. Clearly, on the "experimental" kinetic energy scale the spectra are totally different. The conventional Auger spectrum is produced in the decay of the C 1s ionized state, whereas the deexcitation of the C 1s → 2π excited state produces in general higher energy electrons than the ones seen in the Auger decay. In the following we will adopt the convention to use the term Auger decay for the

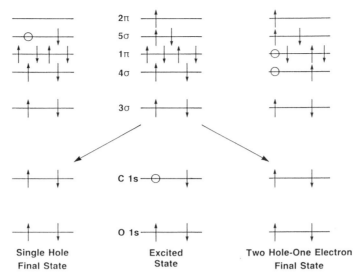

Fig. 3. Schematic illustration of the decay channels for the C 1s → 2π excited state of CO.

decay of a completely ionized system and the term deexcitation for the decay of a neutral core to bound state excited state.

The assignment of the Auger spectrum has been widely discussed in the literature[1]. Here we want to concentrate on the deexcitation spectrum. The electronic configuration of the core electron excited state and the possible decay channels are illustrated in Fig. 3. If the excited 2π electron takes part in the decay, the final state will be a single valence hole state as illustrated on the left. These are the same states that conventionally appear in a photoemission spectrum. On the other hand, if the 2π electron does not participate in the decay and rather stays around as a spectator, then the final state is a two hole-one electron state compared to the molecular ground state. These states also show up in photoemission and are referred to as shake-up states, because the transition is not a simple dipole transition. On the other hand, these states can also be viewed as Auger final states with the presence of a spectator electron.

This simple schematic view suggests a comparison of the various electron emission spectra on a binding energy scale. The binding energy is in general the energy needed to create a certain electronic configuration starting from the ground state of the molecule. Thus for both photoemission and deexcitation the binding energy E_B corresponds to the difference between the photon energy $h\nu$ and the kinetic energy of the measured ejected electron, E_{kin}, like $E_B = h\nu - E_{kin}$. In the case of Auger spectra we have to replace the photon energy $h\nu$ by the core electron ionization energy in this relationship in order to get to a comparable scale.

In Figure 4 we now show the photoemission (PES), deexcitation (DES) and Auger spectrum (AES) of CO plotted on this general binding energy scale. Immediately we recognize that the peaks in the photoemission spectrum and the DES spectrum largely coincide in energy as expected from the schematic in Fig. 3. Thus we get an empirical assignment of the deexcitation spectrum. In the following we want to concentrate on the participator, single hole final states labelled 1-4 on Fig. 4. The assignment of all states, including the spectator configurations, D1-D5, is discussed in more detail elsewhere[2].

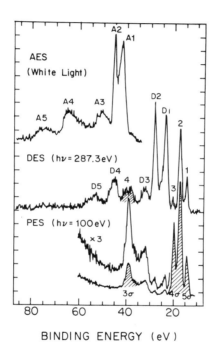

Fig. 4. AES, DES and PES spectra of CO compared on a generalized binding energy scale. Peaks 1-4 are single hole final states, peaks D1-D5 are two hole-one electron final states and peaks A1-A5 are Auger two-hole final states.

We notice that the three outermost single hole states all show up in the DES spectrum, the intensity however is quite different compared to the normal photoemission process[2]. This is due to the different nature of the transition leading to these final state configurations along different pathways. In photoemission the electric dipole field acts globally on the molecular wavefunctions, whereas the DES decay is described by a Coulomb matrix element between the core excited molecular state and the single hole final state of the molecule. Thus in DES the final states are enhanced, which have a large amplitude around the nucleus where the core electron was excited. In DES we have a probe of the localization of electronic wavefunctions within a given molecule. This is similar to the Auger spectrum, however in DES we are dealing with a single hole final state and thus we get the assignment straightforward and unobscured by the Coulomb interaction of the two valence holes in the Auger final state.

The experimental DES spectrum of CO fits rather well to this simple model. The 5σ orbital is a lone pair orbital at the Carbon end of the molecule and the 1π is spread over the whole molecule and actively involved in the screening of the core electron excitation. The 4σ orbital on the other hand, is a lone pair orbital at the oxygen end of the molecule. Thus we do not expect to see a large contribution of the 4σ single hole final state in the C-DES spectrum.

CORE HOLE SPECTROSCOPY ON CHEMISORPTION SYSTEMS

The next step is to try to use the core hole decay spectra in order to learn more about the changes occurring in the electronic structure upon the formation of a chemical bond. Special emphasis will be placed on the single hole states in the deexcitation spectra, since these states are readily assigned empirically by a comparison with the standard photoemission spectrum.

We have chosen to present as example transition metal carbonyls and CO adsorbed on a Cu surface. Transition metal carbonyls have a center of one to several metal atoms. To these metal atoms CO groups are bound to saturation. The predominant CO adsorption geometry is through the C-atom and in a linear "on top" coordination even though bridge and π-bonded CO species may be present in some of the larger carbonyls with more than one metal atom at the center. Thus these molecules resemble closely the chemisorption sites on surfaces but the electronic structure is still discrete.

Absorption on electron yield spectra show that the C 1s → 2π resonance does not vanish upon chemisorption of CO. Accordingly we can select the photon energy to excite this resonance in the chemisorbed state. The resulting electron emission spectrum is then compared to the photoemission spectrum as shown in Fig. 5. We immediately recognize two major differences in the DES spectra of CO in the carbonyls compared to the molecule in gas phase. First, the 5σ and 1π emission is energetically almost degenerate. This is a consequence of the chemisorption bond with the substrate which lowers the energy of the 5σ orbital relative to the other CO orbitals. Second, the intensity of the 4σ emission in DES is almost one order of magnitude larger than in isolated CO. This is caused by the involvement of the 4σ orbital in the screening of the C 1s hole. Due to the formation of the 5σ bond with the substrate, the σ space around the C atom is relaxed and the 4σ electrons are actually involved in screening the C 1s hole. In isolated CO the σ-space around the C-atom is completely taken up by the 5σ electrons and the 4σ electrons remain the O lone pair electrons even when an extra positive charge is created at the C-nucleus.

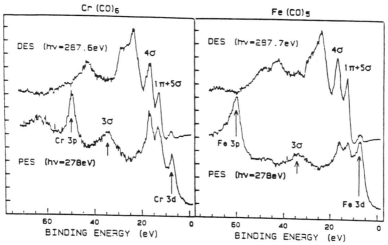

Fig. 5. Deexcitation spectra of Fe(CO)$_5$ and Cr(CO)$_6$ compared to the direct photoemission spectra. The single hole final states are labelled.

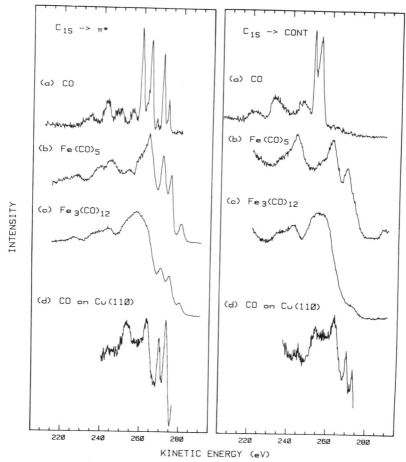

Fig. 6. Dexecitation (left) and Auger electron spectra (right) of free and coordinated CO.

Another piece of evidence for the charge transfer process occurring in the screening of a core hole becomes apparent in the spectra shown in Fig. 6. Here the DES spectra and the Auger spectra are compared for CO, transition metal carbonyls and CO adsorbed on a Cu single crystal surface. The lines appearing on the high energy side of the main Auger peak in the carbonyl spectra and for chemisorbed CO have to be assigned to single hole final state configurations. The presence of these lines in the Auger spectra is a result of the interatomic charge transfer screening process occurring upon the creation of a C 1s hole. An electron has been transferred from the metal into the analog of the molecular 2π orbital of the adsorbate complex in order to screen the C 1s hole. Therefore independent of whether this orbital was filled by direct excitation or via charge transfer screening, at the time of the core hole decay, there is an electron in the "2π" orbital, and the Auger and DES spectra are almost identical for carbonyls and chemisorbed CO.

The Auger and DES spectra for CO chemisorbed on Cu are shown in Fig. 7 on an expanded kinetic energy scale. Both spectra are almost identical and coincide exactly on this energy scale. This is a truly remarkable result which is very important for understanding the dynamics of the core hole decay in adsorbate systems[3]. The core hole excitation spectrum, the normal photoemission spectrum, for CO on Cu is known to exhibit several prominent satellite lines, with an intensity comparable or larger than the

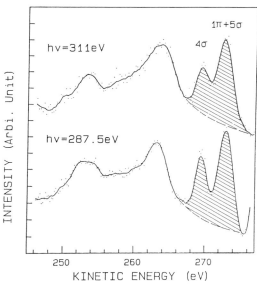

Fig. 7. Comparison of AES (top) and DES (bottom) spectra of CO on Cu(110). The shaded peaks are "single hole" final states. Note the kinetic energy scale.

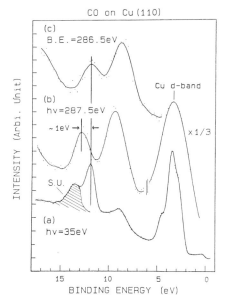

Fig. 8. Comparison of PES (a), DES (b), and AES (c) of CO chemisorbed on Cu(110) on a binding energy scale. The structure labelled S.U. (shake-up) is missing in both the AES and DES spectra.

intensity of the main line[4,5]. These lines spread over a total energy of 7 eV, indicating the corresponding spread in the energy of the core hole excited state. Nevertheless the lines seen in the Auger spectrum are much sharper (FWHM ≤ 2 eV) and there is no evidence of any lines resulting from the decay of the higher core excited states. This means that the higher energy core hole states decay via energy transfer to the metallic substrate into the ground state of the core hole, long before the core hole itself decays. Accordingly the width of the satellite lines is much larger than the width of the fully screened lowest energy core hole state.

Another timing aspect is related to the process itself. Photoemission is a "sudden" perturbation of the system and that is why shake-up side bands are observed in photoemission spectra. However we notice that the 4σ single hole state line in Auger and in DES does not exhibit the same satellite structure as in photoemission (see Fig. 8). This must mean that the two electron transition in AES and DES is substantially slower such that the system has time to settle into its adiabatic ground state.

The third point to be raised in connection with the Auger and DES spectra of chemisorbed CO on Cu is the energy position of the lines. As mentioned earlier, the two spectra show lines which coincide at the same kinetic energy. However this means automatically that in the generalized binding energy scale introduced above there is a discrepancy of 1 eV in the DES spectrum in comparison with both AES and photoemission as shown in Fig. 8. This is a result of the energy difference between the fully screened core hole state (E_i = 286.5 eV) and the C 1s → π^* excitation energy of E_i = 287.5 eV. Apparently the π^* excited state looses 1 eV in energy by converting into the fully screened core hole state before the deexcitation. The energy calibration is explained more carefully in Refs. 2 and 3.

The only difficulty arising from this picture is that in general the π^* excited state is considered to be equivalent to the fully screened core hole state. This however is only true if the spin of the electron is not taken into account. There are two different spin combinations for this configuration. The singlet configuration, which is the only one that can be excited by an optical transition, is about 1.45 eV higher in energy in gas phase CO[6] than the corresponding triplet state. In analogy to the results for isolated gas phase CO we propose that the fully screened core hole state in chemisorbed CO is equivalent to the triplet π^* configuration. Thus the apparent energy loss of 1 eV in DES can be explained by a spin flip of the excited π^* electron via substrate scattering. This again has to occur much faster than the core hole decay, which is consistent with the enlarged width of the π^* transition in chemisorbed CO.

HIGH RESOLUTION SPECTRA

The high brightness soft x-ray beam of an undulator source makes it possible to even detect vibrational substructure in the electronic core hole decay spectra. As example we want to discuss here the DES spectrum of N_2. These types of measurements give insight into the effects due to the nuclear motion and changes of the equilibrium molecular geometry occurring together with the electronic transition[7].

N_2 is an ideal test case for this type of an experiment. Upon core to π^* excitation there is a large (~.065Å) change of the internuclear distance occurring in the molecule and similar changes are observed upon the creation of the single valence hole excited states. However the comparison

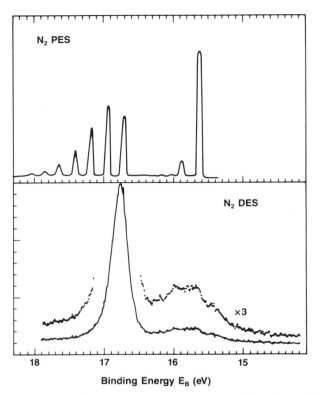

Fig. 9. High resolution photoelectron spectrum of N_2 (top) (from ref. 8) and deexcitation spectrum of N_2 (bottom). Both spectra result in the formation of the $X^2\Sigma_g^+$ and $A^2\Pi_u$ ionic final states of N_2^+. The substructure is due to excitation of vibrations in the molecule simultaneous with the electronic transition.

between the direct photoemission and the deexcitation, leading to identical final state electronic configurations, reveals that the vibrational substructure is completely different.

The top portion of Fig. 9 shows the photoelectron spectrum of N_2 in the range of the $X^2\Sigma_g^+$ ($3\sigma_g^{-1}$) and $A^2\Pi_u$ ($1\pi_u^{-1}$) final states of N_2^+ taken from the literature[7]. The $3\sigma_g$ electron is non-bonding and therefore upon its ionization the equilibrium distance of the two N nuclei does not change significantly (less than 0.02A). Accordingly the 0-0 vibrational transition is observed in the photoelectron spectrum and higher side bands are excited only with a small probability. The $1\pi_u$ electron on the other hand is strongly bonding and its removal changes the mean internuclear separation by +0.077Å. Consequently a large number of higher vibrational excitations are observed in this electronic final state.

The deexcitation spectrum on the other hand shows the opposite behavior. The $X^2\Sigma_g^+$ final state exhibits a lot of vibrational substructure whereas the $A^2\Pi_u$ final state coincides more or less in one peak. Even though several vibrational levels are excited in the primary transition, the deexcitation involves largely transition between vibrational substrates of the same quantum number in the transition to the $A^2\Pi_u$ final state. This means that the system has taken on the eigenfunctions of the core excited state within the lifetime of the core hole. This time is only about 10^{-14} sec and this corresponds to about one period in the vibrational motion of the two nuclei also. Since this time is so extremely short, coupling between the excitation and deexcitation could be observed in other systems even though for N_2 these effects seem to play a minor role.

I hope that I was able to show that a detailed study of the electron emission spectra generated in the decay of neutral or ionized core hole configurations in free and chemisorbed molecules gives new insight into multi electron dynamics in these systems. Especially the single hole states observed in DES give a rather clean picture because the assignment of the states is straightforward by a comparison with theory.

I would like to thank R. Carr, C. T. Chen, R. A. DiDio, W. K. Ford, In-Whan Lyo, H. R. Moser, R. Murphy, R. Reininger, D. Sondericker and especially E. W. Plummer for their help and numerous contributions to these studies. These experiments were carried out at SSRL and NSLS both of which are supported by DOE. These experiments were also partially supported by NSF under contract No. NSF-DMR-8120331.

REFERENCES

1. N. Correia, A. Flores-Riveros, H. Agren, K. Helenelund, L. Asplund, U. Gelius, J. Chem. Phys. 83, 2036 (1985).
2. W. Eberhardt, E. W. Plummer, C. T. Chen, W. K. Ford, Austral. J. of Physics 39, 853 (1986).
3. C. T. Chen, R. A. DiDio, W. K. Ford, E. W. Plummer, W. Eberhardt, Phys. Rev. B32, 8434 (1985).
4. P. R. Norton, R. L. Tapping, J. W. Goodale, Surf. Sci. 72, 33 (1978).
5. W. Eberhardt, J. Stöhr, D. Outka, R. J. Madix, Sol. State Comm. 54, 493 (1985).
6. D. A. Shaw, G. C. King, D. Cvejanovic, F. H. Read, J. Phys. B17, 2091 (1984).
7. R. Murphy, In-Whan Lyo, W. Eberhardt, submitted to Phys. Rev. Lett.
8. J. L. Gardner, J. A. R. Samson, J. Chem. Phys. 63, 1447 (1975).

FRAGMENTATION OF SMALL MOLECULES

FOLLOWING SOFT X-RAY EXCITATION

W. Eberhardt

Exxon Research and Engineering Company
Clinton Township, Route 22 East
Annandale, NJ 08801, USA

Lecture given at the NATO Advanced Science Institute, Vimeiro, Portugal, August 30 - September 12, 1987.

ABSTRACT

The adsorption of a soft X-ray photon, exciting a core electron in a free molecule, leads to a chain of events the final result of which is the fragmentation of the molecule. In low Z elements the core hole decays preferentially via a two electron Auger transition, which depletes the molecule of electrons out of the bonding valence structure such that it becomes unstable and falls apart into ionic and neutral fragments. Here these processes will be discussed step by step taking CO as example.

INTRODUCTION

A detailed knowledge and understanding of the interaction between radiation and matter on the molecular level is needed in many areas of science. These areas span from astrophysics, where these processes are relevant in understanding the formation and existence of molecules in outer space and in high altitude atmospheres of planets, to radiation damage in biological cells. In chemistry radiation, and specifically soft X-rays, could potentially be used to produce radicals or to selectively activate bonds within a molecule. Of course this information also serves as a testing ground for theory, where very complex calculations are needed to get the details of the potential energy curves of the highly excited states we are dealing with here. The fragmentation of molecules also involves similar processes as the desorption from solids or solid surfaces induced by electronic excitations. A comparison between desorption from surfaces and fragmentation of the same molecule in gas phase will give insight into charge reneutralization mechanisms, which suppress largely the desorption.

Obviously any mechanism that leads to a production of a core electron excitation or vacancy will start the same chain of events. Thus fragmentation of the parent molecules can be observed in electron scattering, high energy ion scattering or even in nuclear reactions involving K electron capture. The use of photons as excitation source has the advantage that the initial excitation is well defined by the photon energy. Thus not only the fragmentation after core electron ionization can be studied, but also the fragmentation processes following core electron excitations into empty molecular orbitals or Rydberg orbitals. Also dipole selection rules apply to the excitation spectrum.

Core Electron Excitation and Fragmentation

Here I want to use CO as an example and discuss step by step the events leading to the production of ionic fragments following either C 1s or O 1s electron excitation [1]. The typical absorption spectrum, illustrating the various core electron transitions of the C 1s electron in CO, is shown in Fig. 1. The C 1s electron can be excited into the lowest unoccupied molecular orbital, the 2 π orbital, by absorption of a soft X-ray photon with an energy of 287.3 eV. Alternately the C 1s electron may be ionized completely, which requires an energy of more than 296 eV; or the C 1s electron may be excited into a series of Rydberry states converging towards the C 1s ionization threshold at 296 eV. Various two electron excitations, where together with the core electron a valence electron is excited simultaneously, are also observed, the strongest one is marked "satellite" in Fig.1 and corresponds to an excited state configuration described as $(1s)^{-1}(1\pi)^{-1}(2\pi)^2$ in single particle notation.

As we will see, all these electronic excitations will produce different fragment patterns. This is illustrated in Fig. 2, where TOF (time-of-flight) ion mass spectra are shown recorded at different photon energies. At low photon energies (140 eV) there is a large peak at m/e = 28 observed, which corresponds to the ionized parent molecule. As we enter into the region of the core electron excitations this spectrum changes rapidly. Now more fragments appear with different branching ratios depending upon the exact nature of the primary excitation.

This is a very exciting result, because it opens the possibility to induce the fragmentation process site specific within a given molecule by making use of the chemical shifts in the core levels of inequivalently bound atoms of the same kind. Thus tuning the photon energy to ionize or excite only the core electrons of one atomic species even in a specific chemical environment one can induce local bond breaking around the site, where the initial excitation took place. This idea had been around for quite a while as site specific fragmentation [2,3]; later it was also re-named "memory effect" [4]. All that means is, that the molecules studied fall apart around the site, where the initial excitation took place. As we will see later, this is however as much a consequence of the details of the electronic decay of the core hole excited state as it is due to the local character of the initial excitation.

The peaks in the spectra in Fig. 2 excihibt distinctly different lineshapes. Only the singly or doubly ionized parent molecule species shows sharp peaks. The fragment ions show up as double structures with a considerable width. This is due to the large kinetic energy released in the fragmentation event. That's why this process is also referred to as "Coulomb explosion". The use of a TOF-mass spectrometer as detector for the ions in this type of experiments has thus several advantages over a quadrupole mass spectrometer. We cannot only determine the mass to charge ratio of the ionic species, but also their kinetic energy. Furthermore the trajectories for a TOF spectrometer are easily calculated and thus relative abundances can be determined. With a quadrupole mass spectrometer is it not easy to verify the discrimination for higher mass species and even more difficult to determine the transmission function if these species are produced with large kinetic energies up to 20 eV and more.

The Electronic Decay of the Core Hole as Intermediate Step

So far we have discussed the initial absorption and the final result, the appearance of neutral and ionic fragments. Apart from the somewhat pathological case of HBr [5], the core electron excited and ionized states in molecules are in general short lived but stable configurations. These core excited states decay preferentially via an Auger process into a two valence hole configuration. Thus the molecule gets depleted of the bonding electrons and falls apart according to the potential energy surface of the doubly ionized valence configuration. The electron emission spectra produced in the Auger decay of a neutral (C 1s → 2π) and ionized C 1s core hole state are shown in Fig. 3. The peaks in these spectra can be assigned to various ionic configurations of CO^+ (C 1s → 2π) and CO^{2+} (C 1s → ∞) as discussed earlier [6]. The key experiment to study these fragmentation processes, therefore, is to measure the fragment ions in coincidence with the energy resolved Auger electrons. Thus we can determine individually the species produced from each of these ionic configurations.

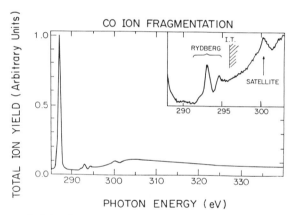

Fig. 1. Total ion yield of CO near the onset of the C 1s excitation. The total ion yield closely resembles the absorption and shows structures due to C 1s transitions into various types of final states.

Now it also is apparent where the "site specific fragmentation" or "memory effect" results from. The Auger matrix element favors electronic configurations with a strong localization around the atomic center where the core hole was created. Thus locally the bonds around this center are depleted of electrons and the molecule breaks apart around this site.

The Auger electron, or the electron observed in DES, carries away most of the energy deposited in the molecule by the absorption of the soft X-ray photon. The energy remaining in the system is equivalent to the "binding energy" defined above in the discussion of the electron spectra generated in the decay of the core hole excited state. From energy conservation we thus

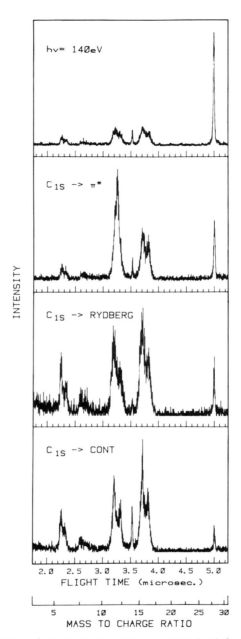

Fig. 2. The CO fragment ion mass spectra measured with a TOF mass spectrometer. The spectra are taken at different excitation energies to reflect the effect of the primary excitation on the branching ratio of the ionic fragments.

can rule out any fragment channel, which requires more energy adiabatically than the energy remaining in the system. These adiabatic energy thresholds can be obtained by adding the various ionization potentials of C and O to the dissociation energy of CO (11 eV). The thresholds derived from the adiabatic energies are shown in Fig. 4 together with the C- and O-Auger spectrum of CO plotted on a binding energy scale. Since the final state of the molecule after the Auger decay is doubly charged, only fragment channels with at least two charges are shown.

Higher charged states may originate from shake-off events in the initial core electron ionization and from Auger cascades. Note here that a hole in the innermost valence orbital of CO, the 3σ orbital, cannot decay via an Auger process. The energy of the $3\sigma^{-1}$ single hole state configuration is about 4 eV below the double ionization potential of $(CO)^{2+}$. Analogously only a $(3\sigma)^{-2}$ configuration can decay again in an Auger cascade, because the $(3\sigma)^{-2}$ double hole state is higher in energy than $(CO)^{3+}$. In single particle notation this

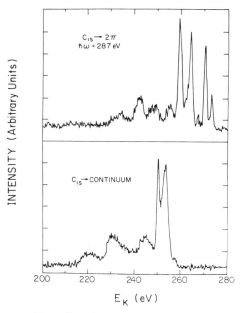

Fig. 3. Electron energy distribution curves of the electrons created in the decay of the primary C 1s core hole excited state in CO. In the top part the decay of the C 1s → 2π excitation is shown, while the bottom part shows the 'standard' Auger spectrum following C 1s ionization.

transition would involve 2 holes and 3 electrons and accordingly the chances for such a transition to occur would be very slim. However, the 3σ hole states are very poorly described as a single hole state and have quite a few multiparticle configurations mixed in. Therefore, this Auger cascade may be more probable to occur. In ion-ion coincidence studies we have observed these events with at least three charges distributed over the two fragments.

It is also interesting to note that the ground state of CO^{2+} is about 5.4 eV higher in energy than the threshold for $C^+ + O^+$. This is due to the Coulomb repulsion between the two holes. Therefore, CO^{2+} is metastable and it depends on the details of the potential energy curves whether its lifetime is long enough to be observed in our experiments.

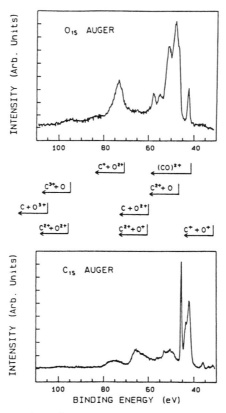

Fig. 4. The CO oxygen and carbon Auger spectra and adiabatic energy thresholds for the production of various ionic fragments.

Auger Electron Ion Coincidence Studies

Now we proceed to actually study the ions in coincidence with specific selected Auger final states populated in the decay of the C and O 1s core holes in CO. This is done by triggering the TDC whenever an Auger electron of a specific kinetic enerty is detected in the CMA electron spectrometer. Then only the ions related to this Auger decay event are recorded. Of course, we have to make sure that the detectors are not swamped by accidental coincidences, when the total count rates are too high. This can be checked by variation of the photon beam intensity and of the ion acceleration voltages as described in more detail in ref. 1.

Fig. 5 shows a series of Auger-electron-ion coincidence spectra of CO taken for various final states of the O 1s core hole decay. The TOF ion mass spectra (c) and (d) are shown at high resolution around the regions where C^+ and O^+ ions (d) or C^{2+} and O^{+2} (c) ions are expected to arrive. These were the only regions where peaks were observed in these spectra. The labels A-E on the TOF ion mass spectra correspond to the equivalent energy set point for the electron spectrometer as marked by the arrows in Fig. 5b. In order to record the ion spectra a static ion extraction field is applied across the interaction region, which causes a slight smearing in the energy distribution curve of the Auger electrons as seen by our spectrometer. Fig. 5a and b show the measured Auger electron energy distribution curves without (a) and with (b) the ion extraction field applied. The main features of the Auger spectrum are still resolved in (b) and, as the coincidence spectra show, we are even

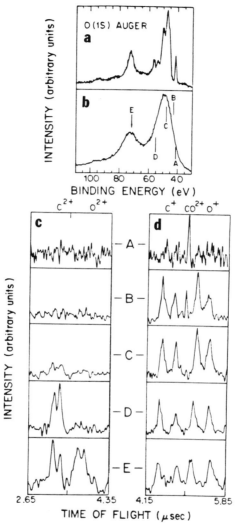

Fig. 5. The TOF ion mass spectra of CO taken in coincidence with oxygen Auger electrons of various kinetic energies as indicated by the arrows A-E in the Auger spectrum (b). A comparison is shown of the Auger spectra taken without (a) and with (b) an ion extractor field applied across the ionization region. The capital TOF mass spectra (c) show the region where C^{2+} and O^{2+} are expected to arrive and spectra (d) show the region containing C^+, CO^{2+} and O^+.

sensitive to the substructure in the major Auger peaks. Otherwise the spectra measured in coincidence with electrons at A would not be different from the one measured at position B in the Auger spectrum.

The sharp peak on top of the Auger spectrum is assigned to the $A^1\Pi$ state of CO^{2+} with a small admixture of the $X^1\Sigma$ molecular ground state of CO^{2+} (7). In single particle notation these states are described as $(5\sigma)^{-1}(1\pi)^{-1}$ and $(5\sigma)^{-2}$, respectively. However, Correia et al. [7] have found that the Σ states are strongly mixed by configuration interaction and are only poorly described in the single particle picture. Nevertheless, we will still use this notation here, keeping the configuration mixing in mind.

In coincidence with these Auger electrons, at position A, we only observe CO^{2+}. This is consistent with the potential energy curves calculated for both these states, which have a well defined minimum at the internuclear separation of the neutral and core excited states. This minimum is bound by a potential barrier of several eV in height [7], such that CO^{2+} is stable for the duration of our measurement. Typical CO^{2+} flight times were 5 μ sec. Within this time the molecule stays intact and does not dissociate into C^+ + O^+ even though it would be favorable on adiabatic energy arguments. The adiabatic energy for this fragmentation channel is about 5.4 eV lower than the minimum of the $X^1\Sigma$ state of CO^{2+} [7,8].

The coincidence spectra taken with electrons of the energies B and C show a new channel which dissociates into C^+ and O^+ fragments. The Auger electrons at this energy correspond to the $(1\pi)^{-2}$ double hole configurations of CO^{+2}. Even though we are not aware of any calculated potential energy curves for these states of Σ and Δ symmetry, these curves have to be strongly repulsive because of the bonding character of the 1π electrons. The observed kinetic energy of the fragments (7 ± 1 eV for C^+ and 5 ± 1 eV for O^+) is consistent with a decay into the lowest C^+ and O^+ channel. This means all the excess energy gets converted into kinetic energy of the fragments. The presence of a CO^{2+} peak at positon B is attributed to the energy spreading in the electron energy distribution when the ion extraction field is applied and indeed, this signal is not seen in curves C.

At position D the Auger final states are now largely triggering an $(4\sigma)^{-1}(1\pi)^{-1}$ and $(4\sigma)^{-2}$ two hole final states. The C^+ and O^+ fragment energies have increased somewhat and additionally there is a large signal of C^{+2} observed, which has to be accompanied by a neutral O partner. The appearance of a C^{2+} fragment as a result of the O 1s core hole decay in CO is somewhat surprising at first. In the single particle picture the 1π and even more so the 4σ hole states, which are the final states at this Auger transition energy, are still localized at the O end of the CO molecule. Only the configuration mixing in these states can account for the "transfer" of charge to the C fragment.

Finally in coincidence with electrons at energy E all possible single and double charged fragments are observed in the spectra. The Auger final states now involve two hole states, where one hole is located in the innermost valence orbital (3σ).

In coincidence with the C 1s Auger decay we get some complementary information since to some extent different hole configurations are observed as the final states of this Auger decay. The fundamental of the undulator was tuned to an energy near 300 eV, consequently also the C 1s → π^* state is excited as seen from the presence of the high energy lines in the Auger spectrum.

The sharp peak at the top of the Auger spectrum corresponds to a transition into the $X^1\Sigma$ ground state of $(CO)^{2+}$. This state was only with a very

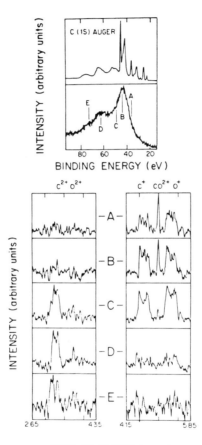

Fig. 6. The TOF ion mass spectra of CO taken in coincidence with carbon Auger electrons (see Fig. 5 for details).

small fraction excited in the corresponding line seen in the O Auger decay. Thus we confirm that both the $X^1\Sigma$ and the $A^1\Pi$ states of $(CO)^{2+}$ are stable on the timescale of our experiment. The presence of a C^+ and O^+ signal at this energy can be explained by the underlying lines of the decay of the C 1s → π^* excitation or to the energy smearing due to the extraction field.

The C^+ and O^+ signals observed at position B and C in the Auger spectrum originate from the $B^1\Sigma$ state, $(4\sigma)^{-1}(5\sigma)^{-1}$ in single particle notation. This state is at a binding energy of 45 eV and below the $(1\pi)^{-2}$ manifold. Consequently the ions have a smaller kinetic energy than the ones observed in coincidence with the O-Auger electrons and the $(1\pi)^{-2}$ double hole configuration. The potential energy curve of the $B^1\Sigma$ state exhibits only a very shallow minimum and the Franck-Condon region for transitions from the core hole excited state into this final state is located on the repulsive part of this potential energy curve. Therefore there is enough energy available to overcome the small dissociation barrier. At position D the result is similar to the O-Auger decay. This is again the manifold of the $(3\sigma)^{-1}(OV)^{-1}$ (outer-valence) final states and all possible fragments are observed.

SUMMARY

In these studies we are able to get a detailed picture of the various processes and pathways leading from the absorption of a soft-X-ray photon by a free molecule to the appearance of neutral and ionic fragments of the molecule. The coincidence studies between the Auger electrons emitted in the core hole decay and the ions reveal the involvement of individual valence electrons in the molecular bond. A comparison with the potential energy curves for the ionic molecular states explains the conversion of electronic energy into kinetic energy of the fragments.

ACKNOWLEDGMENT

I want to thank my co-workers E. W. Plummer, C. T. Chen, I. W. Lyo, R. Reininger, R. Carr, W. K. Ford and D. Sondericker for their help and contributions to these experiments. The experiments were performed at SSRL, which is supported by DOE. This experiment was also partially supported by NSF under Contract No. NSF-DMR-8120331.

REFERENCES

1. W. Eberhardt, E. W. Plummer, I. W. Lyo, R. Reininger, R. Carr, W. K. Ford, D. Sondericker, Austral. J. of Physics, 39, 633 (1986).

2. A. P. Hitchcock, C. E. Brion, M. J. van der Wiel, Chem. Phys. Lett. 66, 213 (1979).

3. W. Eberhardt, T. K. Sham, R. Carr, S. Krummacher, M. Strongin, S. L. Weng, D. Wesner, Phys. Rev. Lett. 50, 1038 (1983).

4. K. Müller-Dethlefs, M. Sander, L. A. Chewter, E. W. Schlag, J. Phys. Chem. 88, 6098 (1984).

5. P. Morin, I. Nenner, Phys. Rev. Lett. 56, 1913 (1986).

6. W. Eberhardt, this volume.

7. N. Correia, A. Flores-Riveros, H. Agren, K. Helenelund, L. Asplund and U. Gelius, J. Chem. Phys. 83, 2036 (1985).

8. A. C. Hurley, J. Chem. Phys. 54, 3656 (1971).

SYNCHROTRON RADIATION APPLIED TO X-RAY FLUORESCENCE ANALYSIS

Pierre Chevallier

LPAN, Université Pierre et Marie Curie, 4 Place Jussieu
75252 Paris Cedex 05
and LURE, Batiment 209C, 91405 Orsay Cedex

I. INTRODUCTION

Whenever physicists develop a new machine it seems that chemists try and make an analytical tool out of it, and sometimes with great success. This has also happened with the application of Synchrotron Radiation (S.R.) to X-ray fluorescence analysis. Since the pioneer work of C. Sparks (1,2) at Stanford a little more than ten years ago this technique has been tested in most Synchrotron Radiation centers and we wish to give a general view of the present status in this field. For simplicity we shall refer to this technique as SRXFA (standing for Synchrotron Radiation X-Ray Fluorescence Analysis) although many other abbreviations may be found in the litterature (SYXFA, SRXRF, ...).

By now there is a growing need, in every human field of activity, for the determination and characterization of elements at trace concentration, that is below one part per million by weight. For qualitative as well as quantitative analyses X-ray fluorescence is often used as a signature of elemental composition. Coupled with other techniques it will be soon able to give interesting chemical information, even at low concentration, thanks to the intense and perfectly well adapted Synchrotron Radiation sources.

In principle X-ray fluorescence analysis is a very simple technique. A given sample is placed in a chamber. A beam strikes this sample in order to create inner-shell vacancies and a detector records the X-ray fluorescence spectrum.

This is a qualitative method of analysis as each X-ray line is characteristic of a given element. It is also quantitiative since the number of recorded X-ray is proportional to the concentration of the emitting element. Moreover it is quite universal, except for very light elements ($Z<12$), because of their too low fluorescence yield, and uneasy to detect characteristic X-ray lines. Another general and quite interesting feature is that it is often a non destructive analytical method.

The variety of beams (photons, electrons, heavy charged particles ranging from proton to uranium) and of detectors now available gives rise to many different combinations and difficulties arise as soon as we wish to choose the best arrangements. In fact as long as we are interested in quantitative determination of elements at the lowest possible concentration, physics considerations should help to choose honestly the best combination. But to beat sensitivity records is not the only interesting thing in research and industry and it is sure that more than one kind of beam -detector combination will always be necessary in order to achieve more precise characterization of the sample. In particular, the distribution of an element in the sample with high spatial resolution, its depth profile, its chemical state (oxydation state, type of surrounding atoms, structure) is more and more often needed. Then there is no universal analytical method using the excited fluorescence in a sample.

II. SEARCH FOR SENSITIVITY

II.1. About minimum detectable limits (MDL)

Whatever the experimental set-up will be, XRFA end's up with a spectrum where the characteristic X-ray lines reflecting the elemental composition of the sample are superimposed over a background. For high sensitivity we wish the fluorescence peak (S) to be as high as possible together with the lowest background (B). In fact it is not the absolute value of this background that really matters, but the statistical fluctuation of this background and they vary as $\sqrt{B}$. Then for MDL estimation we must compare S to $\sqrt{B}$.

In a very simplified way, neglecting absorption, we can write :

$$S_Z = N_Z\, \sigma_{F,Z}\, I_0\, \varepsilon\, t$$

which means that the fluorescence peak area for element Z is proportional to
  (i) the number of atoms Z per unit area (N_Z)
  (ii) the fluorescence cross section of element Z for a given characteristic X-ray line ($\sigma_{F,Z}$)
  (iii) the beam intensity, that is the number of projectile striking the sample per unit time (I_0)
  (iv) the overall detectors efficiency at the considered energy (ε)
  (v) the exposure time (t).

In the same way we can estimate the background to be:

$$B = N_T\, \sigma_B\, I_0\, \varepsilon\, t$$

where N_T represents the total number of atoms per unit area on the sample and σ_B stands for a background production cross section.

For the identification of element Z scientific statistical reason tells us that the ratio $S/\sqrt{B}$ must exceed a given number (n) the exact value of which mainly depends on optimistic consideration. Then MDL for element Z can be computed from the equation

$$\frac{(MDL)_Z}{\sqrt{N_T}} = \frac{\sigma_{F,Z}}{\sqrt{\sigma_B}} \sqrt{I_0\, \varepsilon\, t} > n$$

II.2. How can we improve these limits

From this equation we can derive scientific arguments to attain highest sensitivity.

II.2.1. Sample

For a given sample nothing can be done about $\sqrt{N_T}$ except to try and fix it on the thinest backing as possible. For thin samples (compared with the penetration depth of the projectile) this backing may contribute to a very significant part of the real target atoms and its purity has to be very carefully checked. With photons as incident beam its structure can be of importance for its contribtion to the background through scattering processes.

Liquid samples which can be dried up before examination allow a drastic reduction in the real number of target atoms under analysis. For this reason very low detectable limits are often claimed with this type of samples. Of course care must be taken when looking for volatile elements which can be lost during evaporation.

II.2.2. Detector efficiency

High sensitivity is obtained with detectors of high efficiency. For this reason Si(Li) detectors which allow high solid angle of detection are of universal use. As we shall see they present some drawbacks like poor energy resolution and saturation for high count rate. Wavelength dispersive devices allow high resolution but usually suffer from too low transmission resulting in a waste of information.

II.2.3. Beam intensity

The beam intensity must be maximum. This is often in favour of charged particle beams which are nearly parallel and of small section. Moreover they can be focused rather easily. Technically beams of 1μA (that means some 10^{12} projectile per second) can be focused on small areas. This is of primary importance for the study of small samples and especially for the realization of microprobes. Conventional photon sources, such as X-ray tubes or radioactive sources, cannot compete because of their wide angular diverging beam. The natural collimation of S.R. is then an important advantage and in this sense this mode of excitation compares very well with charged particles. Typically as much as 10^{12} photonss per cm² can be otained in a 1% bandwidth from bending magnets of storage rings, and much higher flux are expected with insertion devices like wigglers and ondulators. Furthermore this natural collimation of S.R. allows to use efficiently refocusing optics for the study of small samples. Although hard X-ray optics have still to progress, microprobes using S.R. are already under investigations.

Table 1.

	Photon		Electron		Proton	
	5 keV	20 keV	20 keV	50 keV	3 MeV	5 MeV
Al (K)	5		50		170	
Cu (K)		14	1900	830	31000	10000
Au (L3)		13	1700	910	71000	21000

Energy left (in keV) in different 1μm thick targets by various projectiles for the production of one characteristic X-ray.

An immediate drawback to high intensity is the ability of the sample to support such intense flux without any damage. This is sometimes a real problem and makes the photon a really interesting projectile. The photon that has not interacted will leave the sample without loosing any energy whereas a charged particle looses its energy in a very short range whether or not it has created a useful ionization.

As an example, table 1 gives the energy left in various 1μm thick pure samples to produce one characteristic X-ray line of this element using different projectiles of different energies. The advantage of monochromatic photons is striking.

II.2.4. *Exposure time*

Long exposure times improve MDL but it is a costly factor. Moreover the great number of samples to be analyzed will set an upper limit to the exposure time. One thousand second analysis which allows the study of about 70 samples per day seems a good compromise. Then it is more important to gain in intensity in order to have the same number of projectiles striking the sample in a shorter time (when the sample can support it).

Finally we must remember that MDL improves only with the square root of detector efficiency, beam intensity and exposure time.

II.2.5. *Fluorescence cross section*

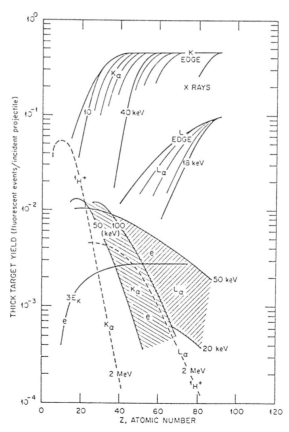

Figure 1. Thick target fluorescence yield for X-rays, protons and electrons (from reference 2).

The fluorescence cross section is the product of the inner-shell ionization cross section, the fluorescence yield of this shell, and the relative intensity of the particular X-ray line under consideration. The last two factors are independent of the projectile which is far from being the case for the ionization cross section.

The inner-shell ionization cross section is much in favour of the photon as it appears in figure 1 (see ref 2), which gives a comparison between photons of various energies and of electron and protons in the usual energy range. Moereover the Z dependence of this cross section is also important.

For photons this cross section scales typically as $Z^4 E^{-3}$.

For charged particles it appears that the product σB^2 where B is the binding energy of the ejected electron, follows a universal curve (different for electrons and heavy particles) which depends on the projectile velocity. As the binding energy varies with Z^2 it appears that charged particle ionization cross sections decrease very rapidly with Z, nearly as Z^{-4}. Furthermore these cross sections rise with the projectile energy up to a maximum when the projectile velocity matches that of the ejected electron. Unfortunately (except for very light elements) this corresponds to an energy much higher than what can be obtained with the small accelerators available for x-ray fluorescence analysis.

Remembering that the universal abundance of elements drops by 12 orders of magnitude from H to U, this Z dependence of the cross section makes the photon a perfect projectile for the identification of trace elements.

II.2.6. *Background and its origin*

Finally, the background production should be the lowest as possible and we must now investigate about its origin.

For electrons, this background is due to bremsstrahlung emitted during the slowing down of the incident electron and of the ejected electrons. As the electron can share half of its energy through a single collision, both contributions are important. This bremsstrahlung spectrum has a continuously decreasing shape from zero energy up to the energy of the incident projectile. This backbround is reduced for low Z matrix samples (or very thin samples). It also decreases with lower energy incident electrons but we must remember that the ionization cross section will also decrease. This bremsstrahlung strongly limits the interest of electron as projectile when speaking of MDL.

With heavy charged particles (protons are mostly used) the background is also due to bremsstrahlung, but the power thus radiated being inversely proportional to the square of the mass of the slowing down particle, its contribution will be much smaller than for electrons. Of course the proton energy is much larger than that commonly uncountered with electrons (about two orders of magnitude) which makes this contribution not completely negligible. The bremsstrahlung due to the ejected electrons is confined to low energy and drops sharply with energy because with heavy projectiles the electrons are mainly ejected with very little kinetic energy. For commonly used proton energy this bremsstrahlung background becomes insignificant over 10 or 15 keV. This very important reduction in the bakcground compared with what is observed for electrons settles the detector limits in the ppm range for most elements.

For photon excitation the background origin is more complex. There is of course a contribution from the bremsstrahlung of the ejected photoelectrons from the whole sample. In this sense one should try to use photons of as low energy as possible in order the photoelectron energy to be minimum. Thus the induced bremsstrahlung is restricted to the low energy part of the spectrum and contributes only very little to the background.

The main source of bakcground with photons is due to scattering (elastic or inalestic) of the incident beam on the whole target. The total scattering cross section is usually much smaller than the photoionization cross section, but there are so many more target atoms than atoms of the element we are interested in that the sample becomes essentially a source of scattered X rays. Through these processes we observe a scattered spectrum which is an image (reduced in intensity) of the incident spectrum. For a continuous spectrum we shall have a continuous background under the fluroescence peaks which strongly limits the sensitivity.

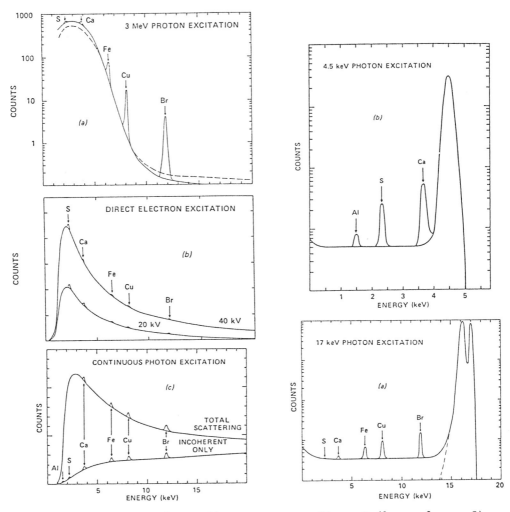

Figure 2 (from reference 3)
(See text for comments)

Figure 3 (from reference 3)

As an illustration of this discussion we present figure 2 calculated spectra which show the relative importance of the fluorescence signal and the background for these typical cases. These calculations are taken from Jacklevic (3). They show the expected spectrum obtained with a 200 eV resolution Si(Li) detector for a target of 250 ng cm^{-2} of Al, S, Ca, Fe, Cu, Br on a 25 mg cm^{-2} carbon backing (10 ppm concentration) under bombardment of 40 keV electrons, 2 MeV protons and a continuous X-ray spectrum respectively. From these pictures we can say that minimum detectable limits seem to be around 100 ppm for electrons, 1 ppm for protons, and 10 ppm for continuous X-ray excitation.

II.3. Interest of Synchrotron Radiation

Sensitivity in trace element analysis can be greatly improved when using Synchrotron Radiation because of some important characteristics of this light. In fact Synchrotron Radiation presents an intense continuous spectrum of highly collimated and polarized light. Each of these characteristics makes S.R. a perfectly well adapted source for X-ray fluorescence analysis. (Details about S.R. will be found in Pr. Brown's lecture of this school and the reader should refer to this text for a full understanding of the interest of this mode of excitation.

II.3.1. *Polarization*

The first advantage comes from the high degree of polarization in the electron orbit plane (horizontal polarization). One can show (see for example ref 4) that in the direction of the incident polarization vector the Rayleigh scattering cross section is zero and the Compton scattering minimum. With new machines and for a well collimated beam centered in the electron orbit plane the horizontal polarization can be as high as 99% and as a consequence there is a considerable reduction in the bakcground if the detector is placed in this plane and at 90° from the incident beam. This is the adopted geometry in every S.R. center.

II.3.2. *Use of monochromatic excitation*

The second improvement comes from the very low vertical divergence of S.R. This divergence (which is of the order of $m.c^2/W$) allows for a very efficient monochromatization of the incident beam, because it is of the same order of magnitude than perfect crystals diffraction pattern.

Considering that the photoionization cross section decreases very rapidly above the absorption edge (typically as E^{-3}) and that the S.R. energy spectrum also drops exponentially with energy (this is true above the critical energy of the machine) we understand that only those photons which have an energy just above this threshold are really efficient for fluorescence analysis. All the others mostly contribute to scatteirng processes and lead to unwanted high count rate in the detector. Moreover the photons with an energy lower than the threshold are totally useless but will contribute to the background under the fluorescence peak of the searched elements. It then appears that by using monochromatic photon excitation (or at least with a broad bandwidth) we may not loose much intensity in the fluorescence signal but the background beneath this signal will be drastically reduced. In fact, the annoying scattered spectrum is now reduced to a peak far over the fluorescence spectrum, which means a considerable reduction in total count rate and background. With Si(Li) detectors the background is essentially due to the low energy tail always accompanying the total absorption peak.

Of course elements whose absorption edge is just under the monochromatic excitation energy will be most efficiently ionized while this excitation mode will be blind to elements with higher absorption edge. With a proper choice of the excitation energy (simply by rotating the monochromator) the experiment can be optimized for the search of a particular element. Even better this tunable energy can be chosen so as to get rid of the too intense fluorescence of major elements if we are interested in the study of lower Z trace elements. These aspects of the benefit of using monochromatic excitation are shown on figure 3 which result from Jacklevic's calculation for the same sample as above, and are to be compared with those on figure 2.

Such monochromatic (or near monochromatic) excitation can be obtained with X-ray tubes after proper filtering of the emitted beam. With special care this has led to very interesting laboratory XRF installation. Nevertheless monochromatic S.R. excitation keeps many advantages over these facilities because of the much more intense flux on the sample, its high degree of polarization and its total and easy tunability.

Another advantage of the natural collimation of S.R. is that refocusing optics can be used quite efficiently. This point is of great interest for the study of small samples (~1 mm^2) but has to be improved to match the requirements of microprobes.

II.4. Minimum detectable limits

MDL can be estimated from equation 1 for thin samples. They depend on the knowledge of cross section but also on beam considerations and one must keep in mind that the total count rate using Si(Li) detectors cannot exceed 10 000 cps. The result depends strongly on the machine as the beam size, its divergence, brilliance, polarization and intensity change drastically from a machine to another.

As a general remark we may say that most S.R. facilities allow analysis for most elements in a range much lower that the ppm and that limitations are mostly due to the detectors saturation.

III. EXPERIMENTAL SET UP

Different kinds of experimental set up can be imagined and are used.

III.1. White beam excitation

This is the simplest case. A pair of slits is used to define a suitable size and shape of the direct indident beam. Such a device is used in Brookhaven for example where it can take advantage of the very high brilliance of the beam delivered by this new machine. It represents the first step in the construction of a much more ambitious project.

To reduce the count rate on the detector a very small beam size is used (typically 50 to 100 µm) and thin foils of aluminium are used as absorbers of the low energy part of the incident beam.

III.2. Monochromatic excitation

Most arrangements use a monochromator before the sample. For simplicity double monochromators in the parallel mode are preferred because they allow a fixed sample position for any incident energy.

Different crystals are used. Perfect crystals like Si or preferable Ge have a small band pass that seldom allow to take full advantage of the incident intensity. This will no longer be a problem with new machines because of the much smaller beam divergence and the improved power delivered by special insertion devices like wigglers. On the other hand this small band pass can be very useful to avoid some interferences in characteristics X-ray identification by a proper choice of the incident beam. For example As(Kα)=10.543 keV cannot be distinguished from Pb(Lα)=10.551 keV, but if the incident energy is chosen between AsK edge=11.867 keV and PbL$_3$ edge=13.035 keV the ambiguity is cleared up most easily.

Special planes, such as Si(111) or Ge(111) are of special interest as they give no second order harmonics. This is very useful when looking at low Z elements in a light Z matrix.

Mosaïc crystals, such as highly oriented pyrolitic graphite, are also widely used, specially with old machines because of their more diverging beam. Higher transmitted intensity allows (at the moment) lower detectable limits but selective excitation is by far not so easy as that with perfect crystals.

Because the experiment is typically located 20 m apart the light source, users are in general allocated 2 m Rad of horizontal beam. This leads to an incident beam size of the order of 40x10 mm, which is perfectly suited for the study of large samples, and a resulting very low energy deposited per unit area.

A last type of monochromator uses a curved crystal in the sagittal plane. This allows a refocusing of the beam in the horizontal plane. This device is not so easy to use as the one with a flat crystal as the focusing point changes with energy. Geometrical optics relation gives:

$$\frac{1}{F_1} + \frac{1}{F_2} = \frac{2 \sin \theta}{R}$$

where θ is the Bragg angle, F$_1$ the beam source to crystal distance and F$_2$ the crystal to sample distance. We already said that F_1 was of the order of 20 m. For convenience F_2 is preferred to be small (<1 m) and this mainly because the magnification is given by F_2/F_1. Thus R, the sagittal radius of the crystal curvature must be small (10 to 20 cm). Pyrolitic graphite again is interesting as it can be pressed to any shape. We then further get the benefit of a large band pass but the splitting of the beam due to the mosaïc spread enlarges the image and reduces somewhat the photon density on the sample. Such a device is quite interesting for the study of small samples but inadequate for reaching the required perfomances of a microprobe.

IV. EXAMPLES OF APPLICATION

We present figure 4 a typical spectrum obtained at LURE (DCI Orsay) with a conventional set up for monochromatic excitation : parallel mosaïc graphite monochromator (σ=0.4°); working conditions : 1.72 GeV, 200 mA, excitation energy 10.8 keV, 0.5 mm Al absorber before the monochromator. The fluorescence spectrum is recorded with a Si(Li) detector 26 mm^2 area 20 cm (in air) from the sample. The sample is a 1 g pellet of cellulose containing 1 ppm of elements V, Cr, Mn, Fe, Co, Ni, Cu, Zn. The recording time is only 100 s.

This spectrum appeals many comments. First of all the Kα lines of all the expected elements show off very easily which gives a good idea of the interest of S.R. for fluorescence analysis. The peak intensity decreases regularly with Z as can be expected from the Z dependence of the cross sections. The relatively high intensity of Fe and Cu suggests a contamination of the cellulose matrix with about 1 ppm of these elements.

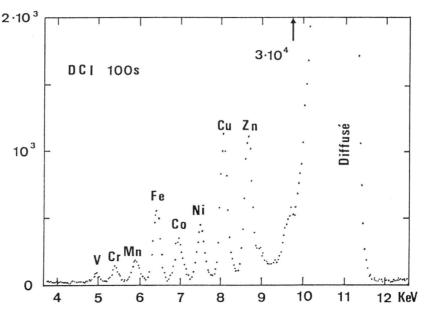

Figure 4. Typical spectrum of 1 ppm V....Zn in a 1 g cellulose matrix. (see text for details).

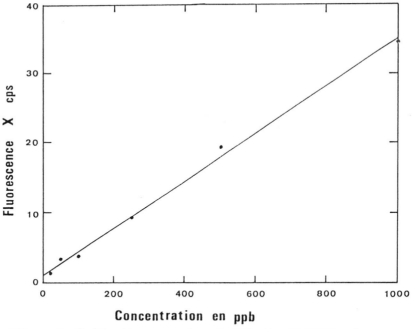

Figure 5. Calibration curve for cobalt in the 10-1000 ppb range.

This gives an idea of the problems encountered when looking at much lower concentrations. For such a multielement sample we see the overlap of the Kβ line of an element with the Kα one of the next element. This adds to the difficulty in computing the area of each peak which is already rendered uneasy by the lack of background reference zone of estimation. Very sophisticated deconvolution programs have to be used to get these areas with good confidence.

But of course the most striking feature in this spectrum is the huge scattered peak which contributes to nearly 99% of the total count rate and whose low energy tail causes the limiting sensitivity background. In fact it is the intensity of this useless peak which imposed to have such an unusually large sample to detector distance. If this scattered peak is reduced (with a higher degree of polarization of the beam, a thiner sample, and a proper filtering of the fluorescence spectrum) then the detector could be placed much nearer of the sample and the fluorescence signal could be higher thus lowering the MDL.

Figure 5 shows the results obtained on the same facility (but with a curved graphite monochromator) with a set of 1 g cellulose pellets containing Co at concentration from 20 ppb to 1 ppm, which corresponds to 170 pg to 17 ng of cobalt in the beam. For a 1000 sec exposure time we can derive a 2 ppb identification limit in this favorable case.

The next spectrum (Figure 6), also taken at LURE-DCI, is that of a single drop of human serum, simply dried on a thin mylar film. An idea of the sensitivity is given by the observation of $SeK\alpha$ whose concentration is of 80 ppb. Synchrotron Radiation proves to be perfectly adapted to the study of medium or high Z elements in low Z matrix, which is typical of biological samples.

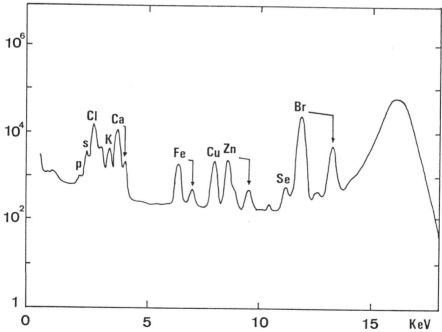

Figure 6. Spectrum of dried human serum (DCI 1000 s.).

Finally figures 8 and 9 show two spectra obtained under white beam excitation in Brookhaven with small size spots of 100 x 100 μm. The first spectrum shows that the sensitivity is lower than that under monochromatic excitation due to a higher background (typically 1 ppm but on very small samples). The second spectrum is that from a deep sea spherule of extraterrestrial origin. It presents an annoying feature (that could become useful in some cases). In the upper spectrum some unidentified peaks appear between PbLα and Fe sum peak. For the lower spectrum the sample

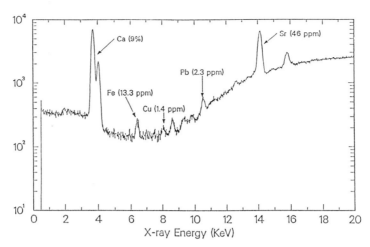

Figure 7. NBS glass standard under white beam excitation (NSLS BNL courtesy of S. Sutton)

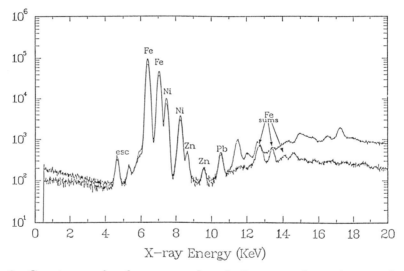

Figure 8. Spectrum of a deap sea spherule for two orientations under white beam excitation (see reference 5).

has been rotated by a few degrees and we notice that the unidentified peaks have disappeared. S. Sutton (5) identified these structures as diffraction peaks on microcrystals in the sample. They could lead to erroneous identifications but on the other hand they can bring more characterization on the structure of the sample. Of course the probability of observation of such peaks is considerably reduced with a monochromatic beam.

V. QUANTITATIVE ANALYSIS

Just like XRF induced by X-ray tubes SRXRA can be quantitative. The same problems, such as homogeneity of the sample, grain size, interelement enhancement ..., will have to be taken into account but the high stability of the beam and the knowledge of the incident radiation spectrum may be an interesting point for S.R.

We consider a beam of $I_0(E)$ photons of energy E arriving with a take-off angle α on the surface of a plane sample. Let $dI_Z(E,x)$ be the number of Kα line of element Z arising from an infinitely thin slab of the target located at the depth x and detected in the direction β.

The sample contains elements $Z_1, Z_2 ... Z_n$ whose concentrations (by weight) are $C_{Z1}, C_{Z2}, ..., C_{Zn}$ ($C_{Z1}+C_{Z2}+... +C_{Zn}=1$). At energy E the mass absorption coefficient of the sample is given by:

$$\mu_m(M,E) = \sum_i C_{Z_i} \mu_m(Z_i,E)$$

where $\mu_m(Z_i,E)$ are the individual mass absorption coefficients for element Z_i and energy E.

The number of incident photons arriving in the slab at depth x is:

$$I_0(E) \, dE \, \exp - (\mu_m(M,E) \frac{\rho x}{\sin \alpha})$$

Out of these $\mu_m(M,E) \rho dx / \sin \alpha$ will interact in the slab. The probability that these interactions will result in a K shell ionization of element Z_i followed by the emission of a Kα line is

$$C_Z \frac{\mu_m(Z_i,E)}{\mu_m(M,E)} \frac{J_K - 1}{J_K} \omega_K(Z_i) \frac{I_{K_\alpha}}{I_{K_\alpha} + I_{K_\beta}}$$

where J_K is the K absorption edge jump and ω_K the fluorescence yield. The three last factors denoted $P(Z,K\alpha)$ only depends on the particular X-ray line of element Z we wish to observe and is independent of the projectile and its energy.

We then must write the probability for the emitted Kα line to leave the target without absorption, that is a factor

$$\exp - (\mu_m(M,Z_{K_\alpha}) \frac{\rho x}{\sin \beta})$$

where $\mu_m(M,Z_{K\alpha})$ is the mass absorption coefficient of the sample for an energy corresponding to the Kα line of element Z.

Further we must introduce the detector solid angle ($\Omega/4\pi$) and efficiency for the Kα line of element Z ($\epsilon(Z_{K\alpha})$);

Finally

$$dI_Z(E,x) = I_0(E) \, dE \, \frac{\rho \, dx}{\sin \alpha} C_Z \mu_m(Z_i,E) P(Z,K_\alpha)$$

$$\exp - ((\frac{\mu_m(M,E)}{\sin \alpha} + \frac{\mu_m(M,Z_{K_\alpha})}{\sin \beta}) \rho x) \frac{\Omega}{4\pi} \epsilon(Z_{K_\alpha})$$

From this relation the number of detected Kα lines of element Z is derived by a twofold integration, first over the targets depth T, then on the incident energy spectrum for all photons with an energy higher than the binding energy of element Z

$$I_Z = \int_{B_K}^{\infty} \int_0^T dI_Z(E,x)$$

As usual two extreme cases are considered:

a) thin targets

$$I_Z = C_Z \frac{\Omega}{4\pi} \frac{\varepsilon(Z_{K_\alpha})}{\sin \alpha} P(Z,K_\alpha) \rho T \int_{B_K}^{\infty} \mu_m(Z,E) I_0(E) dE$$

b) thick targets

$$I_Z = C_Z \frac{\Omega}{4\pi} \frac{\varepsilon(Z_{K_\alpha})}{\sin \alpha} P(Z,K_\alpha) \int_{B_K}^{\infty} \frac{\mu_m(Z,E) I_0(E) dE}{\dfrac{\mu_m(M,E)}{\sin \alpha} + \dfrac{\mu_m(M,Z_{K_\alpha})}{\sin \beta}}$$

These two equations show that there is a direct (although complicated) relation between the concentration and the area of the fluorescence peak on the recorded spectrum.

Many of the parameters entering in these equations, and especially in the case of thin samples, can be obtained by comparison with a standard of known composition. In this ideal case uncertainty is mainly due to the estimation of the area of the considered fluorescence peak and it seems reasonable to claim that, at the ppm level, it should be no more than a few percents if no interfering peak is present.

Reliable estimations can be made by a complete calculation as the shape of the incident spectrum is quite well known when using monochromatic Synchrotron Radiation excitation. In this case uncertainty is mostly due to the various absorption coefficients that enter the equations.

VI. IMPROVEMENTS

Many improvements can be thought off, some of them being already under realization, so as to take full benefit of S.R. for fluorescence analysis. They mainly concern the machine and the X-ray optics, the detection device and finally the kind of research that can be done.

VI.1. Beam considerations

Future or recently developed S.R. facilities present many interesting features for SXRFA.

The electron beam size is of the order of 100 µm instead of a few mm and its divergence very small. As a consequence a much higher degree of horizontal polarization of the photon beam is expected leading to an interesting reduction in background due to scattering. Furthermore the brilliance of the electron source is enhanced by many orders of magnitude. This result is a considerable increase in the number of photons per unit area when using focusing optics and will allow the realization of real microprobes.

The increase of the electron bream energy (typically in the 4 to 6 GeV range) together with new insertion devices like wigglers shifts the X-ray spectrum towards higher energy and experiments up to 100 keV can be thought off. This will allow the study of almost all elements throught their K spectrum which is much easier to observe than the L spectrum usually recorded for elements heavier than Mo or Ag. This possibility has been recently demonstrated at VEPP-4 by Baryshev and al.(6).

Finally an other important gain in brilliance can be expected with ondulators but it seems difficult to expect beams of energy higher than 10 to 15 keV with such devices. Nevertheless they could become very interesting for the study of light element whose fluorescence yield is very low.

VI.2. Detection devices

Up to now Si(Li) detectors are universally used to record X-ray spectrum from SXRFA. On one hand they present many interesting features as they allow a true multielemental analysis with high efficency. On the other hand they present many drawbacks.

First of all their low energy resolution leads to numerous interferences such as $K_\alpha(Z+1)$ with $K_\beta(Z)$ in the transition element region (Cr K_α = 5.414 keV and V K_β = 5.427 keV) or K (Z light) with L (Z heavy) such as As K_α (10.544 keV) and Pb L_α (10.551 keV). Another kind of interference, inherent to these detectors, may appear with the escape peak associated with the total energy absorption peak. As an example Co at trace concentration will be difficult to observe if Zn is also present in notable quantity (Co K_α = 6.929 keV and Zn K_{escape} = 6.897 keV). This problem is even worse with Ge(Li) detectors which one may have to use on wiggler stations. A last kind of interference, due to electronic effects, is linked with the sum peaks that appear at energy $2K_\alpha$, $K_\alpha+K_\beta$ and $2K_\beta$ when a major element with a high fluorescence yield is present in the sample. These spurious peaks are due to pulse pile up in the amplifier and occur when two events are recorded in a time short towards the amplifier shaping time constant. As an example 2 V K_α = 9.902 keV will interfere with Ge K_α = 9.885 keV and Hg L_α = 9.896 keV. Anti-pile up amplifiers are of little help in this case since the two events may be recorded during a single flash of synchrotron light whose duration is typically of 1 ns. Nevertheless pile-up rejection systems have to be used because the tail pile-up observed under S.R. excitation is no longer a continuous background but present peaks that reflect the time stucture of the photon beam (7).

A major drawback of these solid state detectors is their incapacity to support high count rates. In fact their energy resolution degrades as the count rate increases and we can fix an upper limit of 5000 cps. Another point is that live time corrections to compensate rejected pulses become less reliable passed this same limit. Such count rates are already largely attained with existing S.R. machines simply with the elastic and inelastic scattered peak (refer to figure 4 where 99% of the count rate is due to this useless peak) and one has often to reduce the detector solid angle or the incident beam intensity to keep the acceptable total count rate.This results in an important loss of useful information. In fact there is no need for the more intense beam of new machines if another kind of detector is not thought off.

A new device, trying to conciliate the advantages of Si(LI) detectors (namely a multielement analysis with high efficiency and those of wavelength dispersive systems (high resolution to avoid most of the energy interferences between many characteristic X-ray lines), has been tested at LURE with some success. It associates a flat mosaic crystal with a position sensitive detector. The principle is shown on figure 9. We see how such a crystal can refocus the fluorescence spectrum emitted from a point source at characteristic points according to the wavelength. Simple geometrical considerations show that a photon of wavelength λ leaving a point source with an angle α in regard to the central ray (we mean the one reaching the crystal surface with the proper Bragg angle), may find inside the mosaic crystal a microcrystal just tilted enough so that Bragg conditions are met. This photon will be refocused on a point, symmetric of the point source. For a small mosaic spread (α_{max} is typically of the order of 1°) and further small in regard to the Bragg angle the image size can be calculated to be

$$\Delta x = \frac{2D}{tg\,\theta_B} \frac{\alpha^2}{\sin^2 \theta_B}$$

where D is the distance between the point source and the crystal plane. This means a first order focusing in α which conferes a high transmission to the system. This transmission competes quite well with usual solid angle that can be found when using Si(Li) detectors.

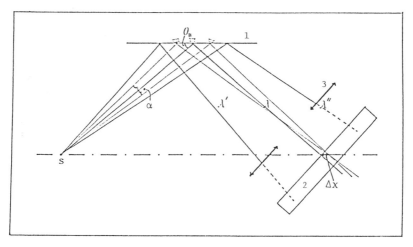

Figure 9. new detector device for X-ray fluorescence analysis. 1) graphite mosaic crystal. 2) Position sensitive detector. 3) Slits.

If a position sensitive detector is placed a the focus and perpendicular to the central ray we see that a large range of Bragg angle can be covered at the same time. Then a multielement analysis (for neighbour element) is still possible. Moreover slits in front of the detector can prevent unwanted wavelengths to reach the detector. This is interesting to get rid of the scattered radiation but also of the fluorescence of major elements and avoid unnecessary saturation of the detector.

We present the spectrum of a spherule of extraterrestrial matter of about 300 µm diameter observed with a Si(Li) detector (figure 10a) and with the new device (figure 10b) under the same conditions. Saturation from the scattered radiation and from the fluorescence of Fe and Ni was avoided and we can observe with high resolution characteristic lines due to Ge and Ga which are only at the 2 ppm level.

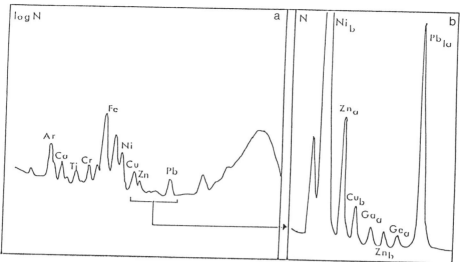

Figure 10. X-ray fluorescence spectrum of an extraterrestrial spherule. a) Si(Li) detector. b) Graphite mosaic crystal plus position sensitive detector.

VII. CONCLUSIONS

In conclusion SXRFA is a very powerful tool for sub-ppm trace element analysis. It already offers unbeatable MDL, even on bulk samples, compared with other fluorescence techniques using X-ray tube protons or electrons.

This technique is still under improvements. Just like PIXE and its association with Rutherford back scattering and Nuclear Reaction Analysis which did not become operational until the arrival of solid state detectors, SXRFA needs new detectors and real X-ray range optics ot take full advantage of its promises.

We wish also to mention that X-ray fluorescence should be more widely associated with other fields of reserach like XANES and EXAFS. With new machines EXAFS spectrum at the ppm level can be obtained in a reasonable time with special detectors such as the one we described earlier. This will bring very interesting characterization of trace elements, an infomation that cannot be reached with other kind of projectiles, and which is essential to understand their chemical behaviour.

Finally the extreme brilliance of new sources offer a real opportunity to build a microprobe with a few µm spatial resolution. Different approaches are in progress (8,9), and due to the very low energy left in the sample through S.R. they promise to achieve very low detection limits (2).

REFERENCES

1. C.J. Sparks Jr, and J.B. Hastings, X-Ray Diffraction and Fluorescence at the Stanford Synchrotron Radiation Project. Report ORNL 5089 (June 1975).

2. C.J. Sparks Jr, X-Ray Fluorescence Microprobe for Chemical Analysis : pp459-512 in Synchrotron Radiation Research, H. Winick and S. Doniach Ed., Plenum Press NY 1980.

3. J.M. Jaklevic, Proceedings of the Energy Research and Development Administration : X- and γ-Ray Symposium (Conf. 760539) Ann Arbor, May 1976.

4. A.L. Hanson, NIM A243 583 (1986).

5. S.R. Sutton, M.L. Rivers, and J.V. Smith, Anal. Chem. 58 2167 (1986).

6. V.S. Baryshev, A.E. Gilbert, D.A. Kozmenko, G.N. Kulipanov, and K.V. Zolotarev, Determination of the Concentration and Distribution of Rare Earth Elements in Mineral and Rock Specimens using the VEPP 4 Synchrotron Radiation, NIM A261 272 (1987).

7. V.B. Baryshev, G.N. Kulipanov, and A.N. Skrinsky, Review of X-Ray Fluorescent Analysis using Synchrotrn Radiation, NIM A246 739 (1986).

8. B.M. Gordon, and K.W. Jones, Design Criteria and Sensitivity Calculations for Multielements Trace Analysis at the NSLS X-Ray Microprobe, NIM B10/11 293 (1985).

9. W. Petersen, P. Ketelsen, A. Knöchel, and R. Pausch, New Developments of X-Ray Fluorescence Analysis with Synchrotron Radiation, NIM A246 731 (1986).

SOFT X-RAY AND X-RAY PHOTOEMISSION STUDIES OF LIGHT METALS AND ALLOYS

L.M. Watson

Metallurgy Department
University of Strathclyde
Glasgow, UK

INTRODUCTION

This paper is not intended to be a comprehensive review of soft x-ray (SXS) and x-ray photoemission spectroscopy (XPS) studies of alloys but is directed rather at an audience who is familiar with solid state physics but not in detail with this particular subject. The objective in all the experimental studies selected for discussion is to shed some light on the behaviour of the valence electrons in metals and alloys; hence the SXS involve transition of electrons in the valence band to some vacancy created in a core level of well defined energy (Fig. 1). The higher the binding energy of the core level the greater is the lifetime broadening, thus to achieve high resolution, spectra in the extreme UV or soft x-ray regions are preferred.

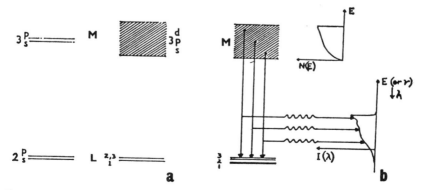

Fig. 1. Schematic of the soft x-ray emission process in 3rd row metals. (a) showing the atomic and solid energy levels, (b) the transitions giving rise to the $L_{2,3}$ emission spectrum.

INTERPRETATION OF SXS

To extract information from the SXS concerning the valence electron distribution a number of approximations are made. A simple one-electron picture is assumed with many body effects being treated separately. The intensity of emission of radiation of energy $h\nu$ is given by:

$$I_{h\nu} = 2 \int_{E_{const}} \frac{d^2k}{\nabla_k(E)} T(E_k)$$

where:

$$\rho(E) = \int_{E_{const}} \frac{d^2k}{\nabla_k(E)}$$

and is the density of states in the valence band of energy E and

$$T(E_k) = \frac{4e^2}{3h^4c^2}(h\nu)^3 \left| \int_{all\,r} \psi_f^*(r)\, r\psi_i(r,k)\, d^3r \right|^2$$

is the dipole matrix element.

The latter is of great significance in SXS because it dictates that x-ray emission can only occur where the initial and final state wavefunctions (ψ_i and ψ_f) overlap in real space. Since the core hole is localised, only the density of states in the region of the emitting atom is sampled. Further, since the matrix element incorporates the dipole selection rules ($\Delta l = \pm 1$, $\Delta m = \pm 1, 0$), only states of 's' and 'd' symmetry make transitions to core hole states of 'p' symmetry (L-spectra), and only states of 'p' symmetry make transitions to core hole states of 's' symmetry (K-spectra). Hence SXS probes the partial density of states in a region of space localised around the emitting atom and this can vary considerably with atom species in alloys and compounds - see Fig. 2.

The most important effect on the SXS due to many-body interactions is the broadening of the low energy limit of the emission band caused by intra-band Auger processes which reduce the lifetime of hole states in the valence band in a progressively increasing manner in going from the Fermi energy to the bottom of the band. This makes the location of the bottom of the band and hence accurate band width measurements extremely difficult. Other many-body effects include the much publicised edge singularity (see for example von Barth and Grossmann, 1982) and very weak low energy plasmon satellites. Both of these are of secondary importance in the study of the band structure of alloys.

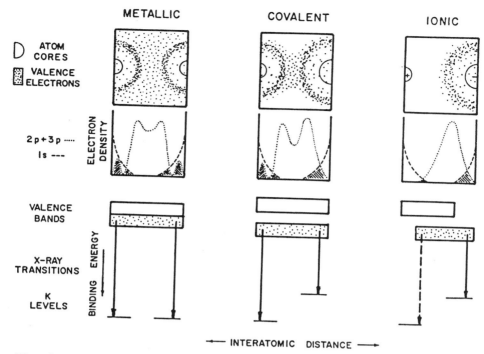

Fig. 2. Schematic representation of change in charge distribution with bond type and the spatial localisation of the inner core level wavefunction (from Nagle, 1970).

INTERPRETATION OF XPS

For the present studies the XPS spectra of interest are those involving the valence band where it is assumed that the final state is high enough above the Fermi energy (~ 1000eV) to be approximated to a plane wave. Therefore, the spectra are taken to be a reflection of the averaged density of states in the valence band modified by the difference in ionisation cross-section between electrons of 's', 'p' and 'd' symmetry (rare earth alloys are not considered here). Inner core level energies are measured to facilitate the location of the Fermi energy in conjunction with the SXS. Much useful information can be derived from the measurement of the asymmetry and satellites of inner core level XPS spectra but will not be discussed here.

EXPERIMENTAL CONSIDERATIONS

In common with all spectrometers, soft x-ray spectrometers consist of a source, an analysing element and a detector. For the ultra-soft x-ray region the analysing element is usually a diffraction grating used at grazing incidence while for the harder x-ray region Bragg reflection from some suitable crystal of known d-spacing is used.

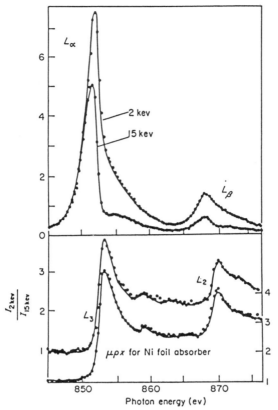

Fig. 3. The anode self absorption spectrum of nickel (from Liefield, 1968).

Obvious experimental pitfalls to be considered are the linearity of reflectivity of the analysing element over the energy region being investigated, linearity of response of the detector, critical absorption by the analysing elements etc. Of particular interest is the problem of self-absorption, i.e. the absorption of x-rays generated within the anode by the anode material itself. Liefeld (1968) elegantly demonstrated that self-absorption can distort considerably the profile of the SXS and to reduce this effect to a minimum, the spectrum should be excited with electrons of the lowest possible energies (threshold excitation). Also the x-ray take-off angle should be normal to the sample surface to minimize the depth through which the x-rays have to travel to escape. He also showed that good use can be made of self-absorption since a plot of the ratio of intensities of two spectra, one taken at a high excitation potential and one take at near threshold excitation produces a profile that compares favourably with the absorption spectrum for the anode material (Fig. 3).

More recently Szasz and Kojnok (1985) have shown that the variation in penetration depth with electron excitation potential can be used in SXS for quantitative depth profiling. Examples of this will be discussed later.

ALLOYS OF LIGHT ELEMENTS

The elements most extensively studied by SXS are aluminium and magnesium because of their good emission intensity, their characteristic profiles and their sharp Fermi edges. Rooke (1968) carried out an in-depth analysis of the SXS of Al and showed that the shape of both the $L_{2,3}$ and K-emission spectra could be explained within the context of the nearly free electron approximation with the exception of an enhanced peak at the Fermi edge of the $L_{2,3}$-spectra which is explained by the Nozieres and de Dominicis (ND) singularity (1969). Watson et al (1968) similarly analysed the magnesium spectrum and reached the same conclusions.

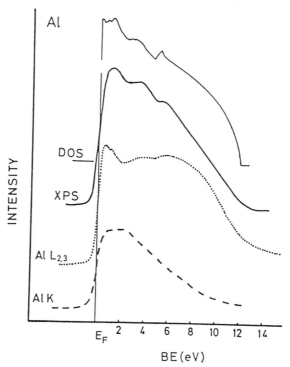

Fig. 4. The density of states (from Rooke, 1968), the XPS valence band spectrum (from Baer and Busch, 1973), the $L_{2,3}$ and the K emission spectra of aluminium. For the L spectrum I/E^3 is plotted to allow for the energy dependence of the transition probability.

Deviations in the shape of the spectra from the nearly free electron-like parabola of the density of states are accounted for by the change in symmetry of the electron states in crossing Brillouin zone boundaries and the consequent effect on the transition probability. The validity of the nearly free electron-like approximation for Al was supported by the XPS measurement of the valence band by Baer and Busch (1973) which showed a parabolic free electron like density of states with a sharp Fermi cut off. Figure 4 shows a comparison of the spectra for aluminium with the theoretical density of states (DOS).

SXS from various Al-Mg alloys have been measured by Dimond (Al and Mg $L_{2,3}$-spectra, 1969) and by Neddermeyer (Al and Mg K- and $L_{2,3}$ spectra, 1973) the latter probably constitutes the most comprehensive SXS study of an alloy system with all possible emission spectra being measured. Both measurements of the $L_{2,3}$ spectra agree extremely well. Figure 5 shows the Al and Mg $L_{2,3}$ spectra as measured by Dimond. At first sight the most obvious conclusions are that the spectral profiles depend more on alloy concentration than on crystal structure and that the valence band width as seen by aluminium atoms is different from that seen by magnesium atoms.

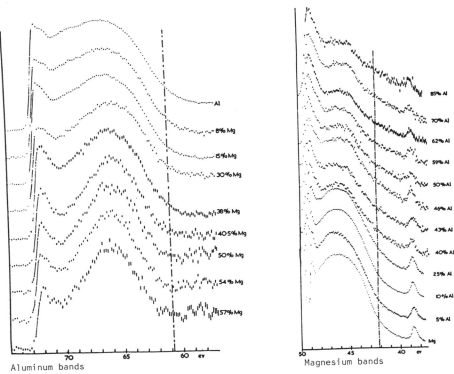

Fig. 5. $L_{2,3}$-emission spectra from a series of aluminium-magnesium alloys.

As the magnesium concentration increases the Al $L_{2,3}$ spectra show an increase in intensity in the low energy region of the band compared with that at the Fermi level, whereas the Mg spectra show an opposite trend. This indicates that the deeper core potentials on the valency three aluminium atoms are pulling s-states to the bottom of the band to screen the excess potential at the aluminium sites at the expense of the s-state density on the valency two magnesium atoms over the same energy region. Also, the valence band width as seen by the aluminium atoms appears to be different from that seen by the magnesium atoms. However, in a clever bit of data analysis Dimond showed that this was not the case. Because instrumental conditions are seldom, if ever, exactly reproducible, direct comparison of the intensities of the spectral profiles from two different samples is not valid. However, if the logarithm of the difference of the intensity of two spectral profiles is plotted against energy it will show a straight line if the two profiles are identical differing only in intensity by a constant factor. Any deviation from a straight line demonstrates a genuine difference in the profiles. Dimond plotted the logarithm of the alloy intensity divided by the pure metal intensity for both the Mg and Al $L_{2,3}$ emission spectra from a 60/40 Al/Mg alloy and these are shown in Fig. 6(a) and (b) respectively, together with the pure metal spectra.

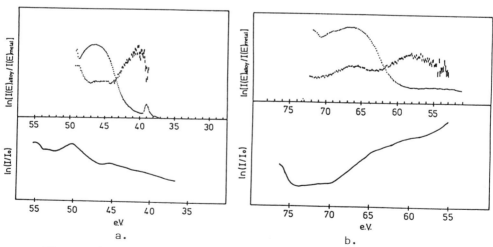

Fig. 6. (a) Mg log (I_{alloy}/I_{metal}) for a 60/40 Al/Mg alloy plotted together with the pure Mg $L_{2,3}$ spectrum (upper) and the absorption spectrum of an Al foil (lower) (from Tomboulian and Pell, 1951). (b) Al log (I_{alloy}/I_{metal}) and the pure Al $L_{2,3}$ spectrum from the same alloy. Lower curve is the absorption spectrum of Mg (from Townsend, 1953).

It can be seen in the log (difference) curve for magnesium that there is an increase in intensity in the alloy spectrum in the low energy region extending the band width of the Mg $L_{2,3}$ spectrum from the alloy to that of the Al emission band. To eliminate the possibility that the increase in the log (difference) curve could be due to absorption of Mg $L_{2,3}$ photons by Al atoms in the alloy, Dimond compared the curve with the absorption spectrum of an Al foil for the same energy region measured by Tomboulian and Pell (1951), shown in the lower part of Fig. 6(a). The decreasing behaviour of the absorption curve over the energy region of interest shows that self absorption is not a contributing factor. The similar treatment of the Al $L_{2,3}$ spectrum from the alloy shows that the general behaviour of the log (difference curve) in this case can be explained by the absorption of the Al $L_{2,3}$ photons by Mg atoms in the alloy (Fig. 6(b)). The absorption spectrum by a Mg foil in this case was measured by Townsend (1953). The conclusion from this study is that at least in concentrated alloys of aluminium and magnesium there exists a common valence band width.

This conclusion was supported by Jacobs (1969) who considered an hypothetical ordered 50/50 Al/Mg alloy. Since the potential well on the aluminium ions is deeper than that on the magnesium ions, the valence electron wavefunctions at the bottom of the band will oscillate on the aluminium sites and will decay exponentially as they penetrate the

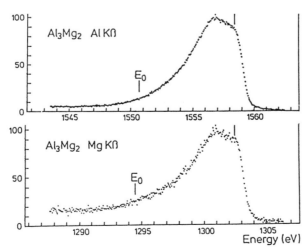

Fig.7. Al and Mg K β spectra from Al_3Mg_2 (from Neddermeyer, 1973).

magnesium sites. However, barrier penetration will occur but the probability of finding an electron in this energy region on the magnesium sites is low, which accounts for the low intensity of the Mg $L_{2,3}$ emission at the bottom of the alloy emission band. In his later study Neddermeyer, with his measurements of both the Mg and Al K-emission spectra from the alloys, showed that the p-states in the valence band are not nearly as sensitive to local potentials as are the s and d electrons as measured by the $L_{2,3}$ spectra but are more delocalised and show an obvious common band width (Fig. 7) - a result predicted by Jacobs (1969).

INTENSITY COMPARISONS AND THE Al-Ag SYSTEM

As has been stated in the previous section, comparisons of the absolute intensities of spectra from different samples is difficult if not impossible. Al $L_{2,3}$ emission spectra from a series of Al-Ag solid solutions were measured by Fabian et al (1971). For presentation they were arbitrarily normalised at their points of maximum intensity. In this form the most obvious feature was the evolution of a secondary peak

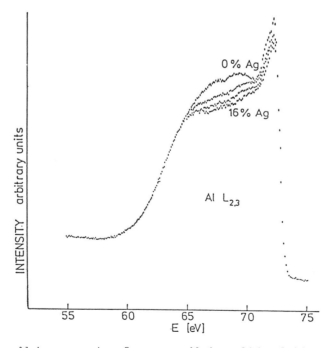

Fig. 8. Al $L_{2,3}$ spectra from some Al-Ag solid solutions with intensities normalised at a point on the low energy back-ground and a point on the tail (from Norris et al, 1974).

in the Al $L_{2,3}$ spectrum at around 65eV which was interpreted by Marshall et al (1969) as being due to the hybridisation of the lower d-bands of silver with the s, p-band of aluminium. Although there may be some truth in this interpretation, a later theoretical study by Kudrnovski et al (1973) of light metal - d-band metal alloys using CPA theory suggested that there would be a depletion of s-states on the light metal atoms over the energy region of the d-bands of the other component. For

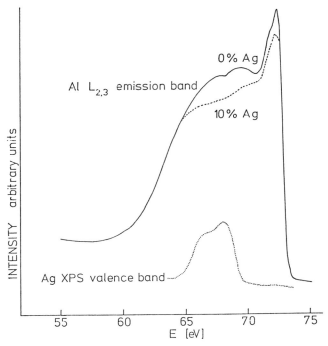

Fig. 9. Al $L_{2,3}$ spectra as in Fig. 8 and the Ag XPS valence band.

comparison with this theory, the spectra were normalised in intensity at two points, one on the low energy tail and one on the low energy background where alloying effects are least likely to be manifest. The resulting spectra are shown in Fig. 8 where it can be seen that the spectra fit extremely well at low energies and again at the Fermi energy; the region in between shows a reduction in intensity which increases with increasing silver content.

Figure 9 shows the $L_{2,3}$ spectrum from pure Al, an Al-10 at%Ag alloy and the XPS spectrum from pure silver. The high intensity part of the XPS spectrum is due to the silver d-bands, the low energy onset of which corresponds to the start of the intensity depletion of the $L_{2,3}$ emission bands in good support of the theory of Kudrnovski et al.

APPLICATIONS OF SXS AND XPS TO BAND STRUCTURE STUDIES

The Mg_2X alloys, where X = Si, Ge or Sn, all crystalize with the antifluorite structure and are all semiconductors with indirect band gaps of 0.73, 0.72 and 0.32eV respectively. Band structure calculations for these alloys have been carried out by Au Yang and Cohen (1969) using an empirical pseudopotential method and were later repeated by Bashenov et al (1978). High resolution XPS spectra of the valence bands of the alloys were measured by Tejeda and Cardona (1976). and reflect the total density of states. These spectra are reproduced in Fig. 10.

The band structures calculated by Bashenov for Mg_2 Si, which are in good agreement with those of Au Yang and Cohen, are shown in Fig. 11 together with the resulting density of states and the XPS spectrum. It can be seen that the theoretical density of occupied states consists of two distinct bands separated by a band gap of almost three electron volts. Tejeda and Cardona considered that the resolution of their spectra was such that a band gap of this magnitude would be much more obvious in their spectrum to say nothing of the absence of any feature corresponding to the singularity at the top of the lower occupied band. Therefore they concluded that the calculations were in error and carried out their own band structure calculations using an empirical tight binding approach. The resulting density of states, both total and the partial density of s-states, are also shown in Fig. 10. The agreement between theory and experiment in this case is much better and suggests that the tight binding approach is better suited to these alloys than the pseudopotential method.

The XPS spectra give confirmation to a certain extent of the accuracy of the total density of states calculation but give little information concerning the symmetry of the states. The author and coworkers have measured the $Mg-L_{2,3}$ and Mg-K spectra of these alloys and are reproduced in Fig. 12. Comparison of the $L_{2,3}$ spectra, which reflect primarily the s-state density on the Mg sites, with the partial density of s-states shown in Fig. 10 shows some agreement in that the intensity increases in the upper band at lower energies.

However the s-state density on the Mg sites in the energy region of the lower band is only significant for Mg_2Sn and even in this alloy it is nowhere near as high as predicted by the calculation. However this does not signify that the calculation is in error since a shift of

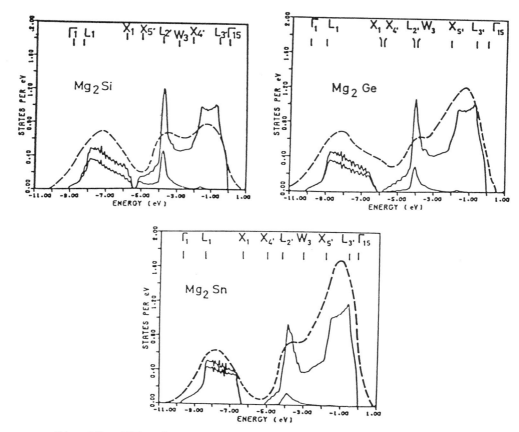

Fig. 10. XPS valence band spectra and the calculated total density of states and partial density of s-states (from Tejeda and Cardona, 1976).

charge in the lower band away from the Mg sites towards the X sites will produce the lowering of intensity in the spectra. Such a shift may well be expected since the valency four X ions will require more screening than the valency two Mg ions. If this is the case the X-$L_{2,3}$ spectra may be expected to show an increase in intensity in this energy region and this does occur for Mg_2Si whose Si-$L_{2,3}$ spectrum was measured by Curry (1968) and is shown in Fig. 13. Thus it would appear that the low energy Mg 3s derived bands are polarised in favour of the X atoms. The K β spectra shown in Fig. 12 and the distribution of s-state density in the calculated N(E) curves show a predominance of p-states at the top of the valence band which are concentrated into a narrower energy region as we go from Mg_2Si to Mg_2Sn. The low energy region of the Mg Kβ spectrum from Mg_2Ge is distorted due to the overlap of the Ge Lβ_3 emission line.

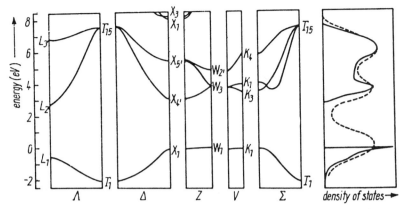

Fig. 11. Band structure and total density of states for Mg_2Si calculated using an empirical pseudopotential method and the XPS spectrum from Fig. 10 (broken line) (from Bashenov et al, 1978).

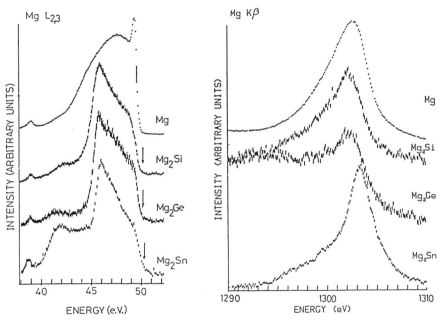

Fig. 12. Mg $L_{2,3}$ and K β spectra from the pure metal and Mg_2X alloys.

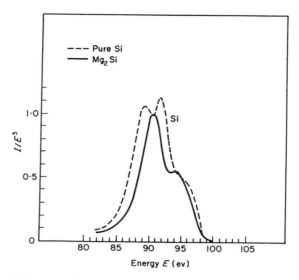

Fig. 13. Si $L_{2,3}$ emission spectra from pure silicon and Mg_2Si (from Curry, 1968).

Conclusions to be drawn from the combined valence band XPS and SXS measurements of these alloys are:

The band structure calculations of Tejeda and Cardona provide a better fit to the experimental data than those of Au Yang and Cohen or Bashenov.

The SXS spectra demonstrate a transfer of charge from the Mg 3s-derived band to the X atom, more strongly for X = Si than for X = Sn.

The p-state density on the Mg sites is concentrated in the upper valence band more strongly for Mg_2Sn than for Mg_2Si and tails off into the lower valence band.

$K\alpha_4$, $K\alpha_3$-INTENSITY RATIOS AS A POSSIBLE MEASURE OF CHARGE TRANSFER IN LIGHT ALLOYS

The $K\alpha_3$ and $K\alpha_4$ satellites are derived from the $K\alpha_1$ and $K\alpha_2$ diagram lines and result from transitions from the $2p_{1/2}$ or $2p_{3/2}$ levels to a vacancy in the 1s level while there is a 'spectator' vacancy in the 2s or 2p levels. In light metals the $K\alpha_1$ and $K\alpha_2$ lines are not resolved and therefore are designated the $K\alpha_{1,2}$ line. Nordfors (1956) was the first to point out the dramatic change in the intensity $IK\alpha_4/IK\alpha_3$ in going from aluminium metal to its oxide. Later Baun and Fischer (1964a,b; 1965) and Fischer and Baun (1967) reported on the variations of the intensity ratios $IK\alpha_4/IK\alpha_3$ of magnesium, aluminium, silicon, their compounds and a number of aluminium binary alloys with 3d transition metals. They confirmed that there were large intensity ratio changes in going from the element to the compound or alloy and showed the intensity ratios varied with alloy concentration.

In order to extract meaningful information from measurement of the $K\alpha_4/K\alpha_3$ intensity ratios, the exact origins of the satellite lines must be known. Demekhin and Sachenko (1967) gave the following assignments for the satellites in the atomic number range $10 < Z \leqslant 14$.

$$K\alpha_3 \quad KL_{2,3} - L_{2,3}L_{2,3} \quad {}^3P - {}^3P$$

$$K\alpha_4 \begin{cases} KL_{2,3} - L_{2,3}L_{2,3} & {}^1P - {}^1D \\ KL_1 - L_1L_{2,3} & {}^3S - {}^3P \end{cases}$$

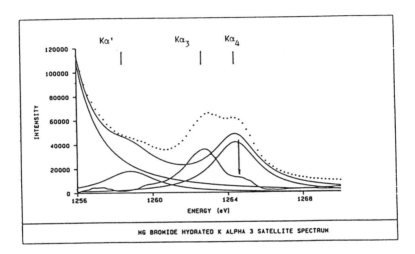

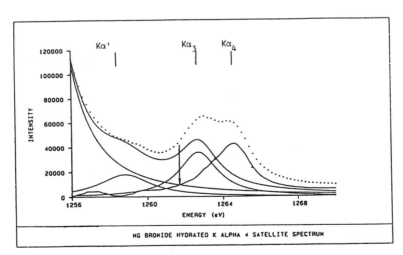

Fig. 14. $K\alpha_3$ and $K\alpha_4$ satellites obtained by subtracting Lorentzians fitted to the other components from the experimental spectrum (dotted curve).

Åberg (1972) also gives the same assignments. Variations in the relative intensities of the satellites with chemical environment is thought to be due to variations in the L_1 hole lifetime brought about by competing $L_1L_{2,3}V$ Coster-Kronig transitions. Shift of charge in the valence band away from the emitting atom will reduce the Coster-Kronig transition rate, increase the L_1 hole lifetime and thus increase the intensity of the $K\alpha_4$ satellite. Unfortunately the correlation between charge transfer and the satellite intensity ratios is not so straight forward. Åberg (1972) points out that the $KL_1 - L_1L_{2,3}$ $^3S - ^3P$ transition gives rise to a contribution to the satellites whose energy position is uncertain and thus may contribute to either the $K\alpha_3$, the $K\alpha_4$ or both. In his Ph.D. work Misra (1986) made an extensive investigation of this subject part of which has been reported by Misra and Watson (1987). Table 1 lists the $K\alpha_4/K\alpha_3$ intensity ratios for a number of magnesium compounds and alloys and illustrates the wide variation with chemical environment and composition. The K-spectra were deconvoluted by fitting Lorentzian curves to the $K\alpha_{1,2}$, the $K\alpha'$, the $K\alpha_3$ and the $K\alpha_4$ lines. A perfect fit in the $K\alpha_3$ and $K\alpha_4$ regions was not achieved. Figure 14 shows the result of subtracting three Lorentzians from the experimental data to produce either the $K\alpha_3$ or the $K\alpha_4$ lines where it can be seen that a distinct component exists in the satellite lines (arrowed in the figure) due to the $KL_1 - L_1L_{2,3}$ transition which varies in shape and position depending upon the valence electron density and L_1 hole lifetime of the magnesium atom in agreement with Åberg.

Table 1.

Compound	$IK\alpha_4/IK\alpha_3$	Alloy	$IK\alpha_4/IK\alpha_3$
Mg-metal	0.80	50 Mg 50 Ag	0.79
Mg-oxide	1.05	75 Mg 25 Ag	0.96
Mg-flouride anhydrous	1.00	60 Mg 40 Cd	0.85
Mg-chloride hydrated	1.38	72 Mg 28 In	0.63
Mg-bromide hydrated	1.17	66.6 Mg 33.3 In	0.58
Mg-iodide hydrated	1.19	60 Mg 40 In	0.67
Mg-chloride anhydrous	1.00	66.6 Mg 33.3 Sn	0.45
Mg-bromide anhydrous	1.22	60 Mg 40 Sb	0.72
Mg-sulphate anhydrous	1.11		
Mg-acetate	0.95		

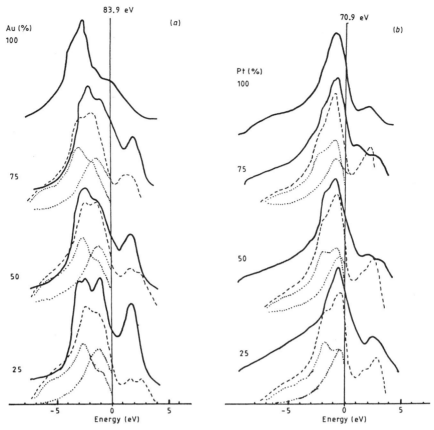

Fig. 15. (a) Au N_6N_7 and (b) Pt N_6N_7 spectra from the pure metals and Pt-Au alloys compared with the theoretical spectra (dashed curves) calculated by Weinberger et al, 1982. The dotted curves at negative energies show the contributions to the overall profiles from the overlap of the N_6 and N_7 spectra. The zero of energy is the Fermi level of the N_7 bands.

THE PLATINUM GOLD ALLOY SYSTEM

The light metals and their alloys lend themselves to investigation by SXS because of their good emission intensity and characteristic profiles. Consequently there have been many investigations of alloys of light metals with transition and noble metals (eg Fuggle et al 1977; Watson et al 1972). Nevertheless SXS can be applied to investigations of the valence electronic structure of alloys of heavy metals. The emission intensity in the soft x-ray region tends to be weak and superimposed on a high Bremsstrahlung background so patience and care are required in accumulating sufficient intensity and in subtracting the background from the spectra.

The Pt-Au system was chosen for study to verify or otherwise the Pt and Au N_6N_7 SXS calculated by Weinberger et al (1982). The N_6 and N_7 spectra from Pt-Au alloys containing 25, 50 and 75 at% Au were measured as well as the spectra from the pure metals (Negm et al, 1987). To superimpose the spectra accurately for comparison, account must be taken of any chemical shift in the N_6 ($4f_{5/2}$) and N_7 ($4f_{7/2}$) inner core levels. These were measured by XPS and the spectra suitably adjusted in energy are shown in Fig. 15 where the zero of energy is the Fermi level for the N_7 spectra. The solid curves are the experimental spectra, the dashed curves are the theoretical spectra and the dotted curves in the negative energy region show the individual contributions from the theoretical N_6 and N_7 spectra to the intensity profiles. Agreement between theory and experiment is quite good given the experimental and computational difficulties for this system. The major discrepancies occur in the N_6 (positive energy) regions of the spectra where, in the Au spectra from the alloys the high intensity in the experimental spectra in this region is due to the overlap of the third order Pt N_4N_6 line. In the Pt spectra the intensity of the experimental spectra in the N_6 region is lower than predicted. This is probably due to the N_6N_7V Coster-Kronig transitions competing with x-ray transitions in filling the N_6 core hole.

SOFT X-RAY DEPTH PROFILE ANALYSIS

The characteristic profiles of valence band SXS from solids and their change with chemical environment suggested to Szasz et al (1985) a novel use for soft x-ray emission spectroscopy in measuring the depth of thin surface layers on solids non-destructively. The energy of the electron beam determines the depth of penetration of electrons into the solid and hence the depth to which x-rays are excited. The depth of penetration is given by the empirical formula (Feldman, 1960; Anderson, 1965)

$$d = 25.6 \frac{\bar{A}}{\rho \bar{Z}^{K/2}} E^K$$

where d is the penetration depth in mm, $\bar{A}$ the atomic mass, E the electron energy in keV, ρ is the density, $\bar{Z}$ the atomic number and $k = 1.2(1 - 0.29 \log \bar{Z})^{-1}$. The equation is valid for compounds or composite substances but $\bar{A}$ and $\bar{Z}$ have to be substituted by averages weighted by the corresponding concentrations. Electrons in their last stages of scattering in the solid will not have sufficient energy to ionise the required inner core levels of the atoms, hence the depth of generation of the x-rays will be slightly less. By taking into consideration the distribution of energy loss of the exciting electrons (see Fig. 16a) and with a knowledge of the spectral profiles of the pure surface layer and pure substrate materials, the ratio of the surface to the substrate contributions to the spectral profiles taken at various excitation energies can be obtained and related to the thickness of the surface layer. This is exemplified for the case of a thin layer of alumina on aluminium in Fig. 16b while Fig. 17 shows the depth profile analysis calculated from these data.

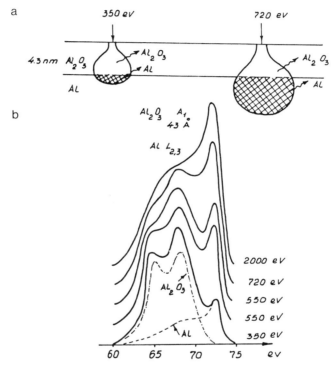

Fig. 16. (a) Electron energy loss distributions for two different energies in a 4.3nm layer of alumina on aluminium. (b) soft x-ray intensity profiles from the alumina on aluminium as a function of excitation voltage (from Szasz and Kojnok, 1985)

SXS APPLIED TO THE STUDY OF METASTABILITY IN AGE HARDENING ALLOYS

Many commercial alloys of aluminium are strengthened by age hardening. In this process the aluminium must be capable of dissolving a certain percentage of the alloying element at elevated temperatures and very little at room temperature in the equilibrium state. Thus it is

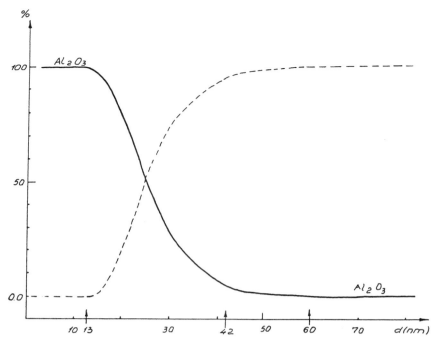

Fig. 17. Depth profile calculated from the data in Fig. 16. (from Szasz and Kojnok, 1985).

possible to form a supersaturated solid solution (SSSS) by rapid cooling from the elevated temperature. On ageing the SSSS at a suitable temperature, one or more intermediate stages of precipitation occurs before the equilibrium phase. These intermediate precipitates can exist in metastable equilibrium indefinitely at room temperatures and form a fine dispersion throughout the lattice which pins dislocations and hence strengthens the metal. The reason for the metastability is not clear (Cohen, 1986).

Although not commercial alloys, dilute silver aluminium-silver alloys are age hardenable. The sequence of precipitates formed from the SSSS by suitable ageing treatments is the following (Kwarciak, 1985): SSSS - spherical Guinier-Preston (GP) zones (η) - disordered (partially redissolved) GP zones (ϵ) - hexagonal intermediate Al_2Ag-type precipitates (γ') - equilibrium Al_2Ag precipitates (γ). The equilibrium precipitates correspond to a minimum in the free energy of the alloy while the metastable phases, although of higher energy, must have associated with them a local minimum in the free energy. Contributions to this minimum could come from the energy of the valence electrons and from entropy considerations.

Following the work of Kertesz et al (1982) and Szasz et al (1985) who showed that age hardening has an effect on the Al $L_{2,3}$-emission spectrum, Negm et al (1988) measured the Al $L_{2,3}$ spectra from a series of Al-Ag alloys in their equilibrium and various intermediate states. Assuming that the distribution of p-states in the valence band does not compensate any energy shift of the $L_{2,3}$ band, they took the mass centre (MC) of the $L_{2,3}$-spectra as a measure of the average valence electron energy in the alloys and plotted this against the various states of the alloys for Al-8, 12 and 16 at% Ag alloys (Fig. 18).

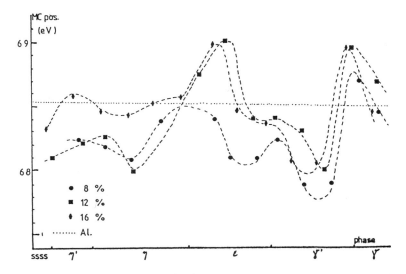

Fig. 18. The mass centre (MC) energies of the Al $L_{2,3}$ spectra from a series of Al-Ag alloys heat treated to produce the precipitation states indicated on the x-axis.

The dotted horizontal line in Fig. 18 represents the MC for pure Al and it can be seen that the MC of the alloy spectra show two distinct minima corresponding to the η and γ' intermediate states suggesting an electronic contribution to the metastability of these states. However in the ε state for the 12 and 16% Ag alloys there is a shift in the MC to positive energies with respect to the pure Al value suggesting that the stabilising influence in this case is configurational rather than electronic. In the equilibrium γ phase the MC energy for the alloy reverts to a value close to that of the pure metal.

REFERENCES

Åberg, T., 1973, X-ray Satellites and their Interpretation, in: "X-ray spectra and Electronic Structure of Matter" A. Faessler and G Wiech, eds., Vol. 1, Munich, 1.

Anderson, C.A., 1965, The Electron Microscope, McKinlay, Heinrich and Wittry, eds., Wiley, New York.

Au Yang, M.Y. and Cohen, M.L., 1969, Electronic Structure and Optical Properties of Mg_2Si, Mg_2Ge and Mg_2Sn, Phys. Rev., 178, 1358.

Baer, Y. and Busch, G., 1973, X-ray Photoemission from Aluminium, Phys. Rev. Letts., 30, 280.

Bashenov, V.K., Mutal, A. and Timofeenko, V.V., 1978, Valence-band Density of States for Mg_2Si from Pseudopotential calculation, Phys. Stat. Sol(b) 87, K77.

Baun, W.L. and Fischer, D.W., 1964a, Influence of Chemical Combination on Aluminium K diagram and non-diagram lines, Nature, 204, 642.

Baun, M.L. and Fischer, D.W., 1964b, Tech. Rep. AFML-TR-64, 350.

Baun, M.L. and Fischer, D.W., 1965, The effect of valence and coordination on K series diagram and non-diagram lines of Mg, Al and Si, Adv. x-ray Analysis, 8, 371.

Cohen, J.B., 1986, Internal Structure of Guinier-Preston zones in alloys, Solid State Physics, 39, 131.

Curry, C., 1968, Soft x-ray emission spectra of alloys and problems in their interpretation, in "Soft x-ray Band Spectra", D.J. Fabian, ed. Academic Press, London & New York, 173.

Demekhin, V.F. and Sachenko, V.P., 1967, Spectral position of K satellites, Bull. Acad. Sci., USSR (Phys. Ser.), 31, 913.

Dimond, R.K., 1969, The soft x-ray emission spectra of aluminium-magnesium alloys, PhD Thesis, University of Western Australia.

Fabian, D.J., Lindsay, G.M. and Watson, L.M., 1971, Soft x-ray emission from alloys of aluminium with silver, copper and zinc, in "Electronic Density of States", L.H. Bennett, ed. National Bureau of Standards (US) Special Publication, 323, p307.

Feldman, C., 1960, Range of 1-10keV electrons in solids, Phys. Rev., 117, 455.

Fischer, D.W. and Baun, W.L., 1967, Effect of alloying on the aluminium K and iron L x-ray emission spectra in the aluminium-iron binary system, J. Appl. Phys., 38, 229.

Fuggle, J.C., Kallne, E., Watson, L.M. and Fabian, D.J., 1977, Electronic structure of aluminium and aluminium-noble metal alloys studied by soft x-ray and x-ray photoelectron spectroscopies, Phys. Rev. B., 16, 750.

Jacobs, R.L., 1969, The soft x-ray spectra of concentrated binary alloys, Phys. Letts., 30A, 523.

Kertesz, L., Kojnok, J. and Szasz, A., 1982, Cryst. Latt. Defects, 9, 219.

Kudrnovski, J., Smrcka, L. and Velicky, B., 1973, Theory of soft x-ray emission in random binary alloys : Effect of the d-resonance in the valence band, in "X-ray Spectra and Electronic Structure of Matter", A. Faessler and G. Wiech, eds., Munich, Vol. 11, p94.

Kwarciak, J., 1985, Kinetics and mechanism of precipitation processes in Al-Ag alloys, J. Therm. Anal., 30, 177.

Liefeld, R.J., 1968, Soft x-ray emission at threshold excitation, in "Soft X-ray Band Spectra", D.J. Fabian, ed., Academic Press, London and New York, p133.

Marshall, C.A.W., Watson, L.M., Lindsay, G.M., Rooke, G.A. and Fabian, D.J., 1969, Interpretation of soft x-ray emission spectra of aluminium-silver alloys, Phys. Letts., 28A, 579.

Misra, U.D., 1986, Origin of $K\alpha'$, $K\alpha_3$ and $K\alpha_4$ satellites in x-ray spectra of light metals : an experimental investigation, PhD Thesis, University of Strathclyde, Glasgow, UK.

Misra, U.D. and Watson, L.M., 1987, $K\alpha$ x-rays from magnesium, some of its compounds and alloys, Physica Scripta, 36, 673.

Nagle, D.J., 1970, Interpretation of valence band x-ray spectra, Adv. in X-ray Analysis, 13, 182.

Neddermeyer, H., 1973, X-ray emission band spectra and electronic structure of alloys of light elements, in Band Structure Spectroscopy of Metals and Alloys", D.J. Fabian and L.M. Watson, eds., Academic Press, London and New York, p153.

Negm, N.Z., Watson, L.M., Norris, P.R. and Szasz, A., 1987, The $N_{6,7}$ soft x-ray emission spectra of platinum-gold alloys., J.Phys. F: Met.Phys, 17, 2295.

Negm, N.Z., Watson, L.M. and Szasz, A., 1988, A soft x-ray investigation of the electronic states of metastable phases in Al-Ag alloys, J.de Physique, in press.

Nordfors, B., 1956, On the K-spectrum of aluminium and its oxide, Ark.Fys., 10, 279.

Norris, P.R., Fabian, D.J., Watson, L.M., Fuggle, J.C. and Lang, W., 1974, Soft x-ray emission and x-ray photoelectron studies of disordered aluminium alloys in relation to CPA theory, J.de Physique, Colloque C4, 35, C4-65.

Norzieres, P. and De Dominicis, C.T., 1969, Singularities in x-ray absorption and emission of metals. Phys. Rev., 178, 1097.

Rooke, G.A., 1968, Interpretation of Al x-ray band spectra I density distribution, J. Phys. C., 1, 767.

Szasz, A., Kertesz, I, Hajdu, J. and Kollar, J., 1985, Aluminium, 61, 515.

Szasz, A. and Kojnok, J., 1985, Soft x-ray emission depth profile analysis : SXDA, Appl. Surface Sci., 24, 34.

Tejeda, J. and Cardona, M., 1976, Valence bands of the Mg_2X (X=Si,Ge,Sn) semiconducting compounds, Phys. Rev. B., 14, 2559.

Tomboulian, D.H., and Pell, E.M., 1951, Absorption by aluminium in the soft x-ray region, Phys. Rev., 83, 1196.

Townsend, J.R., 1953, Solid state absorption spectra of Mg and MgO. Phys. Rev., 92, 556.

von Barth, U. and Grossmann, G., 1982, Dynamical effects in x-ray spectra and the Final-State-Rule, Phys. Rev. B., 25, 5150.

Watson, L.M., Dimond, R.K. and Fabian, D.J., 1968, Soft x-ray emission spectra of magnesium and beryllium, in "Soft X-ray Band Spectra", D.J.Fabian, ed. Academic Press, London and New York, p45.

Watson,, L.M., Kapoor, Q.S. and Hart, D., 1972, The electronic structure of alloys of aluminium with First Transition Series metals studied by soft x-ray spectroscopy, in: "X-ray Spectra and Electronic Structure of Matter", A. Faesler & G. Wiech, eds., Munich, Vol. 11, p135.

Weinberger, P., Staunton, J. and Gyorffy, B.L., 1982, On the $N_{6,7}$ soft x-ray emission spectra of Au_cPt_{1-c}, J. Phys. F:Met.Phys. 12, L199.

X-RAY INELASTIC SCATTERING SPECTROSCOPY AND ITS APPLICATIONS IN SOLID STATE PHYSICS

Nikos G. Alexandropoulos and Irini Theodoridou

Department of Physics, University of Ioannina
P.O. Box 1186
GR-451 10 Ioannina - Greece

ABSTRACT

The x-ray inelastic scattering spectrum is a unique and powerful tool in that its analysis leads in a direct way to the electronic structure of the scattering material. For small scattering angles (Low Momentum Transfer), the spectrum reduces to the Energy Loss Spectrum, directly related to collective excitations of the valence electrons. For Large Momentum Transfer, the individual excitation is dominant and from the Compton spectrum the ground state electron wavefunction can be obtained. For Intermediate Momentum Transfer, the transition from collective to individual excitations occurs, making it the rigorous testing ground for theories and approximations. The discussion includes bulk and surface plasmons, the transition from collective to individual excitations, the one dimensional electron momentum distributions, the x-ray Raman and the x-ray Resonant Raman effects. It also includes a review of the instrumentation used before and after the introduction of γ-rays and synchrotron sources in the x-ray inelastic spectroscopy.

INTRODUCTION

The term inelastic* or incoherent or modified x-ray scattering, refers collectively to a number of interactions

----
*This term is used in the sense that the energy of the electromagnetic wave is not conserved.

of x-rays with condensed matter. The common characteristic of all these interactions is that the frequency of the scattered waves is less than the frequency of the primary radiation. The first and most widely studied of these interactions is Compton scattering or Compton effect and for this reason it is not unusual to consider the term Compton scattering as a synonym for inelastic scattering.

The Compton spectrum differs from most other spectroscopic observations in that, momentum is of main importance rather than energy. However, the most interesting feature of the x-ray inelastic scattering experiments is that the probing particle is a quasiparticle, with energy and momentum related by a multiple valued function, leading to extensive values of energy for each quasimomentum. This is easily understood assuming that an incident photon of energy* ω_1 and wavevector k_1 interacts with an electron of energy ε and momentum p, giving another photon of energy and wave vector ω_2 and k_2, respectively. In the case that energy and momentum are conserved during the interaction, the interaction corresponds to the absorption of a quasiparticle of energy $\omega = \omega_1 - \omega_2$ and momentum $k = k_1 - k_2$, where both ω and k have a functional dependence on ω_1 and the scattering angle θ. In other words, the relation of ω on k depends on the experimental conditions.

To clarify the above, let ω_{cp} and k be respectively the Compton energy shift and the momentum transferred to an initially stationary electron. The relation between ω_{cp} and k is given by:

$$\omega_{cp} = \frac{2\omega_1^2 \sin(\theta/2)}{2\omega_1 \sin^2(\theta/2) + mc^2} = \frac{1.947 k^2 \omega_1}{1.947 k^2 + 511 \omega_1} \qquad (1)$$

where ω_1 is in keV, and k in $Å^{-1}$. Equation 1 is modified, when the scattering electron possesses an initial momentum p to:

$$\omega = \frac{0.986 \, q \, k + 0.261 \, k^2}{0.986 \, q \, k + 0.261 \, k^2 + 68.5 \, \omega_1} \omega_1 \qquad (2)$$

where q is the projection of p on the k direction

----
* $\hbar = m = e = 1$.

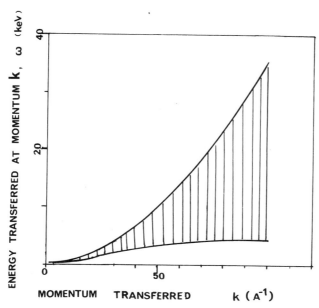

Fig.1: Two dimensional dispersion relation for tne quasiparticle with energy ω=ω₁-ω₂ and momentum k=k₁-k₂.
The shaded area indicates the energies and momentum
available by the right combination of ω₁ and θ.
The area's limits have been drawn on the minimum
(5.4 keV - CrK_α) and maximum (412 keV -¹⁹⁸Au)
values of ω₁, that had been used extensively.

measured in a.u.. It is therefore acceptable to simulate the Compton scattering of a photon ω_1, k_1, by a free electron with the absorption of a quasiparticle* $\omega(k)$ with its dispersion shown in figure 1. To appreciate the total potential of inelastic scattering in the study of condensed matter, it is necessary to associate figure 1 with the fact that in a solid, the energy spectrum of elementary excitations is quite widespread, while their momenta extend only to the boundaries of the Brillouin zone.

When a beam of photons strikes a piece of condensed matter the dominant interaction depends on the relative magnitude of ω and k to ε and p, the respective energies and momenta of the quasiparticle and interacting electron. Figure 2 is a kind of flow chart of all inelastic

----

*It is interesting to mention that although the conservation of energy and momentum leads to inconsistency in the case of absorption of a photon by a stationary electron, this is not the same for our quasiparticle.

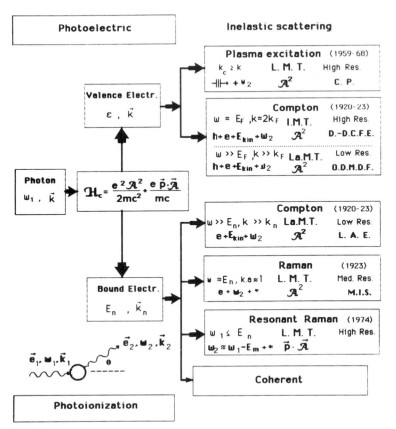

Fig.2: The dominant interactions of a photon of energy ω_1 and momentum k_1 scattered by the electrons of condensed matter. The square for each effect includes, in the left, the kinematical requirement and the products of the interaction, in the middle, informations about the momentum transfer area and the domimant term of coupling Hamiltonian and finally, on the right, the experimental energy resolution requirement for observation and the kind of information that can be derived.

Note: L.M.T. for Low Momentum Transfer, C.P. for Collective Properties, I.M.T. for Intermediate Momentum Transfer, D.-D.C.F.E. for Density-Density Correlation Function for Electrons, La.M.T. for Large Momentum Transfer, O.D.M.D.F. for One-Dimensional Momentum Distribution Function for electrons (reduced to a working method in many research centers) L.A.E. for Localized Atomic Electrons, and M.I.S for Multiparticle Interactions in Solids.

scattering effects with the exception of parametric scattering and the coherent Compton scattering.

There is a large number of publications* on this topic although not evenly spread on all effects, with the most extensive literature on Compton effect; however, the review articles are scarce. Since the epoch-making publication of J.W.M. DuMond[1] in 1933, few other reviews have been published, with the most recent one by M.J. Cooper[2] on Compton Scattering and most probably, only one review article deals with all the effects of inelastic scattering, that of V.A.Bushuev and R.N. Kuz´min[3].

It is probably not out of place to explain, why for many years Compton scattering has been considered as a parasitic effect giving headaches to crystallographers. The history of x-ray inelastic scattering, as any other history, has had its golden age as well as its dark one. There are two golden ages in this history. The first extended from the middle 1920's to the middle 1930's, when some of the most glamorous names in physics were involved in this type of work**. The second started about the early 1960's. The decline in interest during the late 1930's arose because the primitive photon detection systems rendered it impossible to obtain precise experimental data, so the related theories remained for years neither verified nor rejected experimentally. The renewed interest of the 1960's stems from optimistic evaluation of the new detection system which permits low intensity measurements, the various powerful x-ray sources and ultimately the synchrotron radiation in addition to the method's intrinsic superiority in the study of the momentum related effects.

KINEMATICS OF THE INELASTIC SCATTERING

Much useful informations primarily from the peak position of the X-ray inelastic scattering spectrum are derived from the requirement of conservation energy and momentum during the interaction. In the event that the transferred momentum, k, is large enough ($k>>p$) in such a way that the conservation of momentum is applied only between the electron and quasiparticle it is assumed that the interaction is taking place in the Large Momentum Transfer Regime (La.M.T.R.).

----

*The selection of references in this text is made only in order to help the reader to clarify some points and not to pass judgment on the publication's importance.
**e.g. Dirac, Debye, Bloch, Sommerfeld, de Broglie to name few[4].

In the opposite case, where in the momentum calculation is included the momentum of the ensemble where the electron belongs, it is assumed that the interaction is taking place in the Low Momentum Transfer Regime (L.M.T.R.). The limits of these two regimes are well defined only in oversimplified models. In reality, between the two regimes it exists a gray area, which is called Intermediate Momentum Transfer Regime (I.M.T.R.).

In the Large Momentum Transfer Regime, the electrons are treated as though practically free, given rise to the Compton effect, the first and most widely studied form of inelastic scattering effects. Morphologically, the Compton spectrum is characterized by a maximum at a position that depends only on the experimental arrangement and not on the material of the specimen* and a profile, which depends on the specimen**.

Conservation of momentum and energy, which is expressed as: $k = k_1 - k_2 = \hbar^{-1}(p_2 - p_1)$ and $\omega = \omega_1 - \omega_2 = (1/2m\hbar)(p_2^2 - p_1^2)$, respectively provides the relation between the transferred energy and momentum to the electron as:

$$\omega(k) = \frac{\hbar k^2}{2m} + \frac{k \cdot p}{m} \qquad (3)$$

The landmarks of a Compton spectrum shown in figure 3, are reduced from equation 3, as:

$$\omega_c = \frac{mc^2 \omega_1}{2\omega_1 \sin^2(\theta/2) + mc^2} \qquad (4a)$$

$$\omega_{cp} = \omega_1 - \omega_c = \frac{2\omega_1^2 \sin^2(\theta/2)}{2\omega_1 \sin^2(\theta/2) + mc^2} \qquad (4b)$$

$$\omega_2 = \frac{68.5 \, \omega_1}{q \sin(\theta/2) + 68.5(2/mc^2) \, \omega_1 \sin^2(\theta/2) + 68.5} \qquad (4c)$$

----

*In the case that the spectrum is versus wavelength the shift depends only on the scattering angle.
**Since 1945, a new spectral variable has been introduced the electron momentum projection on scattering vector, q, (see equation 10), that makes the Compton spectrum, called the Compton profile, depend only on the material of the specimen and not on the experimental conditions used to obtain it.

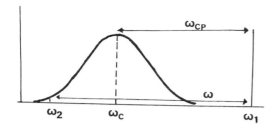

Fig.3: Landmarks of an x-ray scattered spectrum.

$$\omega = \omega_1 - \omega_2 = \frac{q\sin(\theta/2)+68.5(2/mc^2)\omega_1\sin^2(\theta/2)}{q\sin(\theta/2)+68.5(2/mc^2)\,\omega_1\sin^2(\theta/2)+68.5} \quad (4d)$$

The localized core electrons with binding energy $E_n \ll \omega$, are giving a broad Compton spectrum with a shifted* peak.

The reduction of the transferred momentum, (k), by decreasing the scattering angle or the incident photon's energy, leads to the distortion of the Compton profile.

Finally, for** $k \ll k_{T.F.}$, the spectrum consists of a very narrow line resulting from the plasmon creation in the valence band. In this case, the conservation of momentum does not apply to an electron but to the whole electron gas. The position of this line depends on the scattering material and the momentum transfer. In the simplest case of a degenerate electron gas, the dispersion relation between the plasmon energy, ω_p, and the momentum transfer, k, looks like :

$$\omega_p(k) = \omega_p(0) + \frac{3}{10}\frac{v_F^2 k^2}{\omega_p(0)} \quad (5)$$

where $\omega_p(0)$ is the plasmon energy at k=0 and v_F is the electron Fermi velocity. The scattering in this Low Momentum Transfer region, provides a unique technique in the study of the plasmon dispersion relation, specially for materials sensitive to vacuum and high temperature.

Other effects where the conservation of momentum does not hold between the two particles during their collision but between the incident photon and the whole atom to which the electron is bound, are the Raman and Resonant Raman. Although the transferred momentum is quite large compared to the momentum transferred in Compton scattering by free electrons,

----

*Few percent smaller than ω_{cp}.
**k_{TF} is the Thomas-Fermi screening momentum.

it is considered that Raman and Resonant Raman take place at Low Momentum Transfer region. The spectral features in both effects depend only on the specimen's material and not on the experimental conditions* θ and ω_1.

In this section the x-ray inelastic scattering effects have been classified according to the dependence of the energy spectrum dominant features on the experimental conditions, in three categories :

a) The Compton effect, where the spectrum maximum depends only on the experimental conditions ω_1 and θ and not on the specimen's material, b) The Plasma excitation, where the plasmon peak depends both on experimental conditions and on the specimen's electronic structure, and c) The Raman and Resonant Raman,

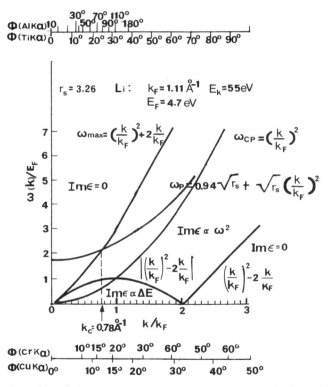

Fig.4: Landmark(ω_{min}, ω_{max}, ω_d, ω_c, ω_p) for the dependence on k/k_F as determined using the R.P.A. for Li and the corresponding scattering angles for four incident radiations.

----

*In the case of x-ray Raman the intensity depends on the scattering angle as $(1+\cos^2\theta)\sin^2(\theta/2)$.

where the energy spectrum discontinuity depends only on the bound electron energies.

Figure 4 shows the landmark for the first two kinds of interaction, Compton and plasmon scattering, according to RPA for the valence electron of Li.

BRIEF REVIEW OF THE SCATTERING THEORY

The most popular theoretical treatment of the inelastic scattering is developed within the weak coupling limit, where the perturbation of the electron of momentum p by the scattering photon is considered as the perturbation by the incident electromagnetic wave, described only by the vector potential, A. The coupling Hamiltonian for this interaction is:

$$H_c = \frac{e^2 A^2}{2mc^2} + \frac{e p \cdot A}{mc} \quad (6)$$

The non-relativistic inelastic scattering cross-section within the limits of the above approximation can be written as:

$$\frac{d^2\sigma}{d\omega d\Omega} = \frac{e^2}{mc^2} \frac{\omega_2}{\omega_1} \sum_f \sum_i |M_{fi}|^2 \delta(E_f - E_i - \omega) \quad (7)$$

where $|i\rangle$ and $|f\rangle$ are the initial and final electronic states and M_{fi} is the matrix element:

$$M_{fi} = \langle f|\exp(ik \cdot r)|i\rangle (e_1 \cdot e_2) +$$

$$+ \frac{1}{m} \sum_n \frac{\langle f|e_2 \cdot p \exp(-ik_2 \cdot r)|n\rangle \langle n|p \cdot e_1 \exp(ik_1 \cdot r)|i\rangle}{E_n - E_0 - \omega_1 - i\Gamma_0} +$$

$$+ \frac{1}{m} \sum_n \frac{\langle f|e_1 \cdot p \exp(ik_1 \cdot r)|n\rangle \langle n|p \cdot e_2 \exp(-ik_2 \cdot r)|i\rangle}{E_n - E_0 + \omega_2 + i\Gamma_0} \quad (8)$$

where $|n\rangle$ is a virtual state and Γ_0 is the lifetime of the hole state created during the excitation.
In the non-relativistic first order perturbation, the dominant term of the coupling Hamiltonian is A^2 and the second and third terms in equation 8 can be neglected in such a way that equation 7 reduces to:

$$\frac{d^2\sigma}{d\omega d\Omega} = \sum_f \sum_i |\langle f|\exp ik \cdot r|i\rangle|^2 \delta(E_f - E_i - \omega) \quad (9)$$

which is the so-called exact Compton cross-section. The error from dropping the term p·A is usually negligible*.

Equation 9 is reduced to a function between Compton profile and the one-dimensional momentum distribution function of the electrons, via the Impulse Approximation**.

Introducing a new variable, q, which is the projection of electron momentum p on the direction of the scattering vector k, ($k = k_1 - k_2$),

$$q = \frac{p \cdot k}{|p||k|} \qquad (10)$$

and is related to the experimental quantities*** $\omega_1, \omega_c, \omega_2, \theta$ as:

$$q \propto \frac{(\omega_c - \omega_2) \omega_1}{\omega_c \omega_2 \sin(\theta/2)}, \qquad (10a)$$

equation 9 is reduced to

$$\frac{d^2\sigma}{d\omega d\Omega} = \sigma_T \frac{\omega_2}{\omega_1} \frac{m}{k} \sum_i J_i(q) \qquad (11)$$

where σ_T is the Thomson scattering cross-section and $J(q)$ is the Compton profile.

The Compton profile is related to the electron's radial density of momentum distribution, $I_i = 4\pi |X_i(p)|^2 p^2$, as:

$$J_i(q) = 2\pi \int_{|q|}^{\infty} |X_i(p)|^2 p \, dp \qquad (12)$$

where $X_i(p)$ is a Fourier transform of the ground state

----

*For an order of magnitude calculation, see ref. 5.
**The physical implication of I.A is that the time that the photon is probing the electron distribution is quite shorter than the time needed by the created hole to alter the potential energy. The I.A has been introduced in the earlier days by Jauncey[6], but for a quantitative foundation of the approximation, the reader can be referred to ref. 5,7,8.
***The introduction of the new variable reduces the data to a universal form, independent of the experimental conditions (ω_1 and θ). The proportionality constant depends on the units. For q in momentum a.u. and $\omega_1, \omega_2, \omega_c$ in keV, the value of this proportionality constant is 68.5.

wavefunction, $\Psi_1(r)$. Equation 12 is one of the most useful relations, responsible for making the Compton spectroscopy a working method in many research centers in the study of the condensed matter. This equation can be easily transformed to:

$$|X_1(p)|^2 = \left|\frac{2}{2\pi q} \frac{dJ_1(q)}{dq}\right| \qquad (13)$$

which permits the determination of the probability of finding an electron with momentum* p.

In 1958 in a ground-breaking approach[9] the spectrum of x-ray inelastic scattering was related directly to the characteristic Energy Loss Spectrum as:

$$\frac{d^2\sigma}{d\omega d\Omega} = -\sigma_T \frac{\hbar k^2}{4\pi^2 e^2 n_0} \, \text{Im} \, \varepsilon^{-1}(k,\omega) \qquad (14)$$

The above relation in conjunction with Kramers – Kronig relation, which permits the determination of Re $\varepsilon^{-1}(k,\omega)$, from Im $\varepsilon^{-1}(k,\omega)$, opens new horizons in the application of the Low and Intermediate Momentum Transfer of the inelastic scattering in the study of solids[10-19]. In other words, it is shown that measurement of the energy distribution of the inelastic scattering is a direct measurement of the dielectric function of the solid, at the frequency and momentum of the energy and momentum transferred to the electron in the solid. Such measurements yield valuable information concerning the role played by the Coulomb interactions between electrons in solids and defines the exact many-body states of the system.

However, in dealing with the quite complicated motion of electrons in solids, it is necessary that one makes several simplifying assumptions. These simplifications are some of the sources of the many unsettled questions on the expeprimentally observed spectral details. Other sources lay on the lack of extensive experimental

----

*For the theoretical calculated as well as the experimentally derived Compton profile, the normalization condition $J(q)dq = N$ (where N is the total number of scattering electrons) is imposed. This normalization process, neglecting the coherent scattered component, introduced an error that according our measurements is not justified e.g. for S and Ti at 59.54 keV the measured ratio for Coherent to Incoherent component is 0.063 and 0.011, respectively.

measurements. The low cross-section that is about two orders
of magnitude less than that of Compton, combined with the
required high resolution and the step-scan mode in the data
collection process, results in low statistics spectra. The
theory and the experiment are in good agreement in the kinematic regime and below the Thomas-Fermi screening vector, $k_{T.F.}$,
or Low Momentum Transfer, where the plasmon dispersion relation
is, according to the free electron model, given in relation 5.
The disagreement appears for Intermediate Momentum Transfer,
where the reported spectra have features not well understood
within the band structure theory. Probably it is fair to say
that many new ideas are expected from the study of condensed
matter in this spectral area, when the experiment reaches the
degree of perfection in resolution and counting that now
is available in other areas of x-rays spectroscopy.

Not too much later after the first experimental observation of plasma excitation[20], a new very exciting feature of the
inelastic scattering spectrum has been observed in 1974, the
Resonant Raman. This is the best example indicating that the
inelastic scattering events are nonlinear events, since no fewer
than two quanta partipate. The effect has been predicted[21] in
terms of the p·A term, in second order of coupling Hamiltonian,
as the absorption of an incident photon of energy below the
absorption threshold of a given level, creating a virtual

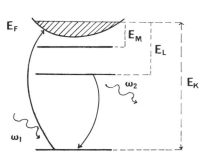 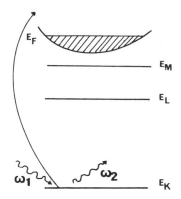

Fig.5a. Energy diagram of Resonant x-ray Raman
scattering, $\omega_1 < E_K$. The
energy $\omega_1 - E_K$ is shared between the ejec-
ting electron and the
emitted photon, ω_2, giving a continuous dis-
tribution for ω_2.

Fig.5b. Energy diagram of x-ray
Raman scattering, $\omega_1 > E_K$.
The energy $\omega_1 - E_K$ is
shared between the
ejecting electron and
the out going photon
ω_2 resulting to a continuous spectrum for ω_2.

ground state hole in the intermediate state to a final state in which there is a higher-shell hole, an electron in the continuum, and an emitted photon, (figure 5a).

In a hand waving argument, it is easy to predict from the second term of the matrix element in equation 8 that the scattered intensity increases as $1/(E_n - \omega_1)$, reaching a maximum which depends on Γ_0. Also, because the ejected electron shares the available energy, $\omega_1 - \omega_m$, where m denotes the final hole state, the Resonant Raman spectrum should be continuum.

The Resonant x-ray Raman has been observed also in γ-ray inelastic scattering[22]. The x-ray Raman effect has created long arguments in the thirties about its observability. The same arguments were repeated again in early sixties and it took several years, for the effect, to be established[23]. Figure 5b shows also the kinematics of the Raman scattering. The incident photon creates a hole in an inner shell, bringing the electron above the Fermi level and emitting simultaneously another photon with energy $\omega_2 < \omega_1 - E_n$. The resulting energy spectrum is a continuum one, because the incident photon's energy is shared between the ejected electron and the outgoing photon. However, the spectrum shows a fine structure near the discontinuity which depends mainly on the density of states in the empty part of the conduction band[24-26] and the inner shell hole potential.

The x-ray Raman spectrum provides similar information on that of the x-ray absorption spectra. The cross-sections of these two effects are related as in the following relation :

$$\left(\frac{d^2\sigma}{d\omega d\Omega}\right)^{inel} \sim \frac{1}{\omega} \left(\frac{d^2\sigma}{d\omega d\Omega}\right)^{abs} \tag{15}$$

INSTRUMENTATION

Almost all kind of detectors spectrometers[27-32] have been used in connection to some inelastic x-ray scattering research. The success or the failure of an experimental work in this area depends on compromising the two contradicting requirements for high resolution and high efficiency, because the common characteristic for all cases of inelastic scattering is the very low scattering cross-section. For Compton scattering, e.g., the cross-section is of the order of a barn and for plasma excitation it is two orders of magnitude less, so only by pushing to the limits the available technology is it possible to extend the frontiers of x-ray inelastic scattering research.

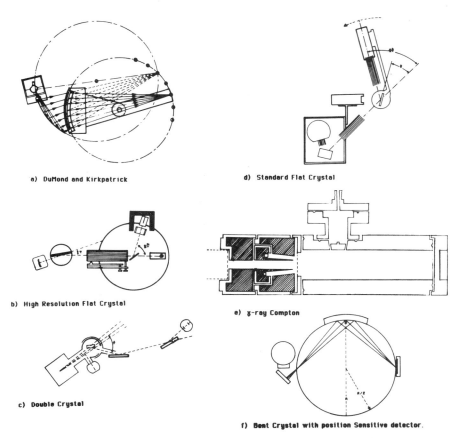

Fig.6. Some of the most popular spectrometers employed in the x-ray inelastic scattering experiments. a) The DuMond – Kirkpatrick multi-crystal (fifty) spectrometer in a focussing arrangement. b) A high resolution flat crystal spectrometer with fine Soller – Slits used in measuring the Low and Intermediate momentum transfer region. c) one of many versions of a double crystal of very high resolution spectrometer, that has been used since late 60's for X-Raman and Plasmon measurements. d) A standard flat crystal arrangement of low resolution extensively used in Compton profile measurement in 70's. e) One of the many versions of γ-ray Compton spectrometers that have been developed for measuring Compton profiles. f) The bent crystal in conjunction with the position sensitive detector, which probably opens a new area in the x-ray inelastic scattering research.

Although the Compton profile and the Raman spectra are not as sensitive to momentum transfer as the plasma excitation, a clean scattering experiment requires a well defined, strong, incident x-ray beam and an appropriate selection of spectrometer and detector. In figure 6 there are shown some of the most popular experimental set ups that have been used in x-ray inelastic scattering experiments.

The DuMond and Kirkpatrick multi-crystal focussing spectrometer (6a) used in conjunction with a specially made tube operating up to 3kW and a photographic plate, rendered the first experimental evidence that the electron is a fermion. The spectrometer consisted of 50 small crystals arranged in an ingenious way to provide high resolution in a focussing geometry. However, the needed time for recording a spectrum even in the most favorable case of the low atomic Z elements, varied between 223 and 1020 hrs. Later, this spectrometer was developed to the one known as DuMond-Kirkpatric and Johansson, using a bent crystal in transmission spectrometer. The flat crystal spectrometer with Soller-Slits has been used in early 60's. One, (6b), specially made with fine slits to provide high resolution, has been used for studying the fine structure on x-ray inelastic scattering. The other, (6d), is a standard flat crystal fluorescent spectrometer, that has been employed extensively in earlier studies of Compton profile. For the plasmon dispersion and Raman measurements, there has been used extensively since late 60's the double crystal spectrometer (6c). In 1971 J. Felsteiner and his co-workers opened a new area in the Compton profile measuring, introducing the energy dispersive spectrometer of Solid State detector. One of many versions of γ-ray Compton spectrometers is shown in figure 6e. The bent crystal spectrometer in step scanning mode, has been used paraller to a double crystal when high resolution was needed. The superiority of the bent crystal to the flat or double crystal spectrometer, lay on the fact that it can be used in conjunction with a high spatial resolution position sensitive detector in an arrangement that combines the high energy resolution of the wavelength dispersive spectrometer with the simultaneous recording of the whole spectrum. The wavelength dispersive spectrometers provide much better energy resolution than that of the energy dispersives*.

----

*The lowest resolution of the wavelength dispersive spectrometer that has been employed in Compton profile measurements was a standard flat crystal fluorescent spectrometer with a LiF crystal. Its resolution was about 310 at 17keV, while at the same energy the resolution of the energy dispersive spectrometer is only 80. A typical resolution of the same instrument at 60 keV, one of the most popular energies used in Compton measurements, is 170. The maximum achievable resolution at 10 keV is the double crystal spectrometer and exceeds the value of 8000. The bent crystal with the position sensitive detector can give resolution near 5000.

The present detection in technology provides position sensitive detectors of spatial resolution better than 100μm and efficiency up to 50%. The combination of position sensitive detectors with bent crystals, in addition to the use of synchrotron radiation, will materialize the dreams of all those who had been working in the past on the very rewarding area of inelastic x-ray scattering. The optimum energy resolution of wavelength dispresive dependence on energy, taking into account all the factors that influence the crystal's rocking curve are discussed elsewhere[33].

Figure 7 shows the energy and momentum resolution as a function of the incident photon energy. For the sake of simplicity, the rocking curve H.W.F.M. is assumed to be independent of the photon energy ; the same simplification is used for the H.W.F.M. of the energy dispersive spectrometer. The above simplifications had as result the disappearance of the maximum that more accurate calculations provide when the dependence of F.W.H.M. on photon energy is taken into consideration.

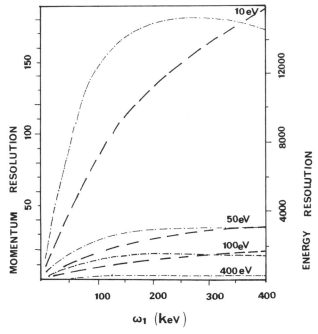

Fig.7. Momentum (right hand scale) and Energy (left hand scale) resolution as a function of incident photon energy. For constant window (FWHM) 10eV, 50eV, 100eV and 400eV.

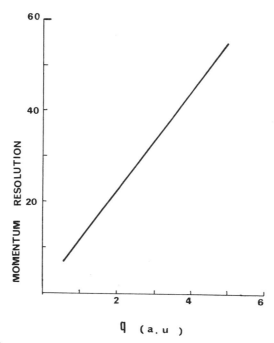

Fig.8. Momentum resolution as function of the projection of electron momentum on scattering wavevector.

Although the momentum resolution is reported* at 2 a.u. for easier comparison between experimental arrangements, it is necessary for a better understanding of the accuracy of the reduced Compton profiles, to use the dependence of the momentum resolution on q, which is as shown in figure 8. This is indicating that in the most interesting case of Compton profile anisotropies (small q), the momentum resolution is very bad. This situation is improved, as mentioned above, in the case of wavelength dispersive spectrometers. However, this is not without drawbacks, because the crystals available for crystal spectrometers are not too many. The optimum energy for Compton profile measurements is around[34] 100keV, where high order reflections are required with low reflectivity. In other areas of the x-ray inelastic scattering research, however, lower energy is required, where the bent crystal spectrometers demonstrate best performance.

----

*The momentum resolution is reported for q=2a.u, an arbitrary value extensively used.

In figure 9 it is shown the spectral area that can be investigated with some of the crystals widely used in x-ray spectroscopy. The upper energy limits can be extended implying high order reflection at the expense of the efficiency.

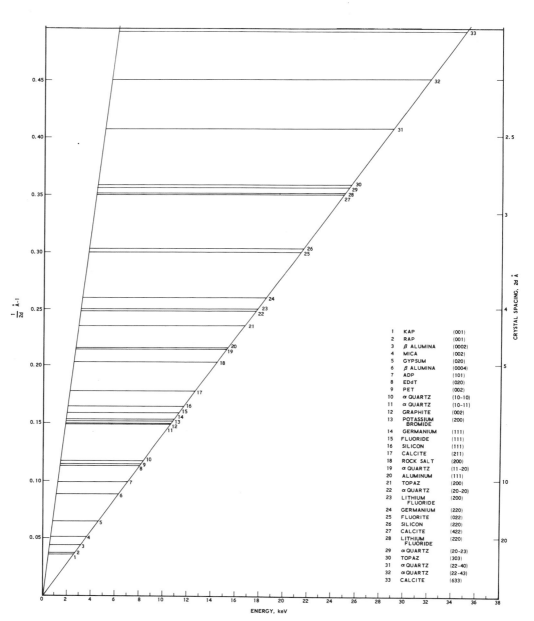

Fig.9. The spectral area that can be investigated by a number of crystals.

METHOD'S ACHIEVEMENTS

The Compton spectra of carbon and beryllium provide one of the first experimental verifications of the Fermi-Dirac distribution of free electrons in Solids, in the early days. The Compton profiles of Li and Na demonstrated, not too long ago, the need to go beyond the one-electron formulation of the conduction electrons momentum density for the simple metals. The same profiles manifested also the role of electron-electron and electron-ion Coulomb interaction in alkali metals. The electron density distribution of transition metals is of tremendous interest, both for the technological applications of their alloys and for the scietific curiosity of the localised nature of d-electron distribution. For that reason extensive measurements of Compton profiles are under way for many years. Extensively also has been studied the Ionic Solids, to determine the importance of negative ions interaction in this group of materials. Finally, as a working experimental method, it has been used in the study of hydrocarbons and in metal hydrogen solid solutions. The most recent application of Compton profile in the study of the condensed matter is the spin-dependent momentum distribution, that became feasible by using sychrotron radiation facilities.

Next in achievements to Compton Profile is the x-ray Plasma scattering, which although up to now by no means can be compared in popularity with Compton scattering, has been used in measuring plasmon dispersion relation and dynamic Structure Factor of simple metals. The reason that this method has not been developed in its full capacity is the inherent disadvatange in the stepping mode of collecting data that rended low accuracy data. It is the authors belief, that this research area will be the main beneficiary of the use of position sensitive detector in conjuction to the focussing geometry.

The x-ray Raman, although cleary can be complementary to x-ray absorption spectroscopy, has not been used yet as a probe in the study of solids; the same is true with the Resonant x-ray Raman.

REFERENCES

1) J.W.M. DuMond, The Linear Momenta of Electrons in Atoms and in Solid Bodies as Revealed by x-ray Scattering, Rev. Mod. Phys. 5: 1 (1933).
2) M.J. Cooper, Compton scattering and electron momentum determination, Rep. Prog. Phys. 48: 415 (1985).
3) V.A. Bushuev and R.N. Kuz'min, Inelastic scattering of x-ray and synchrotron radiation in crystals, coherent effects

in inelastic scattering, Sov. Phys. Usp. 20:406(1977).
4) P. Debye, Phys. Zeits 24: 161 (1923).
   P.A.M. Dirac, Roy. Ast. Soc. 85: 825 (1925).
   L.deBroglie, Ondes et Mouvements, Fasicule 1: 94 (1926)
   F. Bloch, Phys. Rev. 46: 674 (1934).
   A. Sommerfeld, Phys. Rev. 50: 38 (1936).
5) P. Eisenberger and P.M. Platzman, Compton Scattering of X Rays from Bound Electrons, Phys. Rev. 2: 415 (1970).
6) G.E. Jauncey, Quantum Theory of the Intensity of the modified band in the Compton Effect, Phys. Rev. 25:723(1925).
7) R. Currat, P.D. DeCicco, and Roy Kaplow, Compton Scattering and Electorn Momentum Density in Beryllium, Phys. Rev. 3: 243 (1971).
8) R. Currat, P.D. DeCicco and R.J. Weiss, Impulse Approximation in Compton Scattering, Phys. Rev.B 4:4256(1971).
9) P. Nozieres and D. Pines, Electron Interaction in Solids. Characteristic Energy Loss Spectrum, Phys. Rev. 113:1254(1959).
10) Y.Ohmura and N. Matsudaira, Influence of Coulomb Correlation on the Scattering of X-Ray by an Electron Gas. I. RPA Approximation, J. Phys. Soc. Japan 19:1355(1964).
11) A. Tanokura, N. Hirota and T. Suzuki, X-Ray Plasmon Scattering, J. Phys. Soc. Japan 27: 515 (1969).
12) N.G. Alexandropoulos, X-Ray Plasmon Scattering of Lithium, J. Phys. Soc. Japan 31: 1790 (1971).
13) K.C. Pandey, P.M. Platzman and P. Eisenberger, Plasmons in periodic solids, Phys. Rev.B 9 : 5046 (1974).
14) J.C. Asley, T.L. Ferrell and R.H. Ritchie, X-ray excitation of surface plasmons in metallic spheres, Phys. Rev. 10 : 554 (1974).
15) W. Schulke and W. Lautner, Inelastic X-Ray Scattering from Li in the Region of Small and Intermediate Momentum Transfer, Phys. Stat. Sol.(b) 66: 221 (1974).
16) N.G. Alexandropoulos and G.G. Cohen and M. Kuriyama, Evidence of Optical Transitions in X-Ray Inelastic Scattering Spectra: Li Metal, Phys. Rev. Let. 33: 699 (1974).
17) G. Mukhopadhyay, R.K. Kalia and K.S. Singwi, Dynamic Structure Factor of an Electron Liquid, Phys. Rev Let. 34: 950 (1975).
18) G. Barnea, Theory for the complex dynamic structure factor of an electron gas, J. Phys.C: Solid State Phys. 12: L263 (1979).
19) W. Schulke, H. Nagasawa, and S. Mourikis, Dynamic Structure Factor of Electrons in Li by Inelastic Synchrotron X-Ray Scattering, Phys. Rev. Let. 52: 2065 (1984).
20) G. Priftis, A. Theodossiou and K. Alexopoulos, Plasmon Observation in X-Ray Scattering, Phys. Let. 27A: 577 (1968).
21) S. Manninen and P.Suorti, M.J. Cooper, J. Chomilier and G.G Loupias, X-ray resonant Raman cross section and yield in nickel, Phys. Rev.B 34: 8351 (1986).

22) S. Manninen, N.G. Alexandropoulos and M.J. Cooper, Resonant Raman scattering with gamma rays, Phil. Mag. B. 52:899(1985).
23) Smekal, Naturwiss 11: 873 (1923).
B. Davis and D.P. Mitchell, Phys. Rev. 34: 1 (1929).
A. Sommerfeld, Phys. Rev. 50: 38 (1936).
F. Schnaidt, Ann. Phys. (Paris), 21: 89 (1936).
K. DasCupta, Phys. Rev. 128: 2181 (1962).
P.M. Platzman and N. Tzoar, Phys. Rev. 139: A410 (1965).
R.J. Weiss, Phys. Rev. 140: A1867 (1965).
A. Faessler and P. Muhle, Phys. Rev. Letters, 17:4(1966).
T. Suzuki, J. Phys. Soc. Jap., 22: 134 (1967).
Y. Mizuno and Y. Ohmura, J. Phys. Soc. Jap .22:445(1967).
N.G. Alexandropoulos and G.G. Cohen, Phys. Rev. 187: 455 (1969).
24) M. Kuriyama and N.G. Alexandropoulos, On the relationship between x-ray Inelastic Scattering and Absorption spectra, J. Phys. Soc. Jap, 31: 561 (1971).
25) G.G. Cohen, N.G. Alexandropoulos and M. Kuriyama, Relation between X-Ray Raman and Soft x-ray Absorption spectra, Phys. Rev. B, 8: 5427 (1973).
26) T. Suzuki and H. Nagasawa, X-Ray Raman Scattering III. The Angular Dependence of the Scattering Intensity, J. Phys. Soc. Jap., 39: 1579 (1975).
27) J. Felsteiner, R. Fox and S. Kahane, The Electron Momentum Density in Aluminum, Sol. Stat. Com, 9: 61(1971).
28) B.G. Williams, Compton Scattering, MC. Graw Hill, London, (1977).
29) M.J. Cooper, Gamma-ray source properties and Compton Scattering, Nucl. Instr. Meth., 166: 21 (1979).
30) L.V. Azaroff, X-Ray Spectroscopy, Mc. Graw Hill (1974).
31) G. Loupias and J. Petiau, Anisotropic Compton Scattering in LiF using synhrotron radiation, Phys. 41:265 (1980).
32) P. Pattison, H.J. Bleif and J.R. Schneider, Observation of x-ray Raman, Compton and Plasmon scattering using a position sensitive proportional counter, J. Phys.E. Sci. Instr. 14: 95 (1981).
33) N.G. Alexandropoulos, G.G. Cohen, Crystals for stellar Spectrometers, Appl. Spectr., 28: 155 (1974).
34) N.G. Alexandropoulos, Fluorescent sources for an energy dispersive Compton Spectrometer, Nucl. Ins.& Meth.in Phys. Res. A251, 511 (1986).

PARTICLE INDUCED X-RAY EMISSION: BASIC PRINCIPLES, INSTRUMENTATION AND INTERDISCIPLINARY APPLICATIONS

Ede Koltay

Institute of Nuclear Research (ATOMKI) of the
Hungarian Academy of Sciences, Debrecen, Hungary

INTRODUCTION

The name Particle Induced X-Ray Emission refers to a combined process, in which continuum and characteristic X-rays are excited by ion-atom collision events as the consequence of the slowing down of the charged particles involved and the recombination of electron vacancies appearing in the inner shells during their ionisation, respectively.

The process itself is a subject of fundamental research in many respects. As part of this research a number of simplified theoretical models are being composed and tested through a detailed comparison of experimental X-ray emission or ionisation cross sections with data deduced from approximate theoretical calculations. The effect of a multiple ionisation and the chemical state of the atom on the position, width and structure of the lines in the characteristic spectrum represents another broad field of recent investigations. The determination of the angular distribution of characteristic X-rays gives an insight into the mechanism of the ion-atom collision events. The production mechanism of the continuum X-rays which appear as a background of the monochromatic characteristic lines is receiving an increased interest in the present days, as well.

The unique correspondence between the atomic number of the elements excited by the bombarding particles and the energies of the characteristic X-ray lines emitted makes Particle Induced X-Ray Emission a useful method in instrumental analytics.

The elemental analysis has become a tool of great importance of present day fundamental investigations and applied research in a number of fields. Science and technology reached a level in many cases where problems can be solved only on the basis of detailed knowledge of the concentration of all the components - both basic constituents and trace elements - present in the system to be treated. Consequently, there is an increasing need to develop methods of elemental analysis to fulfil the following demands:

- applicability in the widest possible range of atomic number,

- possibility of absolute determination of the amounts of sample components,

- easy methods for sample preparation,

- high sensitivity allowing trace element determination in small volume samples,

- high speed in the preparation and measurements,

- the possibility of obtaining analytical information from microscopic surface areae combined with one- and/or two dimensional scanning for measuring longitudinal and/or lateral distributions for the elemental concentrations, respectively,

- the possibility of obtaining depth distribution curves for the elemental constituents,

- easy way for the automation of the measurements and evaluation at a convenient level of accuracy and precision.

The rapidly developing analytical method based on Particle Induced X-Ray Emission process and referred to hereafter as PIXE meets all the above requirements within certain limits.

In what follows the physical background and basic processes will be briefly reviewed - limited to the practical case of protons as bombarding particles - and details will be presented on the instrumentation for PIXE measurements including the special facilities for microbeam PIXE analysis. Methods of calibration for the determination of relative or absolute concentrations will be treated and mention will be made of the evaluation method for depth profile measurements. The analytical parameters to be achieved by the PIXE method will be discussed.

The continuously broadening field of applications will be illustrated through lists of the main fields of applications in different branches of the science compiled from the literature and mention will be made of the special features of the method from the point of view of the selected fields of application.

The intense development work devoted to bringing the method to a perfection and the wide scope of its application in actual analytical research would make the task hopeless to draw a complete list of important references. Therefore, reference will be made on Conference Proceedings and important review articles to be studied for detailed information and individual papers will be selected somewhat arbitraryly as suggested by the way of our presenting the material to be treated.

PHYSICAL BACKGROUND AND BASIC PROCESSES

The most important articles summarizing the physical background and describing the basic processes underlying PIXE are those written by F. Folkmann[1] and S.A.E. Johansson and T.B. Johansson[2]. More recent papers contribute with details in a number of important fields.

The schematic presentation of the elements which compose an experimental set-up is given in Fig. 1 together with the notations for basic physical parameters to be used throughout this paper. According to the scheme the PIXE procedure can be described as follows: particle beam of energy E_o made diffuse by a scatterer foil and cut by a set of collimators (or micro-focused by a delicate electron-optical system) bombard the sample with an incident angle Θ_i with respect to the normal of sample surface (in some cases the beam arrives to the sample through a thin exit window.). The sample is composed of elements k=1...n with atomic numbers Z_k present with the relative concentrations by weight $C(Z_k)$. The number of bombarding

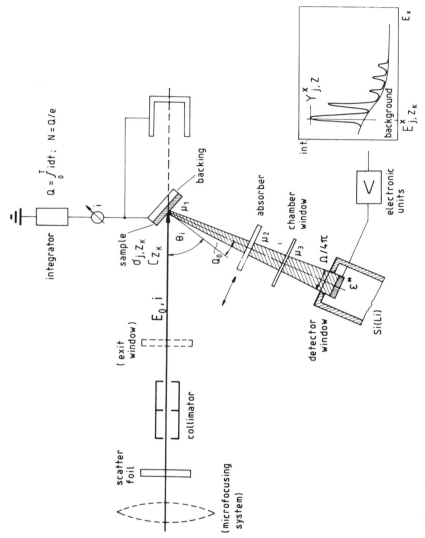

Fig. 1. Schematic of a PIXE set-up and notations for basic physical parameters

particles incident during measuring time T is determined by an integrator device as the time integral of the target current divided by the electron charge. Special care should be taken of an accurate determination of the bombarding beam through avoiding effects by secondary electrons emitted from the sample and from construction elements of the chamber. Characteristic X-ray lines with energy E^x_{j,Z_k}, excited with the cross section $\sigma_{j,Z_k}(E)$ in a differential sample layer which is penetrated by the bombarding particles slowed down to energy E, are detected by a Si(Li) detector. The detector observing the emission with a solid angle $\Omega/4\pi$ is situated at the detection angle θ_o. Absorption terms $\exp(-\mu_1 x_1)$, $\exp(-\mu_2 x_2)$ and $\exp(-\mu_3 x_3)$ take the decreases of X-ray intensity into account, which are caused by the upper target layer, an additional absorber present in or absent from the detection axis and by the chamber window sometimes used to separate beam tube vacuum from the atmospheric pressure or detector vacuum. The detection efficiency ε contains the absorption effect of the detector window, as well. Detector signals processed by standard electronic units and stored in a multichannel analyzer will build up the X-ray spectrum in which characteristic X-lines with intensities Y_{j,Z_k} are superimposed on an X-ray continuum. The continuum is partly due to intrinsic effects during bombardment. A separate component which may appear as the consequence of charging up on targets of insulating character can be easily eliminated by target neutralization. The assignment of X-lines to sample elements and the deduction of elemental concentrations from X-ray intensities require specially developed computer codes.

The main analytical parameter "minimum detectable concentration" is basically determined by the intensity ratio of characteristic lines to continuum background. Optimum conditions of PIXE measurements can be determined on the basis of calculations which take cross section data of the respective components into consideration.

A qualitative description of the process underlying PIXE can be given by the help of Fig. 2 which is composed of five sets of curves reflecting the features of the basic relations.

The characteristic X-ray energies given for K_α, K_β, L_α, L_β and L_γ lines here are unique functions of the atomic number of the emitter atom. K_α and K_β lines for elements from lithium to uranium cover the energy range 0.1-110 keV, while the energy range of L_α L_β and L_γ lines for elements from zinc to uranium is between 1 keV and 21.5 keV [3]. Principally, both K and L lines can be used for detecting any elements present in the sample.

The complex X-ray spectrum emitted by a composed sample under bombardment is observed by the Si(Li) detector in an "energy window" indicated with vertical dotted lines through the figure. The window is set by the strong energy dependence appearing on the detection efficiency curve at energies below 1.3 and above 20 keV. Low energy cut-off is caused by the absorption of soft X-rays in the detector window foil while high energy decrease in efficiency appears as a consequence of the weak absorption of higher energy X-rays in the silicon detector. The efficiency curves in the figure correspond to detector windows made of 12 or 25 μm beryllium foil and silicon thickness of 3 and 5 mm [4]. As clearly demonstrated, by a proper selection of either K or L lines all elements in the interval $11 \leq Z \leq 92$ can principally be detected within the energy window defined by a moderate variation in detection efficiency with varying atomic number. Horizontal dotted line around Z=47 separates with some arbitrariness the regions where K or L lines are preferentially used.

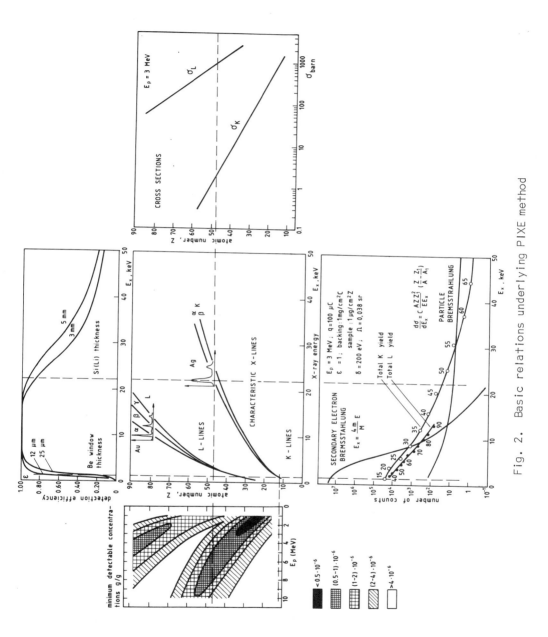

Fig. 2. Basic relations underlying PIXE method

The determinant factor for the intensity of the spectrum lines from the side of the excitation process is the X-ray emission cross section. Contrary to the factors treated above cross sections show a strong dependence on the bombarding energy. In Fig. 2 the dependence of total σ_K and σ_L cross sections on atomic number is given for the proton energy E_p=3 MeV [5,6]. The very high values of the cross sections varying between 1 and 1000 barn indicate that a fast, high yield analytical procedure can be expected with the application of PIXE. A strong decrease of the total σ_K cross section appears with increasing atomic number, while total σ_L cross section remains high for the highest Z values, too. Therefore, the detection of L-lines for atomic numbers above the horizontal dotted line in Fig. 2 motivated by the energy dependence of detection efficiency is highly recommended from the side of cross sections, as well. It can be expected, that in spite of the strong energy dependence of the detection efficiency and X-ray cross sections, the total number of counts for the same concentration will vary within two orders of magnitude only when detecting elements spread in the total $12 \leq Z \leq 92$ interval.

As stated above, characteristic X-ray lines emitted from a sample will be superimposed on an X-ray continuum originating from intrinsic excitation processes. The basic components here are the bremsstrahlung from the slowing of bombarding particles in the thick (sample+backing) layer and the bremsstrahlung from the slowing of secondary electrons released by the bombarding particles from sample and backing atoms. The question on the contribution of other mechanisms mainly important in the case of heavier bombarding particles is being investigated recently [7]. Particle bremsstrahlung shows a slightly decreasing intensity with increasing X-ray energy. The cross section of the emission of bremsstrahlung in the energy interval dE_x around E_x depends on particle energy E as well as on atomic numbers and mass numbers of the stopping and bombarding particles. The spectrum of secondary electron bremsstrahlung shows a sharp cut-off at the high energy side situated around the maximum kinetic energy of the secondary electrons, deduced from the simple kinematic relation for the particle-electron collision process.

The curves shown in the lower part of Fig. 2 presenting data calculated by F. Folkmann [1] reflect the conditions in a clean way. Yields for characteristic and continuum X-rays are shown in the figure for the parameter values given in the figure (δ denotes the energy resolution of the Si(Li) detector). Empty circles and dots are calculated yields for a thin sample of a single element with atomic number indicated, sitting on a thick carbon backing with the thicknesses 1 $\mu g/cm^2$ and 1 mg/cm^2, respectively. The two components of the continuum are shown on the same scale. It is nicely demonstrated for this actual case, that changing for L-rays at atomic numbers higher than 45 compensate for the decrease in the intensities of the K lines as a consequence of the increased total cross section for the L ones.

The relative hights of the characteristic X-ray lines with respect to background level determine the detection limit of the method for a selected element. It is accepted namely that the total number of counts in a line should exceed three times the square root of background counts over the same energy interval, for a safe identification and evaluation of the respective peaks. Consequently, calculations of the above type can give us minimum detectable concentration as the function of atomic number. In the above example the bombarding proton energy was selected arbitraryly to 3 MeV. Due to the strong energy dependence of the cross sections for characteristic and continuum X-rays on the bombarding energy, similar calculations should be performed for a broad energy interval in order of find the best possible conditions for the general use of PIXE. On the left side of Fig. 2 the energy dependence of the minimum detectable concentration levels are given for a broad interval of atomic numbers, as published by S.A.E. Johansson and T.B. Johansson [2].

As a general conclusion from Fig. 2 we claim that PIXE analytical method based on the energy dispersive detection of X-rays with a Si(Li) detector can be best used as a multielemental method around 2 MeV bombarding proton energy with a relatively constant detection limit for a broad range of atomic numbers.

INSTRUMENTAL

The physical conditions for PIXE measurements, schematically presented in Fig. 1 are defined in actual measurements by instruments, which guarantee the prescribed shape, size and current density distribution of the exciting beam, the positioning and changing of the samples in irradiation and detection geometry, the error-free detection of the number of exciting particles, the energy dispersive detection of the emitted X-ray quanta, and the physical modification of the resulting spectra according to the needs of minimizing spectrum background and overload in detection system. Analogue signal processor unit, digital-to-analogue converters, data aquisition and processing systems are applied for converting detector pulses into multichannel spectra to be evaluated by computers in terms of absolute concentrations of sample constituents. In some cases, special electronic and electron-optical units are used for switching off the beam for the periods of signal processing in the analyzer system.

The transport of the bombarding beam and the detection of X-rays take place in the vacuum systems of the beam transport channel and the Si(Li) cryostate, respectively. The bombardment of the sample may be performed in-vacuo or under atmospheric conditions. Consequently, various combinations of transport, irradiation and detection elements can be built up. Internal beam PIXE chambers serve for the irradiation of the sample in the vacuum of the beam transport tube, the Si(Li) detector with its separate vacuum volume looks at the sample through a thin beryllium window on the detector. The detector head either penetrates the vacuum chamber or is situated opposite to a thin exit window on the chamber wall (in-vacuo and through-the-window arrangements, respectively). Sometimes windowless detectors are directly connected to the vacuum of the transport tube. External beam chambers represent an alternative, in which the beam is led to the room atmosphere through a thin exit window and the excitation takes place under atmospheric conditions while X-rays enter detector vacuum through the detector window. Figs. 3.a and 3.b give the schematic presentation of internal beam and external beam PIXE chambers, respectively.

Internal beam PIXE chambers

An internal beam PIXE chamber generally meets the following requirements:

- it provides the possibility of arranging a number of samples simultaneously in chamber vacuum such a way as to facilitate their serial analysis without breaking the vacuum for shifting the samples into irradiation position. In some cases the possibility of a fine continuous sample advance is given along one axis perpendicular to the ion beam to permit the measurement of the radial distribution across a sample spot or the longitudinal distribution along extended samples. In other cases automatic sample change is preferred with accurate positioning of sample center into irradiation spot,

- it defines the beam cross section with a collimator system so as to restrict the irradiation to a given area and to avoid the bombardment of the sample holding frame and/or the target ladder. Elements of the collimator system are connected to different bias voltages,

- it enhances the uniformity of the distribution of beam current density across the sample spot compared to that of the focused beam by getting the beam diffused through a scatterer foil, situated before the collimator system,

- it gives the possibility of measuring the number of bombarding particles through beam current measurements. For thin transparent samples the measurement can be made in a built-in Faraday cup behind the sample, while for thick target measurements the whole chamber block should be insulated from the ground in such a way as to form itself a Faraday cup. The use of such an arrangement together with properly biased collimator elements will help in avoiding misleading effects caused by secondary electrons escaping from sample area or being collected from the region where construction elements outside the cup are under bombardment,

- it cuts the enhanced bremsstrahlung background, which normally appears when the charge of protons deposited in an insulating sample build up an electric field inside the chamber, through the neutralisation of the charge by the electron beam of a built-in electron gun [6]. The gun consists of an emitting cathode filament inside a perforated anode cup. Filament and anode circuits are closed inside the Faraday cup, consequently, the intensity of the electron current does not influence the ion current measured towards the ground. The dramatic effect of charge neutralisation can be seen in Fig. 4 where PIXE spectra taken on human liver sample are shown with electron source switched off and on [9]. Signal-to-background ratio increased by orders of magnitude in the latter case makes the detection of low concentration elements more reliable,

- it permits the application of different total absorption thicknesses between the sample and the detector. As a minimum thickness typically 36 μm thick HOSTAPHAN foil should be used here in order to prevent scattered protons from entering the detector. In case of in-vacuo detector arrangement the beryllium window of the detector represents an additional absorber, while in through-the-window arrangement the window foil of the chamber and an air gap between chamber and detector windows should also be taken into account. The optional insertion of a 36 μm thick aluminium absorber foil makes the detection of elements with atomic number $Z > 20$ easier. Strong continuum X-ray at low energies and high yield of characteristic K-lines for low atomic numbers result in a heavy load of the electronic system in the low energy part of the spectrum and make the detection of heavier elements (mainly present as trace elements in samples composed of light bulk elements) troublesome through pile-up events. The effect of the absorber on the structure of the spectra through depressing the low energy part is clearly demonstrated by the spectra in Figs. 5.a and 5.b taken on thick samples of NBS standard reference material No 1571 with the aluminium absorber removed and inserted, respectively [9].

Electrostatic beam deflection plates and fast electronic driving units also shown in Fig. 3.a serve for on-demand beam bombardment [10,11]. The absence of bombarding beam during signal processing periods achieved by putting high voltage deflection pulses on the deflection plates during the busy period of the electronic system will decrease target deterioration and will help in avoiding spectrum distortions (e.g. additional pule-up) and dead-time correction appearing as a rule with too high counting rates.

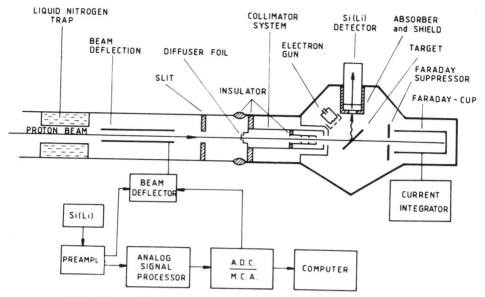

Fig. 3.a. Schematic of an internal beam PIXE system

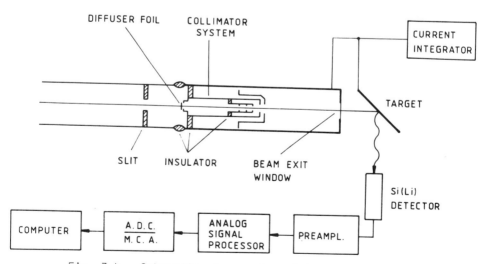

Fig. 3.b. Schematic of an external beam PIXE system

External beam PIXE chambers

Technical difficulties sometimes caused by the exposition of samples to chamber vacuum in internal beam PIXE method can be avoided with external beam configuration, where the bombardment takes place in atmospheric air [12]. The characteristic features of the external beam version can be summarized as follows:

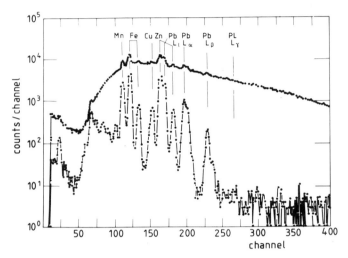

Fig. 4.   PIXE spectra taken on thick human liver autopsy sample with electron source switched off (upper curve) and on (lower curve).

- the method permits the direct irradiation of any type of samples. Samples of liquid form [13], of high volatility and/or heat sensitivity, large size objects, art treasures, etc. may easily be treated. Local heating is intensively reduced by the efficient cooling in the ambient air, charging up is eliminated by the ionized air along the beam. Due to natural air cooling and the absence of target charging the beam intensity can be higher in many cases than in internal beam measurements, consequently, measuring time can be decreased as a rule. It is easy to put the samples in irradiation position and change the samples in serial measurements,

- the beam is led to the air through an exit foil made of beryllium, aluminium or KAPTON [14,15]. High gamma-ray background from metal foils makes the application of KAPTON more appropriate in spite of its limited life time. A foil with thickness of 8 μm withstands air pressure for more than 30 hours when bombarded with 10-20 nA beam intensity. An important task for the safe running of the accelerator in external beam mode is the monitoring of the vacuum in the beam tube connecting to the chamber and the use of a fast acting automatic gate valve which would disconnect the chamber from accelerator vacuum in the case of a rupture of the foil,

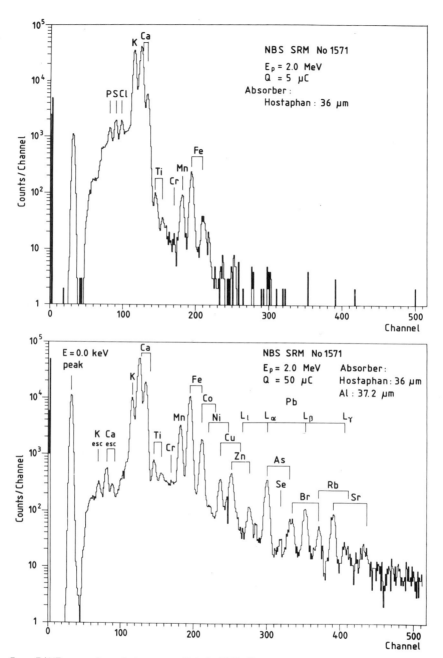

Fig. 5. PIXE spectra taken on thick BNS SRM No 1571 sample without (upper part) and with (lower part) an 37.2 μm Al absorber between target and detector.

- when crossing the air gap towards the sample the beam excites the characteristic X-lines of gases composing air. Nitrogen and oxygen are too light elements to be detected, but argon contributes with an intense component to the background of sample spectra. This structure measured on empty target frame and compared in Figs. 6.a and b [9] with a similar spectrum taken in internal beam mode makes external beam measurements inferior to internal beam ones for elements lighter than Ar. This effect, however, can be cancelled by replacing air to helium in the irradiation unit,

- the accurate measurement of the bombarding beam intensity is made difficult by the emission of secondary electrons from exit window and sample and by the presense of an ionised air gap between these elements. It is conventional to measure the sum of currents on exit window and sample for monitoring the measurements. For a more accurate determination of the number of bombarding particles the intensity of protons elastically scattered from the exit window or the hight of the argon X-ray peak appearing in the spectrum can be used. An oscillating wire beam sensor is proposed for checking beam constancy [16].

PIXE microprobe systems

Electron microscopy developed efficient energy-dispersive X-ray microanalytical systems in which characteristic X-rays excited by the fine--focused electron beam of an electron microprobe [17] is detected by a Si(Li) detector. When setting an energy window on the spectrum at a preselected X-ray energy, - together with the normal optical image deduced from secondary electron intensities - an X-ray image can be constructed which corresponds to a mapping of the concentration levels of the corresponding chemical elements in microareae of the specimen. The application of the electron excitation in X-ray spectrometry limits the potentialities of the method in two respects. On one hand, the characteristic X-lines are superimposed on the continuum bremsstrahlung background, which is orders of magnitude higher than that in the corresponding case of slowing protons in the PIXE process. On the other hand, the optical quality for the X-ray image will be much poorer than for the electron image due to the increased (both in depth and diameter) interaction volume for characteristic X-ray release compared to that for secondary electron emission. Consequently, high bremsstrahlung background will make the minimum detection limit as high as a few tenths of percent for most elements, while the original image resolution of a few tens of nm will be degraded by one or two orders of magnitude in X-ray spectroscopy mode. Original image quality can be maintained for transparent thin samples in transmission electron microscopes.

Focusing a particle beam into microspots of sharply cut current density distribution can be achieved in an electron-optical way. The trajectories of individual particles emitted by a source and accelerated to the interaction energy are to be shaped with a magnetic condensor lens so as to form a beam with small cross over radius at an axial area which serves as the object for an objective lens of short focal length with long object distance. Strong demagnification appearing in such case will result in the desired beam spot provided the proper dimensioning and careful realisation of the optical system minimize aberration effects. Periodic transversal deflections generated in two perpendicular dimensions will permit us to scan the beam over a selected area of the sample placed in the image plane of the objective lens. From a synchronous analysis of electron- and X-ray intensities and spectral distributions by electronic devices two-dimensional electron- and X-ray images can be built up.

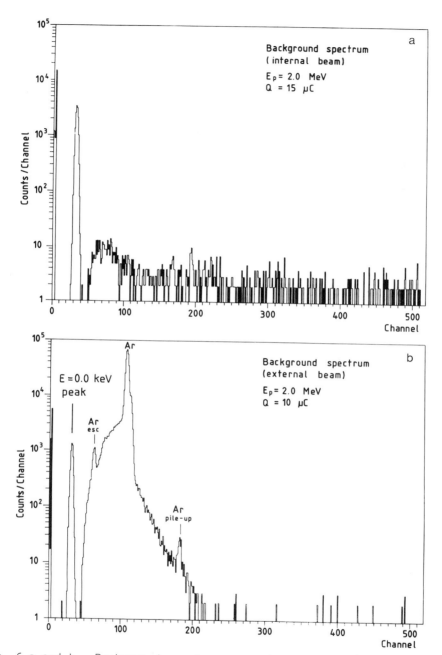

Fig. 6.a and b. Background spectra measured on an empty target frame in internal and external beam mode, respectively.

The low magnetic rigidity of the ten-to-twenty keV electrons used here permits us to shape the electron beams with simple short magnetic lenses, where accurately machined iron pole pieces of rotational symmetry makes high quality focusing easily obtainable. In such lenses electron velocities and magnetic field intensities are nearly parallel to each other, consequently, only the small transversal velocity components result in a weak focusing effect. The schematic of an electron microprobe is given in the left hand side of Fig. 7 together with a typical X-ray spectrum [18] to be obtained with such devices.

The high detection limit in electron microprobe analysers can be lowered in a proton microbeam system in which PIXE excitation mode results in a dramatically reduced background continuum. In a proton microprobe the simple electron source is replaced by a complete electrostatic accelerator followed by a magnetic deflection unit working as a monochromator, while the weak focusing optical system of the electron microprobe is changed to an optical channel which is able to condense and demagnify the proton beam with its magnetic rigidity increased by a factor of $\simeq 400$ compared to that of the electron beam in an electron microprobe. For this purpose classical magnetic lenses can be replaced by superconductive solenoids [19] or a completely new way can be opened by introducing strong focusing elements. Systems of electrostatic or magnetic quadrupole lenses [20] with their strong transversal fields turned out to be the proper tools to be used in ion microbeam facilities; a typical configuration is shown schematically in the right hand side of the figure. As indicated by the symbols of equivalent light optical lenses, a single quadrupole lens posseses focusing and defocusing properties in x and y perpendicular directions, respectively. In order to achieve focusing in both directions a doublet of quadrupoles is used as a condenser system. The optimum configuration for the low-aberration objective lens is given by a specially designed quadrupole quadruplet system [21,22]. Other configurations used in actual facilities are described in original publications [23,24,25,26] and in review articles [27,28,29] covering the field. Beam scanning is achieved by the proper periodic feeding of scanner coils or by periodically excited dipole field components superimposed on the quadrupole field of the lens [30,31].

The characteristic features of the proton microbeam technique can be summarized as follows:

- demagnification factors X_O/X and Y_O/Y generally different from each other amounts to 5-60 for existing microprobes. The practical beam diameter to be achieved depends on the object size cut by the object slits at the cross-over and on the chromatic and third-order spherical aberrations of the objective lens system. Additional beam distortion may be caused by inaccuracies in the quadrupole geometry and misalignments of the singlets composing the objective lens. Minimum spot size varies for different set-ups between 1 and 50 μm,

- the suppression of the beam halo is achieved by the proper construction of the object collimator with scattering rate as low as possible. The transmission efficiency for the collimator for the beam is as low as 10^{-4}. The beam intensity on the sample amounts to 10^{-10} Ampere,

- the method reached the capability of detecting elements in samples at about 10^{-17} g/μm² concentration level,

- proton microprobes may be used both in internal and external beam modes.

From the comparison of the spectra in Fig. 7 taken with electron and proton beam excitation on the same sample [18] the superiority of microPIXE in low detection limits is clearly evident.

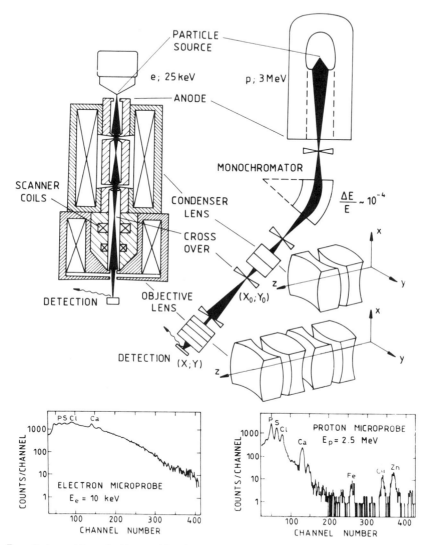

Fig. 7. Schematic of the optical systems of electron and proton microprobes and typical X-ray spectra taken with electron and proton excitation.

CALIBRATION AND EVALUATION

The basic aim of the PIXE method is the determination of (relative or absolute) concentrations of elemental constituents in the sample under investigation. The calibration of PIXE systems, in which sensitivity factors are determined to be used in assigning absolute concentration data to numbers of counts in the X-ray peaks as the function of atomic number, is the subject of a number of articles [32,33,34,35]. The various techniques followed by different authors can be divided into two groups. In the first case the determination is fully based on experimental procedures, in which the sensitivity factors are deduced from measurements performed on standard samples, which are made of pure elements or chemical compounds. The factors are characteristic for the actual detection arrangement used in the analytical procedure. In the second case sensitivity factors are deduced theoretically or in a semiempirical way from calculated cross sections of X-ray excitation and from absorption data for the absorbers present between sample and detector. The detection solid angle and the energy dependent detector efficiency should also be determined.

The calibration procedure and the underlying formalism for the determination of concentrations from X-ray yields are influenced by the sample thickness through three effects. For a sample of finite thickness

- a continuous slowing of the bombarding particles appears along the depth coordinate in the sample. Consequently, X-rays will be excited in layers at different depths with different cross section values for the respective particle energies,

- absorption length for emitted X-rays between points of emission and detection varies with depth coordinate,

- an enhancement of X-rays appears through secondary X-ray fluorescence of those elements with absorption edges just below the emission energies of the dominant constituents of the sample. The contribution of secondary fluorescence increases with increasing thickness of the sample.

It is worth mentioning that all the above effects are matrix dependent, i.e. for a quantitative evaluation use should be made of the concentration values for the main constituents in the sample matrix. On the other hand, they disappear with decreasing sample thickness, therefore it is practical to treat the case of differentially thin samples free of the above effects independently [36].

Thin samples

For a thin uniform homogeneous target, with negligible energy loss of the bombarding proton and no absorption of the X-rays in the sample, the yield of each X-ray peak can be calculated from the formula

$$Y(Z) = \frac{N_{av} \cdot \sigma_Z(E_o) \cdot \omega_Z \cdot b_Z \cdot \varepsilon_Z}{A_Z} M_a \cdot N = K(Z) M_a \cdot N, \qquad (1)$$

where N_{av} denotes Avogadro's number, $\sigma_Z(E_o)$ is the ionization cross section at proton energy E_o [37,38,39,40,41,42], ω_Z and b_Z denote fluorescent yield [43] and branching ratio [44,45], respectively, ε_Z is the detection efficiency [46,47] composed of solid angle $\Omega/4\pi$, absorption factors and absolute detector efficiency ε_Z^* for the bare detector [34], Z and A_Z are atomic number and mass of the element with areal density M_a, respectively, while N stands for the number of bombarding particles. Calibration means the determination of thin target sensitivity factor K(Z) as the function of atomic number.

Thin calibration samples of known areal density prepared by the vacuum evaporation method turned out to be well suited for experimental calibration procedure, in which thin target sensitivity factors for K or L lines are directly measured on a series of standard foils. For theoretical calibration method fundamental parameters are available from the publications referred to above. Here systematic errors may appear as a consequence of using approximate models during the calculations. Additional error may be caused by the use of approximate data for the physical parameters of the system. The careful comparison of the theoretical and experimental calibration methods [34,35] and the agreement found between corresponding data prove that the determination of elemental concentrations can be done with an overall accuracy of 3-5 %, this number includes the errors appearing in spectrum evaluation, as well.

Thick samples

For a thick homogeneous target in which the bombarding protons are completely stopped and the emitted X-rays are subjected to absorption in the source, the yield of each X-ray peak can be written as follows:

$$Y(Z) = \frac{N_{av} \cdot \omega_Z \cdot b_Z \cdot \varepsilon_Z}{A_Z} C_Z N \int_{E_o}^{o} \frac{\sigma_Z(E) T_Z(E)}{S(E)} dE \, , \qquad (2)$$

where C_Z is the concentration of element with atomic number Z, S(E) denotes the stopping power and

$$T_Z(E) = \exp \left[-\left(\frac{\mu}{\rho}\right) \frac{\cos\Theta_i}{\sin\Theta_o} \int_{E_o}^{E} \frac{dE}{S(E)} \right] \qquad (2')$$

gives the photon attenuation in the sample. Data for absorption coefficients at various X-ray energies can be taken from the works [3,48] and [49]. For a sample composed of more than one bulk elements both resultant μ/ρ and S(E) can be deduced by summing over the corresponding data for the basic constituents.

A direct approach for calibration in thick target case is to use thin target sensitivity factors K(Z) and to calculate thick target corrections I(Z) on the basis of the formula [4,50,51]

$$Y = N \cdot C_Z \cdot K(Z) \cdot I(Z) =$$

$$= N \cdot C_Z \cdot K(Z) \int_{E_o}^{o} \frac{\sigma_Z(E) T_Z(E) dE}{\sigma_Z(E_o) S(E)} \qquad (3)$$

obtained by rewriting Eq.(2). The method requires information on sample composition and target thickness. Most commonly used methods are the use of separate thick single element standards for each element to be analyzed with $C_Z = 1$, the addition of internal standards to the sample to be investigated and the application of multielemental standards. Finally formulae (2) and (2') are well applicable for thick sample calibration based on the theoretical methods of fundamental parameters. In the first case the concentration C_Z writes

$$C_Z = \frac{Y(Z)}{Y_s(Z)} \cdot \frac{I_s(Z)}{I(Z)}, \qquad (4)$$

where s stands for standard. In the internal standard method the matrix effect is well accounted for and it does not require the use of absolute beam current, but the homogeneous mixing of the standard material added to the sample is an important requirement. The use of multielemental standard reference material for mass calibration is relatively fast and simple if errors of 10-15 % are acceptable. However, for better accuracy, the composition of both standard and sample has to be known to make the necessary correction for differences in the matrices.

The methods listed above are directly applicable in the case of samples, the matrix of which is composed of bulk elements of known concentrations, and the task is the determination of trace elements which do not contribute effectively to the slowing, absorption and secondary fluorescence behaviour of the matrix. In the case of bulk analysis (e.g. measuring alloys or mixtures) where the above effects are mainly determined by the bulk components of unknown concentrations to be analyzed, an iterative procedure is needed which starts from initial data of the respective concentrations.

One important matrix effect leading to an overestimation of concentrations deduced from measured X-ray yields in thick target analysis is the secondary fluorescence. The error may be significant for some combinations of elements where energy difference between the primary radiation and absorption edge of the lighter constituent is small. Different numerical and analytical procedures resulting in corrections more or less appropriate for ending up with the exact concentration data are treated in the references [5,52,53,54,55,56]. Further work is going on aiming at the development of a unique matrix correction model.

The conditions for homogeneity and uniformity of the sample are not fulfilled in a number of practical cases, therefore the above simplified treatment must be corrected. For taking the actual types of non-uniformity and inhomogeneity into consideration a number of correction methods were developed and published in the literature. The influence of surface irregularities on the sample, which usually are greater than the proton range, can be estimated in special models approximating the surface by a sawtooth function [57] or by a wavy structure [55]. Calculations performed for actual cases resulted in correction functions with values around 10 % for soft X-rays. A special sort of inhomogeneity is observed by metallurgists in some binary alloys where clusterisation of the minor component results in the appearance of grains of different size. The conventional treatment of the data tends to everestimate the concentration of the constituent. A generalized version of Eqs. (2) and (2'), which describe the sample an a sheet composed of parallel layers with grains of given size inside distributed randomly, was found to be appropriate for deducing reasonable concentration values. The formalism also offers a way for calculating the effective grain size by comparing spectra taken at two different bombarding energies [58]. The presence of surface layers on a homogeneous sample represents an important case to be investigated because of the regular appearance of oxides on the surface of metallic samples. Calculated ratios of metal X-ray intensities produced from oxidized samples to those obtained on unoxidized ones, given for different metals and different oxide thicknesses in the work [59] offers the possibility of measuring the thickness of oxide using PIXE if one assumes a certain stoichiometry for the oxide.

In-depth distribution of minor constituents present in a bulk is a question of practical importance in material science. On one hand, depth intervals covered by the range of protons used in PIXE analysis may contain interesting information when sample is subjected to corrosion, diffusion or annealing processes. On the other hand, in some technological processes metals are coated with thin layers to improve their properties; the structure of these layers could be checked in PIXE, as well. In the presence of a layer of a minor constituent with its concentration profile varying with depth coordinate the concentration C_Z in Eq. (2) is to be replaced by the function $C_Z(x)$. Rewriting Eq. (2) with variable x total detected X-ray yield appears in the form

$$Y(Z) = \kappa \cdot N \int_0^{R\cos\theta_i} C_Z(x) \sigma[E_0, E(x)] \exp(-\mu x/\cos\theta_0) dx \qquad (5)$$

where κ is a constant and R is the proton range, $E(x)$ is the proton energy in depth x, $\sigma[E_0, E(x)]$ is the local mean value of the cross section. The application of deconvolution algorithms published in papers [60,61] offers a technique to deduce concentration profiles $C_Z(x)$ varying over the proton range (around 20 μm) from a set of experiments in which the incident bombarding energy is systematically varied over an appropriate energy interval. The critical analysis of the potentialities of depth profiling with PIXE published in paper [61] led authors to the conclusion that, with a properly chosen algorithm for evaluation, the method became quite general and reliable.

Spectrum analysis

The analytical information on the presence and concentration of separate elements in a sample is contained in a complex spectrum of characteristic X-ray peaks sitting on an energy dependent continuum (see e.g. Figs. 5.a and b). Qualitative analysis can be made by assigning atomic numbers to the single peaks or groups of peaks, while quantitative data on concentrations can be obtained from Eqs. (1), (2), (3), (4) or (5) on the basis of Y(Z) values to be deduced from the spectra. The deconvolution of complex PIXE spectra can be made by the use of computer codes based on least square fitting procedure optimizing all the parameters present in the fitting model. A number of software packages for PIXE analysis find widespread application in practical work [35,62,63,64,65]. A recent compilation of PIXE spectrum processing programs carefully tested in intercomparison runs was published in paper [66]. Some of the codes yield corrected Y(Z) peak areae only, for data reduction to elemental masses or concentrations additional codes are needed. The modification aiming at increased flexibility and improved accuracy is going on in most cases.

The typical functions performed by a spectrum processing program can be listed as follows:

- the energy calibration of the spectrum is made on the basis of the energies of two reference X-ray peaks of known energy, deliberately selected from among the strongest known peaks in the spectrum,

- calibration parameters are determined from the measured width of a reference peak to be used for calculating peak widths for any peak energy,

319

- background parameters to be used in the selected background function during spectrum deconvolution are obtained from fitting to a simple spectrum,

- energies and relative intensities are stored in a library file for different X-ray lines of each element covered by the analysis,

- ratios between different X-ray lines of any single element are corrected for the change caused by energy dependent filtering effect of the absorbers present,

- approximate centroid locations, heights and widths are stored for each peak seen in the spectrum,

- Gaussian shapes are used for the description of peaks and correction is performed for the deviation from this simplified shape,

- corrections are made for escape peaks and pile-up peaks appearing in the spectrum,

- least square fitting is performed for optimizing the parameters and deducing net peak areae for all the lines emitted by the sample,

- analytical procedures are performed following any of the above treated calibration techniques to deduce quantitative analytical information (absolute mass or concentration) from peak areae.

Analytical capabilities of PIXE

Criteria for the evaluation of the analytical capabilities of a technique and for a comparison with other methods of instrumental analysis can be given in terms of a number of parameters. They also may help one in choosing the most appropriate method for solving a given analytical task. The most important quantitative parameters in this respect can be listed as follows:

- the sensitivity, defined as the slope of line intensity versus elemental concentration, normally expressed in units (counts/time)/($\mu g/cm^2$),

- detection limit, defined as the minimum concentration at which signal above background equals to three times the standard deviation of the background for a given unit of time, normally expressed in units $\mu g/g$,

- accuracy, expressed by the deviation of the deduced concentration from the true value,

- precision, measured by the scatter of concentration data taken on a number of identical samples.

Other important features of a method are speed, selectivity, range of applicability, forms of the samples requested by the method, non-destructive character, etc. These parameters and features change from element to element even within the same technique and depend on the character of the samples, too. Therefore a general decision on the merits of a selected technique can only be made on the basis of careful systematic exploration.

Papers [2,3] gave a detailed treatment of these questions and summarized the results of early investigations. Tests on the accuracy typical for PIXE were performed by a number of authors in measurements on carefully prepared standard samples [66,67,68]. Some works were devoted to comparative meas-

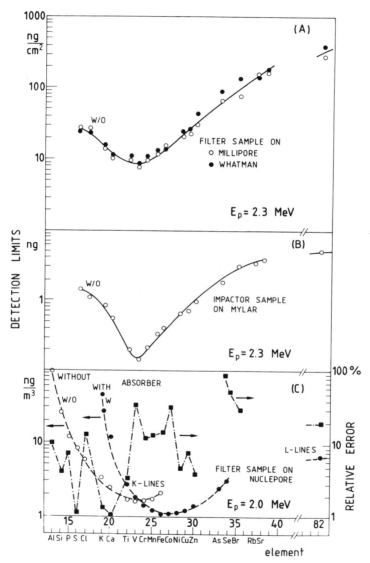

Fig. 8. Thin target detection limits and relative errors deduced from measurements [73,75] on atmospheric particulate spectrum. Curves W/O and W show the effect of Al-absorber (37.2 μm) on detection limits

urements, where parallel analyses were performed with PIXE and other instrumental methods on samples of unknown composition [69]. Among others, PIXE was compared with photon induced X-ray emission analysis in terms of sensitivity and detection limit [70,71,72]. Attainable precision was investigated in serial measurements on sets of identical samples [73].

The tests performed on thin multielemental samples composed of air pollution particulates represent an important special case. On one hand, the strict validity of Eq. (1), i.e. the absence of any matrix effects here offers a clean picture on basic characteristics of the method. On the other hand, PIXE finds a very important application in the analysis of air particulate matter, therefore a careful characterization of the method in this field is of high practical importance. Representative data taken from papers [73] and [75] are shown in Fig. 8 where detection limits expressed in areal density and absolute mass units µg and µg/m^3-air are presented as the function of atomic number for different sample backings. Part A and B were measured [73] without (W/O) additional filtering, while curves W/O and W in part C [75] demonstrate the effect of filtering with an aluminium filter 37.2 µm thick on improving minimum detection limits for higher atomic numbers. The shape of curves W/O with a minimum around $Z = 23$ nicely reflects the qualitative features expressed by the detection limit contours in Fig. 2. From areal density data in part A and measured sample thickness the lowest detectable concentrations of 1-2 ppm by weight can be obtained, which is in accordance with the data calculated in papers [1,2]. Detection limits and statistical error for part C are calculated [35] from measured spectra. Relative errors strongly influenced in the case of some trace elements by their low concentration give a guess on accuracy and precision here.

As shown in paper [34] thin sample calibration constants deduced in a careful experimental and evaluation procedure reach the estimated accuracy of 2-3 % between $22 \leq Z \leq 30$ and 5 % for $11 \leq Z \leq 20$, while analytical accuracy and precision determined on actual thin samples are influenced by sampling technique and concentrations of different constituents.

As it follows from the definition, the detection limit depends on the total number of counts acquired in the spectrum, through increased statistical accuracy in longer runs. Therefore minimum detection limits and speed of the method are correlated parameters. To give a typical example, curves in parts A and B were taken with total number of counts 500 000 during a measuring time of 1000 sec.

Good selectivity of PIXE is disturbed for some pairs of elements by interference between K and L or M lines, emitted with energies close to each other from the lighter and heavier members of the pairs, respectively. An example of practical importance is given by the interference between sulphur K X-rays and lead M X-rays. However, known intensity ratios of K, L and M lines, taken into consideration in unfolding spectra for all the elements within the energy window of the detector make the resolution of the method reliable in most cases [76].

An important contribution to the determination of general features in thick target PIXE is contained by paper [67]. The calibration curves for different rare earth analytes were found to be linear for concentrations lower than 0.3 weight percent down to the detection limits amounting to 0.5-5 ppm, with a precision between runs around 3-5 %. The range of applicability without matrix corrections is claimed to be between 2 % and detection limit. Within this interval no distortion was caused by specific interelement effects from other rare elements present with concentrations below 2 %, even in the case when the concentration of additional constituents exceeds the concentration of analyte by a factor of 100. Significant interelement effects are expected at concentrations higher than 5 %.

PIXE method is principally non-destructive. During the analytical process samples are only subjected to a moderate heat load equivalent to the loss of beam energy in the sample. With bombarding intensity kept on the conventional 1-10 nA level this load does not cause an appreciable modification of the state and composition of the sample in most cases. In external beam version objects to be analyzed can be mostly put into analyzing position in their original form, however, depending on the task of the analyst the need of preparing samples of appropriate form and state may appear. In internal beam mode the additional requirements of sample stability against more intense warm-up and other vacuum-induced effects should be taken into consideration. Due to the physical simplicity of the case of differentially thin targets it is advisable in many cases to prepare samples in such form.

Special care should be taken of the basic rules governing the total error introduced by the entire sampling operation [77] aiming at the preparation of representative samples.

Because of the wide-spread application of PIXE in many interdisciplinary fields a broad variety of sample forms and sample preparation techniques have been developed. A critical compilation of the sample preparation methods specially developed for X-ray emission elemental analysis has been published in the work [77].

The inherent detection limits determined by the physical rules outlined above can be improved if sample preparation is combined with preconcentration methods aiming at the increase of concentrations of analytes in the target with respect to those in the original samples. A complete review and critical evaluation of preconcentration procedures published for the case of water samples is given in paper [78].

INTERDISCIPLINARY APPLICATIONS

Proton Induced X-ray Emission method, built up on the basic physical rules and technical tools outlined in the previous parts of the paper became, during its relatively short history, a well developed versatile and useful method of instrumental analytics. While the improvement of the method is still going on to meet all the requirements of accurate routine analysis of large series of samples, it finds a great variety of applications in many different fields of science and practice.

The increase of activity in development and application is well documented by the rich material on International Conferences on Particle Induced X-ray Emission and its Analytical Applications [79,80,81,82]. On the other hand, results based on analyses performed with PIXE method are being published in an increasing number of papers appearing in journals of the respective fields of science, too. Therefore, the only realistic aim within the limited frame of the present paper is to give a few examples arbitraryly selected from the huge amount of information to illustrate the broad scope of applications in some of the most important fields.

Atmospheric aerosols and environmental samples

The increasing pollution of air by the emission in heavily industrialized areae called the attention of scientists working in the field of environmental research to the importance of measuring the composition and particle size distribution of atmospheric aerosols. Ecological and toxical effects are mainly governed by the concentrations of different components, while the atmospheric residence time and transport distances from the source to the points of deposition depend on the size distribution of the particulate matter.

Both natural and antropogenic sources contribute to the origin of aerosols. The first component appears as a result of erosion of soils and spraying sea water. In order to be able to separate man-made component accurately, the determination of natural atmospheric level of aerosol constituents is of basic importance. Therefore a world-wide activity has been set up to make multielemental analysis on atmospheric samples either containing all sizes of particulate matter or separated into particle size classes. A comparison of the composition of samples with those from most abundant crustal rocks and sea water may facilitate the separation of the components.

The same measurements can be used for investigating physical processes of the atmosphere itself. Long range transport of air masses and its influence on aerosols of remote areae, aging of pollutants during transport, tracing air masses through characteristic trace elements, meteorological effects are the most important questions to be investigated here.

Health related aerosol investigations are mainly performed in work environment, deposition of aerosols in respiratory tract is an important research field in this respect, too.

Analytic capabilities of PIXE are well fitted to aerosol research, where multielemental analyses with low detection limits are requested to be performed on thin samples of small absolute mass. The characterisation of individual airborn particles is also possible by the use of microbeam facilities. The great variety of topics investigated by the PIXE method is illustrated by a list of fields and selected references given in Table 1.

Sampling for aerosol analysis with PIXE method is fully based on physical processes. Accurately measured air volume is pumped through filtering or inertial equipments [113] designed for producing thin layers of deposited aerosols containing particulates without any discrimination in their size or separated according to their aerodynamic diameters. Integral and size resolved sampling can be combined with time resolved collecting of the particles in order to obtain time resolved integral samples or time sequence samples separated into particle size fractions. For integral sampling high purity fine pore filters (preferably Nuclepore) are used in filter holders. Size fractionation is obtained by inertial separation of the different components. In inertial spectrometers [114] a continuous variation of the size of deposite particulates is achieved along the longitudinal coordinate on a stripe of filter - a size fractionated filtering is performed by the sampler. In inertial cascade impactors [111,115] air flows through a number of classification stages in series, with higher velocities in each successive stage. Each stage is transparent for particles travelling with the air flow if their diameter is smaller than a critical value defined by the aerodynamic parameters of the stage. Particles not fulfiling the condition will impact on an impaction plate contained by the stage and serving as sample backing in PIXE analysis. In such a way particles will get fractioned into size intervals, one sample will be built up in each stage. Time sequence filtering is made by a streamer in which the air flowing through an inlet channel deposits particulates to a surface element of a filter disc. With the slow continuous or stepped rotation of the disc around an axis parallel to the air flow the time variation of aerosol composition will appear as a variation of the composition of deposited layer

Table 1. PIXE method in aerosol research

MAIN FIELDS OF APPLICATION	REFERENCES
NATURAL COMPONENT	
- soil derived and sea spray components	83,84
- aerosols in rural and remote areae	85,86
- dust storm events	84,87
- volcanic activity	84,88,89
ANTROPOGENIC COMPONENT	
- separation of antropogenic and natural components, by enrichment factors	75
- global and regional monitoring air quality programs	90,91,92
- pollution sources, polluted air	87
- source identification	93
- biomass burning	83,87
- dust particulates and exhaust by car traffic	94,95
- coal fly ash and particulate emission from thermal power stations	96
- urban aerosols	
- characterisation of individual airborn particles	28,97
- time dependent and sequential measurements on aerosols	87,98,99
TRANSPORT OF AEROSOLS	
- long range transport of air masses	100,101,102
- ageing and transformation of antropogenic pollutants during transport	103
- regional signatures as tracers in long range transport, local and distant effects	75,104,105,106
- regional concentrations and meteorology	87,107
MEDICAL PROBLEMS	
- work environmental aerosols, occupational safety and health	108,109
- in-plant aerosols, mining environment	110
- deposition and retention of dust in the respiratory tract	111,112

with the angular coordinate of the deposited spots. Multistage streamers, rotating impactors and DRUM samplers [116] combine inertial separation in impactors with rotating impaction surfaces for time resolving.

Biomedical research

The variation in the elemental concentration of consituents of blood components, tissues, body fluids and hair taken from humans are the subject of detailed investigation in the present days related to a number of physiological and pathological processes. On the basis of the results obtained in these investigations one can expect that in the near future the determination and control of trace element concentrations in human organism will play an important role in the everyday praxis of diagnosis and therapy of human diseases.

Constituents of the living material are usually divided into bulk elements, essential trace elements and toxic elements. The group of bulk elements is composed of H, C, N, O, Na, Mg, P, S, Cl, K, Ca, while elements F, Si, V, Cr, Mn, Fe, Co, Ni, Cu, Zn, Se, Mo, Sn, and I are considered as essential trace elements. Some authors claim the importance of other trace elements, too. Trace elements are essential for life because of their role as key components of enzyme systems and of proteins. For man and domestic animals there are very narrow permitted concentriation intervals of the different trace elements necessary to the normal life functions. Both deficiences and excesses may result in a number of disorders. This is why trace element analysis became highly important as a research and diagnostic tool in medicine. Few other elements such as Be, Hg, Pb, U and transuran elements are known to be toxic even at extremely low concentration.

Except for the light bulk elements H, C, N, O, Na and toxic element Be all the other constituents of living material can be analyzed with about the necessary detection limits by the help of PIXE method. Microbeam facilities also offer the possibility of detecting local effects in microscale down to the size of individuals cells.

According to the diversity of biomedical problems to be investigated a number of special samples can be used in PIXE analyses. Three important groups are formed by blood tissue and hair samples.

A strong variety in the concentration of the constituent elements of blood samples taken from different individuals makes trace element analysis of blood of central importance in view of its possible use in diagnostics. Comparative measurements performed on whole blood as well as on blood serum and plasma of the same individual makes diagnostic more specific and permit to clear up the mechanism of different processes resulting in changes of blood composition.

Determination of the concentrations of elemental constituents in human and animal tissues are of special importance because of the well established effect of the accumulation of some elements by different organs in malignant diseases and other disorders. E.g. abnormal concentrations appear in liver tissue samples in the case of deteriorated liver functions, deviation can be found in spleen and liver in diabetic patients.

Hair analysis is considered to be a unique tool of diagnostic value in medicine. Beside the typical bulk constituents more than twenty trace elements can be identified with concentrations between 0.1 and 100 ppm in hair samples by present day techniques. Fine resolution measurements on the elemental distribution along hair samples to be achieved by shifting hair samples in small longitudinal steps accross the fine proton beam typical for PIXE measurements may have a high additional value in getting information on the history of medical events. Length variation of the elemental composition are correlated with the metabolic changes of the respective elements in the past. Such a way environmental effects, the course of poisoning by heavy metal, changes in nutritional conditions and physiological changes in the patient can easily be followed in the function of time.

An illustrative list of fields and selected references is given in Table 2.

Table 2. PIXE method in biomedical research

MAIN FIELDS OF APPLICATION	REFERENCES
ENVIRONMENTAL FACTORS	
- nutrition; excess and deficiences caused by dietic uptake, malnutrition	117,118
- environmental carcinogenic factor, role of selenium in cancer morbidity rate	119
- toxicology	
- effects of aerosols on respiratory tract	111,112,120,121
EXCESS OR DEPLETION OF ELEMENTS CORRELATED WITH PATHOLOGICAL CONDITIONS	
- trace elements in neurologic disease	122
- regional distribution in human brain	123
- arteriosclerosis, elements in arteries	124,125
- changes and localised effects in liver necrosis and cirrhosis	123,126,127
- diabetes and changes in trace elements	128
- infectious diseases and trace elements	
- renal insufficiency, variation during hemodialysis	
- diagnostic use of elemental correlations	129
CANCER DIAGNOSTICS AND THERAPY	
- trace elements in cancer tissues	130,131,132
- elemental profile in malignant tissues and erythrocytes	133
- variation of concentrations in radio- and chemotherapy	134
- diagnostics and therapy	135,136
- hematological malignancies	132
GYNECOLOGY AND OBSTETRICS	
- elemental concentrations during pregnancy	128,137,138
- nutritional and toxic effects on developing fetus, immature and low weight birth	118
- maternal-fetal interaction through placenta	118
- diabetic pregnancies	128
- elements in colastrum and breast milk	118,139
- trace elements in hair during pregnancy	118
ODONTOLOGY	
- structure of tooth enamel and cementum	140,141,142
HAIR ANALYSIS	
- history of events from longitudinal scanning	143
- trace element migration from lateral scanning	144,145
- trace elements in hair correlated with carcinoma	146
- reiiability of hair analysis	143,147

Table 3. PIXE method in other research fields

MAIN FIELDS OF APPLICATION	REFERENCES
MATERIAL SCIENCE, MATERIAL TESTING	
- surface layers on silicon substrates	148, 149
- surface oxid layer	59,
- depth profiling by PIXE	60, 61
- location of foreign atoms in host lattice from PIXE and channeling	150, 151
- composition of alloys	58, 152
- sputtering yield measurements	
- impurities deposited from plasma discharge in nuclear fusion device	153, 154
- analysis of insulating glass	155
- analysis of lubricating oil	156
ENVIRONMENTAL RESEARCH	
- trace elements in water samples	157, 158
- suspended particulates in river, lake and sea	159, 160, 161
- emission from thermal power station studied in tree rings	162
- trace elements in coal	163
ARCHEOLOGICAL RESEARCH	
- historical documents studied by PIXE on paper and inks	164, 165
- provenance study of pottery and obsidian artefacts	166, 167
- color additives in glass	168
- elemental analysis of coins	169
- forgery of jewels detected by elemental concentrations	170
- bones from mummy analyzed by PIXE	171
GEOLOGICAL RESEARCH	
- major and trace elements in mineral samples	172, 173
- formation of rocks studied by elemental analysis	
- solar system studied by the analysis on meteorites and cosmic dust	174, 175
- trace elements in lunar rocks	176
MISCELLANEOUS APPLICATIONS	
- forensic science	177

Due to the diversity of possible samples in biomedical analysis a number of physical and chemical preparation methods are to be used aming at homogenisation, pulverisation and preconcentration of the original samples. Ashing procedures may be applied for getting rid of bulk elements with the purpose of ending up with samples of low background continuum. Special care should be taken of avoiding contaminations and loss of elements during sampling and preparation. Details on sampling and preparation technique are treated in paper [77].

Other applications

While the overwhelming majority of PIXE applications is being concentrated on aerosol and biomedical research the method finds increasing interest in a number of other fields, too. Without trying to give a systematic description of all the directions in which PIXE contributes with new possibilities expanding the potentialities of conventional methods of the disciplines, we illustrate the diversity of successful applications with the list given in Table 3.

From the point of view of analytical procedure the application of thick targets is characteristic for the above applications except for the investigation on water samples. It also means, that the introduction of the method to the regular use on a set of similar samples demands the determination of thick target calibration factors including all special corrections for matrix effect. In many cases use is made of the fact, that PIXE can be performed on the material to be investigated in its original form (in material science and archeology) without the need for any sample preparation. Non-destructive character of the method permits the application on precious archeological objects. The application of microbeam facilities may reveal special structural effects on a microscopic scale.

SUMMARY

Proton Induced X-ray Emission method is developed to a level, where basic processes are well understood, instrumental questions are mainly solved, reliable methods are available for calibration and spectrum evaluation. A number of PIXE groups are active in accelerator laboratories all over the world.

With the proper evaluation of the analytical potentialities of the method and with the application of fast and reliable sample preparation techniques well fitted to the selected field of application PIXE finds its own role in the competition with present day methods of instrumental analytics and contributes with new results to the interdisciplinary research based on sensitive multielemental determination of elemental constituents in a number of fields.

REFERENCES

Items marked with an asterisk are Conference Proceedings and review articles containing a number of individual papers and extensive bibliographies, respectively.

1.* J. Folkmann, J. Phys. E: Sci. Instr. 8:429 (1975)
2.* S. A. E. Johansson and T. B. Johansson, Nucl. Inst. and Meth. 137:473 (1976)
3. J. Leroux and T. P. Think; "Revised Tables of X-Ray Mass Attenuation Coefficients", Corporation Scientifique Claisse, Inc., Quebec, (1977)
4. Silicon (Li) X-Ray Detectors in "EG&G ORTEC Instruments for Research and Applied Science" Catalog (1979)
5. R. K. Gardner and T. J. Gray, Atomic Data and Nuclear Data Tables 21:515 (1978)
6. T. L. Hardt and R. L. Watson, Atomic Data and Nuclear Data Tables 17:107 (1976)
7. K. Ishii and S. Morita, Nucl. Instr. Meth. B3:57 (1984)
8. M. Ahlberg, G. Johansson and K. Malmquist, Nucl. Instr. Meth. 131:377 (1975)

9. I. Borbély-Kiss and Gy. Szabó: private communication
10. K. G. Malmqvist, E. Karlsson and K. R. Akselsson, Nucl. Instr. Meth. 192:523 (1982)
11. X. Zeng, and X. Li, Nucl. Instr. Meth. B22:99 (1987)
12. A. Katsanos, A. Xenoulis, A. Hadjiantoniou and R. W. Fink, Nucl. Instr. Meth. 137:119 (1976)
13. A.H. Khan, M. Khaliquzzaman, M. Husain, M. Abdullah and A. A. Katsanos, Nucl. Instr. Meth. 165:253 (1979)
14. J. Räisänen and A. Anttila, Nucl. Instr. Meth. 196:489 (1982)
15. A. Anttila, J. Räisänen and R. Lappalainen, Nucl. Instr. Meth. B12:245 (1985)
16. B. Hietel, F. Schulz and K. Wittmaack, Nucl. Instr. Meth. B3:343 (1984)
17.* D. Vaughan, ed. "Energy dispersive X-ray Microanalysis. An Introduction". KeVex Corporation, Foster City, California, 1983
18.* F. Bosch, A. El Goresy, W. Herth, B. Martin, R. Nobiling, B. Povh, H. D. Reiss and K. Traxel, Nucl. Sci. Appl. 1:1 (1980)
19. H. Koyama-Ito and L. Grodzins, Nucl. Instr. Meth. 174:331 (1980)
20. P. W. Hawkes, "Quadrupoles in electron lens design", Academic Press, New York, London 1970
21. A. D. Dymnikov, T. Fishkova, and S. Ya, Yavor, Sov. Phys. - Techn. Phys. 10:340 (1965)
22. J. A. Cookson and F. D. Pilling, Report AERE - R 6300 (1970)
23. F. Watt, G. W. Grime, G. D. Blower and J. Takács, IEEE Tr. NS-28, No.2:1413 (1981)
24. G. Bonani, M. Suter, H. Jung, C. Stroller and W. Wölfli, Nucl. Instr. Meth. 157:55 (1978)
25. R. Nobiling, K. Traxel, F. Bosch, Y. Civelekoglu, B. Martin, B. Povh and D. Schwalm, Nucl. Instr. Meth. 142:49 (1977)
26. W. M. Augustyniak, D. Betteridge and W. L. Brown, Nucl. Instr. Meth. 149:665 (1978)
27.* T. A. Cahill, Ann. Rev. Nucl. Part. Sci. 30:211 (1980)
28.* K. Traxel and U. Wätjen, Particle-induced X-ray emission analysis (PIXE) of aerosols in "Physical and chemical characterisation of individual airborne particles", K. R. Spurny, ed., Ellis Horwood Limited Publishers, Chichester, 1986
29.* K. Traxel, Nucl. Instr. Meth., to be published
30. G. W. Grime, J. Takács and F. Watt, Nucl. Instr. Meth. B3:589 (1984)
31. E. Koltay and Gy. Szabó, Nucl. Instr. Meth. 35:88 (1965)
32. G. I. Johansson, J. Pallon, K. G. Malmqvist and K. R. Akselsson, Nucl. Instr. Meth. 181:81 (1981)
33. K. R. Akselsson, S. A. E. Johansson and T. B. Johansson, in "X-Ray Fluorescence Analysis of Environmental Samples", T. G. Dzubay, ed., Ann Arbor Science Publ. Inc., Ann Arbor, 1977. p.175
34. W. Maenhaut and M. Raemdonck, Nucl. Instr. Meth. B1:123 (1984)
35. I. Borbély-Kiss, E. Koltay, S. László, Gy. Szabó and L. Zolnai, Nucl. Instr. Meth. B12:496 (1985)
36. J. L. Campbell and J. A. Cookson, Nucl. Instr. Meth. B3:185 (1984)
37. G. Basbas, W. Brandt and R. Laubert, Phys. Rev. A17:1655 (1978)
38. W. Brandt and G. Lapicki, Phys. Rev. A20:465 (1979)
39. W. Brandt and G. Lapicki, Phys. Rev. A23:1717 (1981)
40. O. Benka and A. Kropf, At. Data and Nucl. Data Tables 22:219 (1978)
41. J. S. Lopes, A. P. Jesus and S. C. Ramos, Nucl. Instr. Meth. 164:219 (1978)
42. A. P. Jesus, T. M. Pinheiro, I. A. Niza, J. P. Ribeiro and J. S. Lopes, Nucl. Instr. Meth. B15:595 (1986)
43. M. O. Krause, J. Phys. Chem. Ref. Data 8:307 (1979)
44. I. H. Scofield, At. Data and Nucl. Data Tables 14:121 (1974)
45. Md. R. Khan and M. Karimi, X-Ray Spectrometry 9:32 (1980)
46. J. Pálinkás and B. Schlenk, Nucl. Instr. Meth. 169:493 (1980)
47. E. C. Montenegro, A. Oliver, F. Aldape, Nucl. Instr. Meth. B12:453 (1985)

48. V. McMaster, M. Delgrande, J. Mallet and J. Hubbel, University of California Report UCRL-50174 (1969)
49. E. C. Montenegro, G. B. Baptista and P. W. E. P. Duarte, Atomic Data and Nuclear Data Tables 22:131 (1978)
50. M. R. Khan, Appl. Phys. Lett. 33:676 (1978)
51. H. C. Kaufmann and J. Steenblik, Nucl. Instr. Meth. B3:198 (1984)
52. W. Reuter, A. Luriox, F. Cardone and J. F. Ziegler, J. Appl. Phys. 46:3194 (1975)
53. B. Van Oystaeyen and G. Demortier, Nucl. Instr. Meth. 215:299 (1983)
54. F. W. Richter and U. Wätjen, Nucl. Instr. Meth. 181:189 (1981)
55. Z. Smit, M. Budnar, V. Cindro, M. Ravnikar and V. Ramsak, Nucl. Instr. Meth. B4:114 (1984)
56. Z. Smit, M. Budnar, V. Cindro, M. Ravnikar and V. Ramsak, Nucl. Instr. Meth. 228:482 (1985)
57. M. Ahlenbery and R. Akselsson, "Proton Induced X-Ray Emission in the Trace Analysis of Human Tooth Emanel and Dentine", LUNP 7512 (1975)
58. M. A. Respaldiza, G. Madurga and J. C. Soares, Nucl. Instr. Meth. B22:446 (1987)
59. J. Rickards, Nucl. Instr. Meth. B12:269 (1985)
60. J. Végh, D. Berényi, E. Koltay, I. Kiss, S. Seif El-Nasr and L. Sarkadi, Nucl. Instr. Meth. 153:553 (1978)
61. I. Brissaud, J. P. Frontier and P. Regnier, Nucl. Instr. Meth. B12:235 (1985)
62. P. Van Espen, H. Nullens and F. Adams, Nucl. Instr. Meth. 145:579 (1977)
63. G. I. Johansson, X-Ray Spectrom. 11:194 (1982)
64. E. Clayton, Nucl. Instr. Meth. 218:541 (1983)
65. E. Bombelka, W. Koenig, F. W. Richter and U. Wätjen, Nucl. Instr. Meth. B22:21 (1987)
66. U. Wätjen, Nucl. Instr. Meth. B22:29 (1987)
67. R. P. H. Garten, K. O. Groeneveld and K. H. Koenig, Fresenius Z. Anal. Chem. 307:97 (1981)
68. L. E. Carlsson, Nucl. Instr. Meth. B3:206 (1984)
69. P. S. Z. Rogers, C. J. Duffy and T. M. Benjamin, Nucl. Instr. Meth. B22:133 (1987)
70. E. Bombelka, F. W. Richter, H. Ries and U. Wätjen, Nucl. Instr. Meth. B3:296 (1984)
71. J. A. Cooper, Nucl. Instr. Meth. 106:525 (1973)
72. J. V. Gilfrich, P. G. Burkhalter and L. S. Birks, Anal. Chem. 45:2002 (1973)
73. M. S. Ahlberg and F. C. Adams, X-Ray Spectrom. 7:33 (1978)
74. V. Valkovic, R. B. Liebert, T. Zabel, H. T. Larson, D. Miljanic, R. M. Wheeler and G. C. Phillips, Nucl. Instr. Meth. 114:573 (1975)
75. I. Borbély-Kiss, L. Haszpra, E. Koltay, S. László, Á. Mészáros, Gy. Szabó and L. Zolnai, Physica Scripta, in press
76. Xin-Pei Ma, G. R. Palmer and J. D. MacArthur, Nucl. Instr. Meth. B22:49 (1987)
77.* V. Valkovic, "Sample preparation techniques in trace element analysis by X-ray emission spectroscopy". IAEA-TECDOC-300. International Atomic Energy Agency, Vienna, (1983)
78.* R. Van Grieken, Analytica Chimica Acta 143:3 (1982)
79.* S. A. E. Johansson (ed.), Proc. 1st Int. Conf. on Particle Induced X-ray Emission and its Analytical Applications, Nucl. Instr. Meth. 142:1 (1977)
80.* S. A. E. Johansson (ed.), Proc. 2nd Int. Conf. on Particle Induced X-Ray Emission and its Analytical Applications, Nucl. Instr. Meth. 181:1 (1981)
81.* B. Martin (ed.), Proc. 3rd Int. Conf. on PIXE and its Analytical Applications, Nucl. Instr. Meth. B3:1 (1984)

82.* H. H. Andersen and S. T. Picraux (ed.), Proc. 4th Int. Conf. on Proton Induced X-Ray Emission and its Analytical Applications, Nucl. Instr. Meth. B22:1 (1987)
83.* J. W. Winchester, Nucl. Instr. Meth. 142:85 (1977)
84.* J. W. Winchester, Nucl. Instr. Meth. 181:367 (1981)
85. R. E. Van Grieken, T. B. Johansson, K. R. Akselsson, J. W. Winchester, J. W. Nelson and K. R. Chapman, Atmospheric Environment 10:571 (1976)
86. F. Adams, M. Van Craen, P. Van Espen, D. Andreuzzi, Atmospheric Environment 14:879 (1980)
87.* J. W. Winchester, Nucl. Instr. Meth. B3:455 (1984)
88. M. Darzi, Nucl. Instr. Meth. 181:359 (1981)
89. J. P. Quisefit, G. Robaye, P. Aloupogiannis, J. M. Delbrouk Habaru and I. Roelandts, Nucl. Instr. Meth. B22:301 (1987)
90. C. Q. Orsini, P. A. Netto and M. H. Tabacnicks, Nucl. Instr. Meth. B3:462 (1984)
91. R. A. Eldred, T. A. Cahill and P. J. Feeney, Nucl. Instr. Meth. B22:289 (1987)
92. B. A. Bodhaine, J. J. Deluisi, J. C. Harris, P. Houmere and S. Bauman, Nucl. Instr. Meth. B22:241 (1987)
93. E. Svietlicki, H. C. Hansson and B. G. Martinsson, Nucl. Instr. Meth. B22:264 (1987)
94. M. Öblad and E. Selin, Physica Scripta 32:462 (1985)
95. S. Amemiya, Y. Tsurita, T. Masuda, A. Asawa, K. Tanaka, T. Katoh, M. Mohri and T. Yamashina, Nucl. Instr. Meth. B3:516 (1984)
96. E. S. Gladney, J. A. Small, G. E. Gordon and W. H. Zoller, Atmospheric Environment 10:1071 (1976)
97.* K. R. Spurny, (ed.), Physical and Chemical Characterisation of Individual Airborn Particles, Ellis Horwood Ltd., Chichester (1986)
98.* H. J. Annegarn, T. A. Cahill, J. P. F. Sellschop, A. Zucchiatti, Physica Scripta, in press
99. P. Metternich, R. Latz, J. Schader, H. W. Georgii, K. O. Groeneveld and A. Wensel, Nucl. Instr. Meth. 181:431 (1981)
100. H. O. Lannefors, T. B. Johansson, L. Granet and B. Rudell, Nucl. Instr. Meth. 142:105 (1977)
101. P. Metternich, H. W. Georgii and K. O. Groeneveld, Nucl. Instr. Meth. B3:475 (1984)
102. M. Öblad and E. Selin, Atmospheric Environment 20:1419 (1986)
103. R. A. Eldred, T. A. Cahill, L. L. Ashbaugh and J. S. Nasstrom, Nucl. Instr. Meth. B3:479 (1984)
104. K. A. Rahn, Atmospheric Environment 15:1457 (1981)
105. K. A. Rahn and D. H. Loewenthal, Science 23:132 (1984)
106. H. Lannefors, H. C. Hansson and L. Granat, Atmospheric Environment 17:87 (1983)
107. S. E. Bauman, R. Ferek, E. T. Williams and H. L. Finston, Nucl. Instr. Meth. 181:411 (1981)
108. K. G. Malmqvist, G. I. Johansson, M. Bohgard and K. R. Akselsson, Nucl. Instr. Meth. 181:465 (1981)
109. K. G. Malmqvist, Nucl. Instr. Meth. B3:529 (1984)
110. H. J. Annegarn, A. Zucchiatti, J. P. L. Sellshop and P. Booth-Jones, Nucl. Instr. Meth. B22:325 (1987)
111.* Particle Size Analysis in Estimating the Significance of Airborne Contamination, International Atomic Energy Agency, Vienna (1978)
112.* J. Heyder, Physica Scripta, in press
113.* D. Y. H. Pui, Physica Scripta, in press
114.* K. R. Akselsson, Nucl. Instr. Meth. B3:425 (1984)
115. V. A. Marple and K. Willecke, Atmospheric Environment 10:891 (1976)
116. H. J. Annegarn, T. A. Cahill, J. P. F. Sellschop and A. Zucchiatti, Physica Scripta, in press
117. M. Barrette, G. Lamoureux, R. Lecomte, P. Paradis, S. Monaro and H. A. Menard, J. Radioanal. Chem. 52:153 (1979)

118. G. S. Hall, N. Roach, M. Naumann and U. Simmons, Nucl. Instr. Meth. B3:332 (1984)
119.* V. Valkovic, Nucl. Instr. Meth. 142:151 (1977)
120. R. L. Walter, R. D. Willis, W. F. Gutknecht and R. W. Shaw, Nucl. Instr. Meth. 142:181 (1977)
121. J. W. Winchester, D. L. Jones and Mu-tian Bi, Nucl. Instr. Meth. B3:360 (1984)
122. H. A. Van Rinsvelt, R. W. Hurd, J. W. A. Kondoro, J. M. Andres, J. P. Mickle, B. J. Wilder, W. Meanhaut and L. De Reu, Nucl. Instr. Meth. B3:377 (1984)
123. W. Meanhaut, J. Vandenhaute, H. Duflou and J. De Reuck, Nucl. Instr. Meth. B22:138 (1987)
124. M. Simonoff, Y. Llabador, G. N. Simonoff, P. Besse and C. Conri, Nucl. Instr. Meth. B3:368 (1984)
125. T. Cichocki, D. Heck, L. Jarczyk, E. Rokita, A. Strzalkowski and M. Sych, Nucl. Instr. Meth. B22:210 (1987)
126. D. Heck, A. Ochs, A. Klempnow, K. P. Maier and C. Kratt, Nucl. Instr. Meth. B22:196 (1987)
127. F. Watt, G. W. Grime, J. Takács and D. J. T. Vaux, Nucl. Instr. Meth. B3:599 (1984)
128. S. Gődény, I. Borbély-Kiss, E. Koltay, S. László and Gy. Szabó, Intnat. J. of Gynecology and Obstetrics 24:201 (1986)
129. H. A. Van Rinsvelt, R. D. Lear and W. R. Adams, Nucl. Instr. Meth. 142:171 (1977)
130. E. Johansson, U. Lindh, H. Johansson and C. Sundström, Nucl. Instr. Meth. B22:179 (1987)
131. M. Uda, K. Meada, Y. Sasa, H. Kusuyama and Y. Yokode, Nucl. Instr. Meth. B22:184 (1987)
132. G. Weber, G. Robaye, B. Bartsch, A. Collignon, Y. Beguin, I. Roelandts and J. M. Delbrouck, Nucl. Instr. Meth. B3:326 (1984)
133. E. Johansson and U. Lindh, Nucl. Instr. Meth. B3:637 (1984)
134. É. Pintye, Z. Dézsi, L. Miltényi, I. Kiss, E. Koltay, Gy. Szabó and S. László, Strahlentherapie 158:739 (1982)
135. P. J. Chang, C. S. Yang, M. J. Chou, C. C. Wei, C. C. Hsu and C. Y. Wang, Nucl. Instr. Meth. B3:388 (1984)
136. B. Hietel, F. Schulz and K. Wittmaack, Nucl. Instr. Meth. B3:343 (1984)
137. I. Borbély-Kiss, E. Koltay, S. László, Gy. Szabó, S. Gődény, and S. Seif El-Nassr, J. Radioanal. and Nucl. Chem. 83/1:175 (1984)
138. S. Gődény, I. Borbély-Kiss, E. Koltay, S. László and Gy. Szabó, Intnat. J. of Gynecology and Obstetrics 24:191 (1986)
139. P. M. Parr, IAEA Bulletin, 25; No.2:7 (1983)
140. H. J. Annegarn, A. Jodaikin, P. E. Cleaton-Jones, J. R. F. Sellschop, C. C. P. Madiba and D. Bibby, Nucl. Instr. Meth. 181:323 (1981)
141. M. A. Chaudhri and T. Ainsworth, Nucl. Instr. Meth. 181:333 (1981)
142. M. A. Chaudhri, Nucl. Instr. Meth. B3:643 (1984)
143. J. Bacsó, L. Sarkadi and E. Koltay, Intnat. J. of Appl. Rad. and Isotopes 33:5 (1982)
144. A. J. J. Bos, C. C. A. H. Van det Stap, V. Valkovic, R. D. Vis and H. Verheul, Nucl. Instr. Meth. B3:654 (1984)
145. N. Limic and V. Valkovic, Nucl. Instr. Meth. B22:163 (1987)
146. X. Zeng, H. Yao, M. Mu, J. Yang, Z. Wang, H. Chang and Y. Ye, Nucl. Instr. Meth. B22:172 (1987)
147. V. Valkovic and N. Limic, Nucl. Instr. Meth. B22:159 (1987)
148. R. Mann, C. Bauer, P. Gippner and W. Rudolph, Journal od Radioanal. Chemistry 50:217 (1979)
149. F. M. El-Ashry, M. Goclowski, L. Glowacka, J. Jaskola, J. Marczewski and A. Wolkenberg, Nucl. Instr. Meth. B22:450 (1987)
150. K. H. Ecker, Nucl. Instr. Meth. B3:283 (1984)
151. R. S. Bhattacharya and P. P. Pronko, Appl. Surface Sci. 18:1 (1984)
152. J. Räisänen, Nucl. Instr. Meth. B22:442 (1987)

153. S. Amemiya, A. Asawa, K. Tanaka, Y. Tsurita, T. Masuda, T. Katoh, M. Mohri and T. Yamashina, Nucl. Instr. Meth. B3:549 (1984)
154. B. L. Doyle, L. T. McGarth, A. E. Pontau, Nucl. Instr. Meth. B22:34 (1987)
155. I. Borbély-Kiss, M. Józsa, Á. Z. Kiss, E. Koltay, B. Nyakó, E. Somorjai, Gy. Szabó and S. Seif El-Nasr, Journal of Radioanal. and Nuclear Chemistry 92/2:391 (1985)
156. B. Babinski, M. Goclowski, M. Jaskola, M. Kucharski and L. Zemlo, Nucl. Instr. Meth. 181:523 (1981)
157. K. M. Varier, G. K. Mehta and S. Sen, Nucl. Instr. Meth. 181:217 (1981)
158. R. Cecchi, G. Ghermandi and G. Calvelli, Nucl. Instr. Meth. B22:460 (1987)
159. W. C. Burnett and G. T. Mitchum, Nucl. Instr. Meth. 181:231 (1981)
160. K. Maeda, Y. Sasa, M. Maeda and M. Uda, Nucl. Instr. Meth. B22:456 (1987)
161. J. Kleiner, G. Lindner, E. Recknagel and H. H. Stabel, Nucl. Instr. Meth. B3:553 (1984)
162. M. Nagj, J. Injuk and V. Valkovic, Nucl. Instr. Meth. B22:465 (1987)
163. J. R. Chen, H. Kneis, B. Martin, R. Nobiling, K. Traxel, E. C. T. Chao and J. A. Minkin, Nucl. Instr. Meth. 181:151 (1981)
164. B. H. Kusko and R. N. Schwab, Nucl. Instr. Meth. B22:401 (1987)
165. B. H. Kusko, T. A. Cahill, R. A. Eldred and R. N. Schwab, Nucl. Instr. Meth. B3:689 (1984)
166. P. Fontes, I. Brissaud, G. Lagarde, J. Leblanc, A. Person, J. F. Saliege and Ch. Heitz, Nucl. Instr. Meth. B3:404 (1984)
167. P. Duerden, J. R. Bird, E. Clayton, D. D. Cohen and B. F. Leach, Nucl. Instr. Meth. B3:419 (1984)
168. S. J. Fleming and C. P. Swann, Nucl. Instr. Meth. B22:411 (1987)
169. Z. Smit and P. Kos, Nucl. Instr. Meth. B3:416 (1984)
170. G. Demortier, B. Van Oystaeyen and A. Boullar, Nucl. Instr. Meth. B3:399 (1984)
171. M. Cholewa, W. M. Kwiatek, K. W. Jones, G. Schidlovsky, A. S. Paschoa, S. C. Miller and J. Pecotte, Nucl. Instr. Meth. B22:423 (1987)
172. K. G. Malmqvist, H. Bage, L. E. Carlsson, K. Kristiansson and L. Malmqvist, Nucl. Instr. Meth. B22:386 (1987)
173. L. E. Carlsson and K. R. Akselsson, Nucl. Instr. Meth. 181:531 (1981)
174. D. S. Woolum, D. S. Burnett, T. N. Benjamin, P. S. Z. Rogers, C. J. Duffy and C. J. Maggiore, Nucl. Instr. Meth. B22:376 (1987)
175. R. D. Vis, C. C. A. H. Van der Stap and D. Heymann, Nucl. Instr. Meth. B22:380 (1987)
176. H. Blank, A. El Goresy, J. Janicke, R. Nobiling and K. Traxel, Nucl. Instr. Meth. B3:681 (1984)
177. S. Sen, K. M. Varier, G. K. Mehta, M. S. Rao, P. Sen and N. Panigrahi, Nucl. Instr. Meth. 181:517 (1981)

ELECTRON ENERGY LOSS SPECTROSCOPY IN REFLECTION GEOMETRY

Falko P. Netzer

Institut für Physikalische Chemie
A-6020 Innsbruck, Austria

1. Basic Concepts

The inelastic scattering of electrons provides a powerful technique to study the primary processes of electronic excitation and to obtain information on the electronic structure of materials. The energy loss ΔE suffered by an incident electron originally at energy E_p directly defines a fundamental electronic excitation of the target system. Electron energy loss spectroscopy (EELS) of condensed phases may be carried out either in the transmission mode, in which fast electrons ($50 < E_p < 300$ keV) pass through thin films ($\lesssim 100$ nm) of solids while undergoing mainly bulk excitations, or in reflection mode, in which slower electrons $20 \text{ eV} \lesssim E_p \lesssim 2$ keV suffer both electron loss and elastic backscattering from the surface. Here the effective sampling depth is directly dependent on the primary energy E_p, and both bulk and surface excitations will be present. In the early days of EELS the technique was often called "characteristic electron loss spectroscopy" thereby pointing at the fact that element specific structure was observed in the spectra, whose origin was then still unrevealed.

Inelastic scattering processes involve two main types of electronic excitations: (a) low-energy excitations of valence electrons in the $0 < \Delta E < 30$ eV range, which may be collective in character (plasmons) or one-electron transitions; (b) higher-energy core excitations, which may at least schematically be regarded as one electron excitations.

In many respects electronic excitation by inelastic electron scattering is similar to photon excitation in optical absorption spectroscopy. The similarities are most readily visualised by considering electrons as a virtual radiation source of "white light" character as illustrated in Fig. 1. When a fast electron of velocity v passes an atom at a distance a, an inner core electron experiences a sharply pulsed electric field F(t) in the direction perpendicular to the electron path. Following the arguments of Brion et al.[1] Fourier transforming F(t) gives frequency resolved field components F(ω), which are approximately uniform up to a limiting frequency related to the speed and impact parameter of the incident electron. Over a wide frequency range the core electron may "mistake" the electric field variation that it sees for that of a white radiation source. This is why fast electrons with high impact parameter, and hence low scattering angle, scatter inelastically according to radiative dipole selection rules. In this regime an electron gun may be regarded as a virtual radiation facility in the spirit of a "poor man's synchrotron".

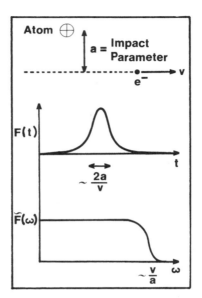

Fig. 1. The electric field sensed by a core electron as a fast electron passes by.

However, electrons are much more than a poor photon substitute. In optical absorption it is impossible to vary independently the energy and the momentum of a photon, since Maxwell's equations fix one of these once the other is determined. Therefore, in energy-momentum space excitations by photon absorption are represented as a single line [2]. In a scattering experiment, on the other hand, the energy and momentum transfer may be varied over a wide range, and an entire region in energy-momentum space becomes accessible. This ability of electron scattering together with superior and constant energy resolution over a wide spectral range in electron loss experiments make EELS one of the most versatile of all spectroscopies for the study of electronic structure. However, it is important to notice that electron impact excitation is not a resonant process. With an electron beam of energy E_p all excitations with energies ΔE up to a substantial fraction of E_p are generated with reasonable cross sections. Electron beam excitation is therefore less specific than resonant photon excitation, unless a coincidence experiment is performed in which both the inelastically scattered and the electron emitted in the deexcitation process are detected. Then, the information content is identical to that of a photon excited experiment [3].

Inelastic electron scattering has been applied for many years to the investigation of electronic properties of gases and solid targets, and several comprehensive reviews have appeared in the recent literature - see e.g. those of Celotta and Huebner [4], Bonham [5], King and Read [6], Trajmar [7], Kuppermann et al. [8], Schnatterly [2] or Raether [9]. EELS in reflection mode to probe the electronic structure of solid surfaces has been somewhat neglected in the past, but review articles of Froitzheim [10], Avouris and Demuth [11] or Netzer et al. [12,13] may provide some guidance for the interested reader.

It is useful to compare electron scattering in atoms and free molecules with the case of scattering from solid surfaces. In atoms and molecules there is a clear distinction between excitation to a discrete state and ionisation. In metals at least all one electron transitions to above the Fermi level (the highest filled level at OK) lie in a continuum and technically correspond to ionisation. However, in some systems it is possible to reach excited states in the continuum that retain some discrete atomic or molecular character, and these may be thought of as localised resonances that may autoionise into the continuum or undergo direct recombination. The socalled giant resonances of the 4d shell in the rare earths are a particularly clear example of this phenomenon, and this will be discussed in some detail in section 5. In solids collective oscillations of the outer electrons (plasmon) tend to be prominent in electron induced excitations below 30 eV giving the valence electron loss region a rather different character from the case of most atoms and molecules (see section 4).

EELS in reflection geometry is an inherently surface specific technique. It has controlable surface sensitivity since the energy of exciting primary electrons and so their inelastic mean free path can be varied over a wide range. The inelastic mean free path of electrons λ, that is the distance over which electrons can carry messages about characteristic energies, varies typically between 0.3 and 5 nm for electrons of energy 20 - 2000 eV. This is shown in Fig. 2 where a socalled "universal λ versus E" curve is reproduced from the poineering work of Seah and Dench [14]. Note the broad minimum between 50 - 150 eV - the region of highest surface sensitivity - and that λ for most elements tends to scatter around this curve. If EELS is performed on electrons reflected from a solid surface the inelastic scattering process has to be either proceded or followed by an elastic backscattering event so that the loss electron can reach the detector. The elastic backscattering event is necessary because inelastic scattering cross sections are dominated by small angle scattering processes (see section 2); thus the momentum necessary for the loss electron to be turned around in order to escape from the surface has to be provided by the whole crystal. As shown by Ibach and Mills [15] for small-angle scattering two alternative single scattering consecutive events - loss-before-diffraction (LD) or diffraction-before-loss (DL) - represent the dominant contributions to the backreflected current, higher order processes being of only minor importance. In this two-step model the elastic reflectivities of the surface at the particular primary and loss energies have to be considered. In practice it is advantageous to emphasise DL processes in the experiment by choosing a primary energy at which the elastic reflectivity is at maximum [16-18]. In DL processes the backreflected current in the loss spectrum is a direct measure of the inelastic scattering cross section, whereas for LD processes the inelastic scattering cross sections of the various excitations are modulated by the energy dependent elastic reflectivity.

Energy loss peaks always relate to the elastically reflec-

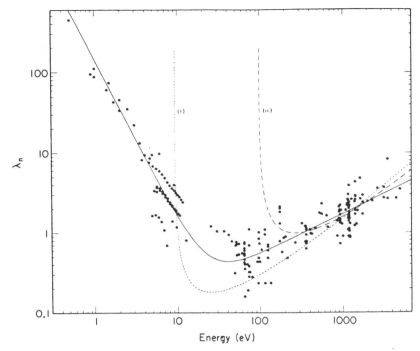

Fig. 2. Compilation of inelastic mean free path measurements of elements (λ in nm). From Seah and Dench [14].

ted primary peak at E_p in the spectrum by a <u>constant energy separation</u>, the loss energy ΔE; they shift therefore with the primary peak if E_p is varied. In this way variation of primary energy allows energy loss peaks to be distinguished from other features in the secondary electron spectrum, e.g. Auger peaks, which occur at <u>constant kinetic energy</u>.

2. Some Elements of Theory

The inelastic scattering of electrons by atoms and solids may be treated in two basic regimes according to whether or not the velocity of the incident electron v is large compared to the orbital speed of the electron being excited. In the former case the first Born approximation may be applied which means that the interaction between beam and sample is weak and averaged over the unperturbed beam and sample. Accordingly, the results give information about excitations from the unperturbed ground state of the sample. Experimentally in the case of single atom inelastic scattering, the Born approximation is known to be accurate for inelastic scattering from He for electron energies as low as 250 eV [2], but high Z materials require higher beam energies. Within the validity of the first Born approximation Bethe [19] derived his rigorous scattering theory with the very important result that the differential cross section for scattering depends directly on the momentum transferred in the collision as we will discuss below. If the incident electron energy is not much greater than the binding energy of the electrons being excited a much more complicated pattern of interaction emerges [4,6,8]. In this "slow" regime selection rules for excitation break down and new scattering channels become important. The ideas inherent in the Born-Bethe approach survive resonably intact in the intermediate energy regime between "fast" and "slow", and the Born-Bethe approximation forms a good basis for the interpretation of most (but not all as we shall see) EELS experiments on solids. The main results of Bethe's analysis are summarised in the following, but for deeper insight the articles of Inokuti [20], Inokuti et al. [21], Bonham [5], Schnatterly [2], Leapman [22] and Inokuti and Manson [23] are recommended.

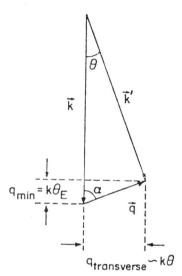

Fig. 3. Momentum conservation in energy loss. The scattering diagram shows the incident electron momentum $\underline{k}$ and the momentum transfer $\underline{q}$ for scattering angle θ.

2.1 Born-Bethe Formalism

Let us consider now the interaction of a fast electron of energy $E_p = 1/2 \, k^2 = 1/2 \, v^2$; k is the incident wave vector, v the velocity and we shall adopt Hartree atomic units, setting relative to SI units $e = \hbar = m_e = 4\pi\varepsilon_o = 1$; the atomic unit of energy is then 27.2 eV, and the unit of distance the Bohr radius 0.0529 nm. If the electron loses energy ΔE in being scattered into a final state with wave vector $\underline{k}'$ momentum conservation yields the momentum transfer (Fig. 3)

$$\underline{q} = \underline{k} - \underline{k}' \qquad (1),$$

and energy conservation for small scattering angles θ gives

$$\underline{q}^2 = k^2 \, (\theta^2 + \theta_E^2) \qquad (2)$$

where $\theta_E = \Delta E/2E_p$; $q_{min} = k \, \theta_E = \Delta E/(2E_p)^{1/2}$ is the minimum momentum transfer for a given ΔE and its direction is parallel to $\underline{k}$. To describe inelastic scattering we require the differential cross section with respect to ΔE and $\underline{q}$. The cross section may be written according to Bethe [19]

$$\frac{d^2\sigma}{d\Omega\, d(\Delta E)} = 4\, \frac{k'}{k}\, \frac{1}{q^4}\, S(\underline{q}, \Delta E) \qquad (3)$$

where $d\Omega$ is an element of solid angle and $S(\underline{q}, \Delta E)$ is a dynamical form factor for inelastic scattering. For an N electron system $S(\underline{q}, \Delta E)$ takes the form

$$S(\underline{q}, \Delta E) = \left| \langle f | \sum_{j=1}^{N} \exp(i\underline{q} \cdot \underline{r}_j) | i \rangle \right|^2 \qquad (4)$$

with $\underline{r}_j$ the position vector of the jth electron. In equ. 4 the form factor is expressed in terms of matrix elements for transitions from initial states $|i\rangle$ to final states $|f\rangle$. This is appropriate for core excitations, but for valence excitations a different formulation is more useful as we shall see below.

The merit of the Bethe formalism of equ. 3 is its partitioning of the cross section into two parts, a term which describes the kinematics of the scattering process and a term which describes the dynamics, that is the excitation properties, of the target system. The term $4(k'/k)(1/q^4)$ is the Rutherford cross section for scattering of one free electron by another at rest, so that $S(\underline{q}, \Delta E)$ represents the ratio of scattering from a system with electronic structure to the idealised "structureless" electron case.

In order to obtain angular distributions of inelastic scattering cross sections it is necessary to relate the momentum transfer $\underline{q}$ to the scattering angle θ. When $E_p \gg \Delta E$ near forward scattering dominates (θ small, $k'/k \approx 1$) and with use of equ. 2 the differential scattering cross section (equ. 3) reduces to

$$\frac{d^2\sigma}{d\Omega d(\Delta E)} = \frac{4}{k^2}\, \frac{1}{\theta^2 + \theta_E^2}\, \frac{1}{q^2}\, S(\underline{q}, \Delta E) \qquad (5).$$

This is the relevant formula for EELS performed at high angular resolution. Equ. 5 implies that at low scattering angles the main angular dependence comes from a Lorentzian factor $(\theta^2 + \theta_E^2)^{-1}$ with the characteristic angle $\theta_E = \Delta E/2E_p$ representing the "half width" of the angular distribution. The way in which the spectrometer aperture in relation to θ_E affects experimental loss spectra in reflection geometry will be discussed further with an example in section 4.

For a core excitation between a localised initial state $|i\rangle$ and any kind of final state $|f\rangle$ form factor may be written

$$S(q, \Delta E) = |\langle f | \exp(i\underline{q} \cdot \underline{r}) | i \rangle|^2 \qquad (6)$$

where a one electron approximation has been adopted. The operator governing the transition may expanded in a Taylor series as

$$\exp(i\underline{q}\cdot\underline{r}) = 1 + iq(\hat{\mathcal{E}}_q\cdot\underline{r}) - \frac{q^2}{2}(\hat{\mathcal{E}}_q\cdot\underline{r})^2 + \ldots \quad (7)$$

where $\hat{\mathcal{E}}_q$ is a unit vector in the direction of $\underline{q}$. The first term vanishes since $|i\rangle$ and $|f\rangle$ are orthogonal. The second term has exactly the same form as the dipole matrix element for photon absorption, where $\hat{\mathcal{E}}_q$ is replaced by the polarisation vector of the electric field. Therefore, provided that $\underline{q}\cdot\underline{r} \ll 1$, i.e. $q \ll 1/r_c$ with r_c the orbital radius of the core electron, the operator reduces to the second term and supports transitions of the dipole form. In this regime of high incident energy and small scattering angles (in the limit $q \to 0$) electron loss and optical absorption are strictly analogous and govered by dipole angular momentum selection rules, where $\Delta l = \pm 1$. For finite momentum transfer q the third term of the operator expansion comes into play and allows monopole and quadrupole transitions [2] ($\Delta l = 0, \pm 2$). Such transitions are, of course, regarded as "forbidden" in optical spectroscopy. Higher terms in the expansion of $\exp(i\underline{q}\cdot\underline{r})$ cause higher order multipole transitions. By scanning the electron loss spectrum at different values of q different multipole transitions can therefore be observed. Thus even within the Born-Bethe regime monopole, quadrupole and octopole transitions may be prominent.

A closely related quantity to the form factor introduced above is the generalised oscillator strength (GOS) [20], which is defined as

$$\frac{df(q,\Delta E)}{d(\Delta E)} = \frac{2\Delta E}{q^2} S(\underline{q},\Delta E) \quad (8).$$

It can be shown that the GOS is related to the optical oscillator strength by the relation[24]

$$\frac{df(q,\Delta E)}{d(\Delta E)} = \frac{df_o}{d(\Delta E)} + q^2 \left(\frac{df_o}{d(\Delta E)}\right)' + \frac{q^4}{2}\left(\frac{df_o}{d(\Delta E)}\right)'' + \ldots \quad (9)$$

where $df_o/d(\Delta E)$ is the optical oscillator strength, and the primed brackets are, respectively, the first and second derivatives with respect to q^2 of the GOS evaluated at $q^2 = 0$. In the limit $q \to 0$ the higher terms in the expansion vanish and the GOS becomes equal to the optical oscillator strength, which in turn is proportional to the total photon absorption cross section.

The energy and momentum dependence of the GOS may be displayed in a twodimensional plot known as the Bethe surface - see Inokuti [20] for detailed discussion. The important message of these plots in the present context is the fact that the maximum of the GOS in the optical regime, i.e. at $E_p \gg \Delta E$, occurs at small q, that is scattering is dominantly in the forward direction [20,25]. This forms the physical basis of the two-step model in reflection geometry, which has been discussed in section 1.

2.2 Dielectric Formalism

The Born-Bethe formalism is most easily applied to single atoms or gaseous targets, where the GOS can be calculated as a function of loss energy and momentum transfer using an atomic model. It may be useful also for describing inelastic scattering taking place from inner-atomic shells in a solid. For outer-shell scattering, however, collective effects involving many atoms become important, and it is advantageous to describe the interaction of electrons with the solid in terms of a dielectric response function $\varepsilon(q, \Delta E)$ [9]. The structure of the basic result (equ. 3) with the partitioning into a Rutherford cross section and a form factor depending on momentum transfer still remains intact, but the GOS may now be reinterpreted and, for small q, takes the form

$$\frac{df(q, \Delta E)}{d(\Delta E)} = \frac{2 \Delta E}{\omega_p^2} \text{Im}\left(-\frac{1}{\varepsilon(q, \Delta E)}\right) \quad (10)$$

where the plasmon frequency $\omega_p = (4\pi n)^{1/2}$, with n the effective electron density in the solid. $\text{Im}(-1/\varepsilon)$ is termed the bulk dielectric loss function. The complex dielectric response function is defined as

$$\varepsilon(q, \Delta E) = \varepsilon_1(q, \Delta E) + i\varepsilon_2(q, \Delta E) \quad (11),$$

and for $q = 0$ the dielectric constant ε is related to the optical constants, the refractive index n and the extinction coefficient k, by

$$\varepsilon(0, \Delta E) = (n + ik)^2$$
$$\varepsilon_1 = n^2 - k^2 \qquad \varepsilon_2 = 2nk \quad (12)$$

In the dielectric loss function as given by

$$\text{Im}\left(-\frac{1}{\varepsilon}\right) = \frac{\varepsilon_2}{\varepsilon_1^2 + \varepsilon_2^2} \quad (13)$$

the denominator is responsible for differences that can occur between EELS and optical absorption spectra; the latter are proportional to ε_2 [26]. For electron energy losses $\lesssim 50$ eV (i.e. losses due to valence electron excitation), $\varepsilon_1^2 + \varepsilon_2^2$ is usually appreciably different from unity and there is a large difference between energy loss and optical absorption spectra. For energy losses greater than about 100 eV (i.e. losses due to core electron excitation) $\varepsilon_1 \approx 1$, $\varepsilon_2 \ll 1$ so that

$\text{Im}(-1/\varepsilon) \approx \varepsilon_2$. Thus, EELS spectra are similar to X ray

absorption spectra. Also, since for free atoms $\varepsilon_1 \approx 1$ and $\varepsilon_2 \ll 1$ the loss function for inner-shell excitations is generally similar to the corresponding functions for free atoms and molecules. Since ε_2 is proportional to a matrix element as defined in equs. 6 and 7 the connection between Born-Bethe theory and dielectric formalism is most readily made for the inner-shell region of solids.

Maxima occur in the energy loss function when

$$(\varepsilon_1^2 - \varepsilon_2^2)\left(\frac{d\varepsilon_2}{d(\Delta E)}\right) - 2\varepsilon_1 \varepsilon_2 \left(\frac{d\varepsilon_1}{d(\Delta E)}\right) = 0 \quad (14)$$

and $d^2[\text{Im}(-1/\varepsilon)]/d(\Delta E)^2$ is negative [26]. Maxima in $\text{Im}(-1/\varepsilon)$ then occur near energies for which there are maxima in ε_2 (excitation of interband transitions) or $\varepsilon_1 = 0$, ε_2 small

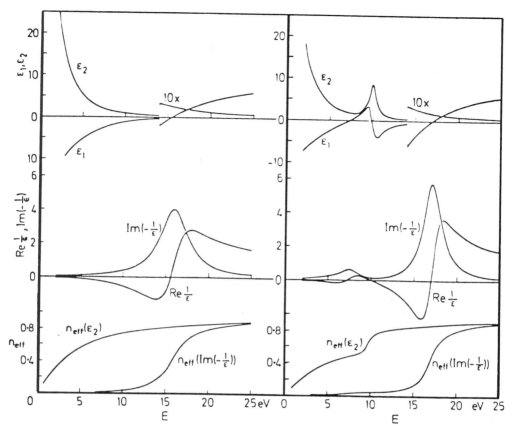

Fig. 4. Dielectric properties of a free-electron gas with $E_{BP} = 16$ eV and $\Delta E_{BP} = 4$ eV. On the right hand side, interband transitions have been added at $\Delta E = 10$ eV (Daniels et al. [35]).

(excitation of volume plasmons) - see Fig. 4.

The plasmon, a quantum mechanical quasiparticle, may be regarded as a set of coupled oscillators interacting with each other and with a passing electron via electrostatic forces or, in a different but related picture, as a longitudinal charge density fluctuation of quasi free valence electrons. In the simplest situation of a free-electron gas the plasmon energy is given by

$$\omega_p = (4\pi n)^{1/2} \qquad (15)$$

(or $\omega_p = (\dfrac{ne^2}{\varepsilon_0 m_e})^{1/2}$ in S.I. units, where m_e is the electron rest mass). Plasmon poles in the energy loss function occur for $\varepsilon_1 = 0$ and ε_2 small and of negative slope as shown in Fig. 4. This is the condition for a volume plasmon. At the surface of a solid additional longitudinal waves of charge density appear; these travel along the surface but decay rapidly in the direction perpendicular to the surface [9]. For a free-electron metal the surface plasmon at the metal - vacuum interface is

$$\omega_s = \frac{\omega_p}{\sqrt{2}} \qquad (16)$$

and conditions for surface plasmon occurrence are $\varepsilon_1 = -1$ and ε_2 small and decreasing. More generally, the surface loss function is given by [9]

$$\text{Im}\left(-\frac{1}{\varepsilon+1}\right) = \frac{\varepsilon_2}{(\varepsilon_1+1)^2 + \varepsilon_2^2} \qquad (17).$$

This expression takes into account the terminating boundary of the solid, but not specific features of the electronic structure of the surface such as surface states or the like.

Besides collective excitations electrons in the valence region of solids can undergo single particle excitations from occupied states below the Fermi level to unoccupied states above the Fermi level. In the dielectric loss function interband transitions are characterised by peaks in the ε_2 function (see Fig. 4, right hand side panels). The transition rate is determined by the details of the electronic band structure, and usually reflects peaks in the joint density of occupied and unoccupied states (JDOS). Peaks in the JDOS occur where initial and final state bands are approximately parallel. If both initial and final states of the transition are localised, the transition energy is not just the sum of one-electron energies of the states involved, but excitonic effects due to the electron-hole Coulomb interaction also become important. Excitonic effects are particularly relevant in semiconductors and insulators, if the final states of one-electron transitions from the valence band are located in the band gap. Interband transitions in the energetic vicinity of plasmon excitations induce shifts of the plasmon energy [9,25] by modifying ε_1 and ε_2 as indicated in Fig. 4. As a rule of thumb the loss peaks tend to shift so that their energy separation is increased.

Insights into which electrons in the solid are contributing to which loss features can be gained by examining the differential oscillator strength $df(0, \Delta E) / d(\Delta E)$. Integrating with respect to ΔE gives

$$n_{eff} = \int_0^{\Delta E} \frac{df(0, \Delta E')}{d(\Delta E')} \, d(\Delta E') \qquad (18)$$

where n_{eff} goes to N, the number of electrons per atom as $\Delta E \to \infty$ (the Thomas-Kuhn-Reiche sum rule).

2.3 Beyond Born-Bethe

As E_p approaches ΔE the incident and target electrons move at similar speed, and so the simplicity of the Born-Bethe analysis collapses since there is no longer a simple functional dependence of cross section on momentum transfer. Total inelastic cross sections for dipole processes build up from threshold to a maximum at a few times threshold, and roll off slowly with increasing E_p. Monopole and quadrupole cross sections peak closer to threshold, and fall off more rapidly with energy at high E_p[27]. Near threshold the exchange between incoming electron and target electron assumes great importance. It is then possible to change the spin of the target e.g. from a singlet to a triplet state, giving rise to transitions which are not

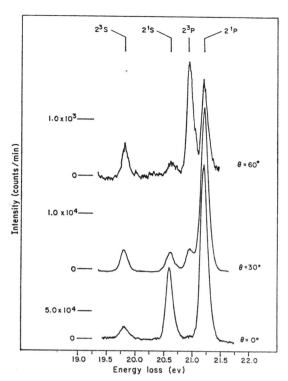

Fig. 5. Energy loss spectra of He at 36 eV impact energy for various scattering angles θ (Rice et al.[28]).

allowed in the simplest Born-Bethe approach. Now total excitation cross sections peak very close to threshold and fall off very rapidly as E_p increases. However, at high angles of scattering transitions involving change of spin have cross sections much higher than transitions involving no spin change. A combination of energy and angle dependence of loss intensities can therefore be used to identify transitions involving spin change. In order to illustrate the variation of scattering cross sections of different excitations with scattering angle, Fig. 5 from Rice et al.[28] compares electron loss spectra of He at 36 eV impact energy as a function of scattering angle.

For $\theta = 0°$ the dipole allowed $1^1S \rightarrow 2^1P$ transition dominates the spectrum, but at $\theta = 60°$ the spin-forbidden $1^1S \rightarrow 2^3P$ transition becomes the most intense feature in the spectrum. Therefore, the spin-forbidden excitation gains prominence at higher scattering angles compared to the spin-allowed transition.

2.4 Adsorbates

Let us now mention briefly the case of atoms and molecules adsorbed on a solid surface. Ibach and Mills[15] have developed a two-layer model for dipole scattering within the framework of the dielectric formalism: an adsorbate layer of thickness d and dielectric constant ε_s rests on a substrate with dielectric constant ε_b. In the limit $q_\shortparallel \cdot d \ll 1$ ($q_\shortparallel$ is the momentum transfer parallel to the surface), which is satisfied for $\theta \lesssim \theta_E$, the differential scattering cross section may be separated into bulk σ_b and surface σ_s contributions

$$\frac{d^2\sigma}{d\Omega\, d(\Delta E)} = \frac{d^2\sigma_b}{d\Omega\, d(\Delta E)} + \frac{d^2\sigma_s}{d\Omega\, d(\Delta E)} \qquad (19)$$

$$\frac{d^2\sigma_b}{d\Omega\, d(\Delta E)} \propto \mathrm{Im}\left\{-\frac{1}{\varepsilon_b+1}\right\} \qquad (20)$$

$$\frac{d^2\sigma_s}{d\Omega\, d(\Delta E)} \propto q_\shortparallel \cdot d\, \mathrm{Im}\left\{-\frac{1}{\varepsilon_s}\left(\frac{\varepsilon_b^2 - \varepsilon_s^2}{(\varepsilon_b+1)^2}\right)\right\} \qquad (21).$$

The presence of ε_b in the surface expression (equ. 21) shows that the charge fluctuations in the adsorbed layer excite the substrate, so that the field seen by the electron is that produced by both the excitation in the adsorbate and its image in the substrate. The excitation cross section in the adsorbate is thus determined by the total transition dipole moment, which in classical terms is the vector sum of the real dipole and its screened image dipole in the substrate. Note also that the surface scattering cross section has an extra $q_\shortparallel \cdot d$ term, which causes the surface loss intensity to decrease less rapidly than bulk losses when observed at progressively higher inelastic scattering angles.

Since the adsorbate transition dipole moment includes contributions from the underlying substrate, surface selection rules appear which allow in certain cases the determination of molecular orientation at surfaces. As noted by Rubloff [29] the dipole selection rule which is widely used in interpreting vibrational EELS data [15] may be extended to electronic EELS if ε^2 is large at the energy of the electronic transition. This is the case for a number of metals and semiconductors up to an energy of 5-10 eV. Accordingly, electrons scattered in the specular direction, that is in the dipole scattering regime, cannot excite those transition dipoles of the adsorbate which are oriented parallel to the surface, but can excite with maximum efficiency those that are perpendicular to the surface. As the direction of a transition dipole with respect to the surface is determined by the molecular orientation, structural information is possible by comparing relative intensities of valence excitations of molecules in the gas phase and in the adsorbed phase [30,31].

3. Instrumentation

Every electron energy loss experiment requires three components: a source of electrons, a target system and a spectrometer unit for analysing the energy and the momentum of scattered electrons. Depending on the geometry of the experiment the various components have to be arranged in different ways, and particular designs have been developed for transmission and reflection EELS. Apart from the different geometrical arrangement of electron source, target and electron energy analyser in transmission and reflection mode, the primary energy E_p of the incoming electrons forms the major difference between the two experimental set-ups. In reflection EELS the primary

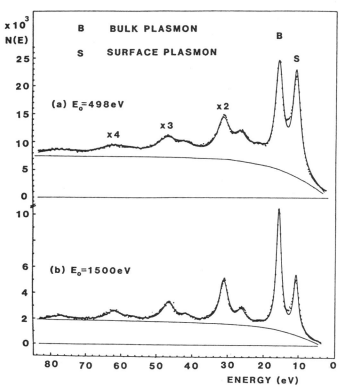

Fig. 6. EELS spectra of Al (Van Attekum and Trooster [33]).

energy ranges from a few eV to several keV, in accord with the desired depth sensitivity, whereas in transmission electron loss E_p is of the order of several tens to several hundreds of keV. High E_p's are necessary in transmission mode to ensure weak interaction of electrons with matter and therefore the penetration through (thin slices of) the probing material. Information on instrumentation of transmission EELS is reviewed in the book of Egerton [25], and the basic design of electron spectroscopy experiments on surfaces is discussed in the article of Riviere [32].

4. Valence Excitations

Inelastic electron scattering cross sections in the region of valence excitations in simple metals are dominated by plasmon losses. This is shown for polycrystalline Al metal in Fig. 6, where two different primary energies have been used [33]. The spectra consist of a series of doublet lines corresponding to multiples of the bulk plasmon at 15.6 eV and of the surface plasmon at 10.7 eV. The experimental values are close to those calculated for a free-electron model using three valence electrons per Al atom. Up to five multiple plasmon losses can be recognised in Fig. 6, and their intensities decrease according to a Poisson distribution of the probability for multiple plasmon excitation [2]. Note that the surface plasmon peak at 10.7 eV is considerably more intense at E_p = 498 eV than at E_p = 1500 eV, and this is as expected from the higher surface sensitivity at the lower primary energy.

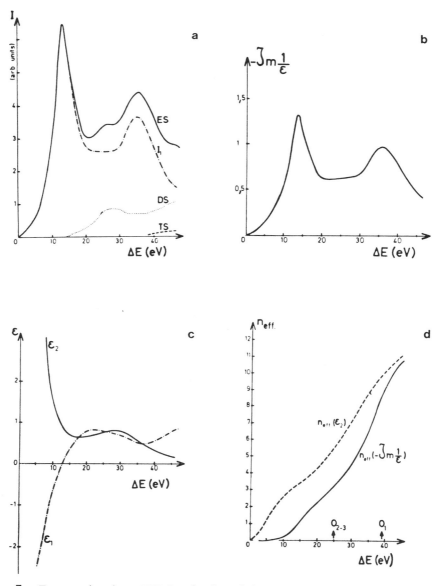

Fig. 7. Transmission EELS of Tb: (a) uncorrected loss spectrum; (b) Im$(-1/\varepsilon)$; (c) $\varepsilon_1(\Delta E)$ and $\varepsilon_2(\Delta E)$; (d) $n_{eff}(\Delta E)$. After Colliex et al. [34].

Fig. 7 shows experimental EELS in the valence region of the rare earth metal Tb (a) together with the loss function (b), the real and the imaginary parts of the dielectric constant ε (c) and n_{eff} (d) [34]. The experimental data have been recorded in transmission mode, but the figure demonstrates nicely the data analysis which has been performed by Colliex et al. [34]. The raw loss spectrum (panel a) has been corrected for instrumental aperture effects as well as multiple scattering to give Im $(-1/\varepsilon)$ (panel b), which undergoes a Kramers-Kronig transformation (see Daniels et al. [35]) to give ε_1 and ε_2 (panel c). A bulk plasmon peak is recognised at $E_{BP} = 13.3$ eV

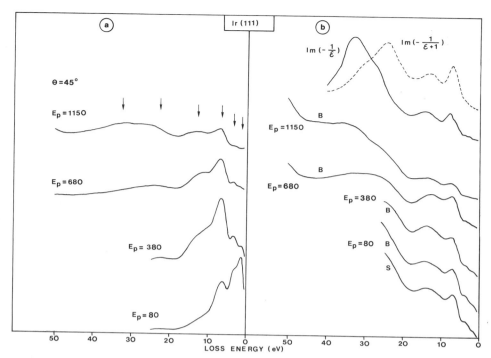

Fig. 8. Angle resolved EELS of Ir(111) in the specular direction at various E_p. (a) Raw data; (b) spectra normalised by the spectrometer aperture weighting function. After Matthew et al.[36].

with ε_1 and ε_2 behaving qualitatively free-electron like, but no surface plasmon is observed in accordance with the bulk character of the experiment. Converting the loss function to differential oscillator strength according to equ. 10 and integrating with respect to ΔE gives $n_{eff}(\Delta E)$ (panel d). In Tb n_{eff} remains low to just below the plasmon loss and builds up to a value $n_{eff} \approx 3$ by the 5p (O_{23}) threshold (three valence electrons contributing to the plasmon). This is an indication that the compact 4f shell of binding energy 2-10 eV is playing little part in loss events at this stage. The 5p oscillator strength is largely dissipated in the strong 35 eV resonance loss as evidenced by the rapid rise of n_{eff} from near 3 to 9 between the 5p and 5s (O_1) threshold.

In transition metals interband transitions and plasmon excitations are intimately mixed resulting in complex spectral behaviour. In many cases collective and one-electron excitations cannot be separated unambiguously, and the plasmon energies may differ significantly from free-electron values. An example of this are the loss spectra of Ir, where the consequences of the angular acceptance of the spectrometer in relation to the characteristic angle Θ_E are also demonstrated. Fig. 8(a) shows EELS of an Ir(111) single crystal surface, recorded in specular reflection geometry for E_p = 80 - 1150 eV with an analyser of high angular resolution (spectrometer acceptance angle $\Theta_s = \pm 1°$)[36]. At E_p = 80 eV structures at ΔE = 1.4 eV and 3.5 eV are dominant, but at E_p = 1150 eV those features are barely discernible and higher energy losses prevail. These changes in EELS profile resulting from variation of E_p are complicated by the way in which the characteristic loss profile angular breadth θ_E varies relative to the spectrometer acceptance angle θ_S. If θ_S is fairly small in absolute terms the differential scattering cross section in equ. 5 may be integrated over the solid angle of the spectrometer aperture to yield for specular reflection[36]

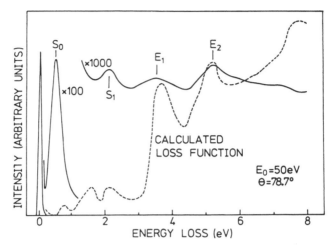

Fig. 9. EELS of a cleaved (111) 2x1 silicon surface.

$$\frac{d\sigma}{d(\Delta E)} \propto \ln\left(1 + \frac{\theta_S^2}{\theta_E^2}\right) \text{Im}\left(-\frac{1}{\varepsilon(q_{min}, \Delta E)}\right) \quad (22)$$

In Fig. 8(b) the spectra have been normalised by the weighting factor $\ln\left(1 + \frac{\theta_S^2}{\theta_E^2}\right)$. The spectra are now very similar in shape for all primary energies, and show good correspondence with the bulk and surface loss function Im $(-1/\varepsilon)$ and Im $(-1/\varepsilon+1)$ derived from optical data. Accordingly, the broad structure around 30 eV loss energy has primarily plasmon character, but below $\Delta E = 20$ eV one-electron and collective excitations are difficult to disentangle.

On semiconductor surfaces one-electron transitions involving localised filled and empty surface states are particularly attractive features for EELS investigation. As an example Fig. 9 shows a high resolution energy loss spectrum of a cleaved (111) 2x1 silicon surface recorded by Froitzheim et al.[37]. The transitions labelled S_0 and S_1 involve dangling bond surface states as evidenced by their sensitivity to adsorbates or surface order, while E_1 and E_2 are transitions between bulk bands.

In Fig. 10 the build up of an alkali layer (Na) on a single crystal surface of Ru has been followed by EELS[38]. The for-

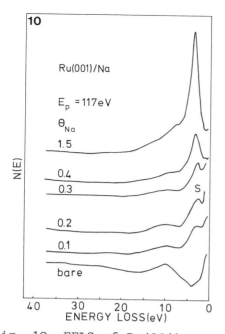
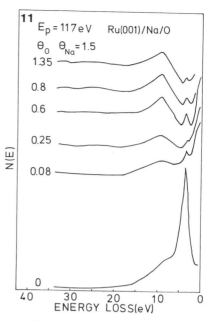

Fig. 10. EELS of Ru(001) covered with various amounts of Na. A Na monolayer corresponds to $\Theta_{Na} \approx 0.45$ (Surnev et al. [38]).

Fig. 11. EELS of Ru(001) precovered with ~3 ML of Na and subsequently exposed to oxygen (Surnev et al. [38]).

mation of the Na layer is characterised by a loss transition between 3-4 eV, which is interpreted at low Na coverages ($\Theta_{Na} < 0.5$) in terms of a one-electron transition between hybridised Na 3s - Ru 4d derived states, but is ascribed to a surface plasmon once a continuous overlayer ($\Theta_{Na} > 0.5$) has been formed. This surface plasmon of the Na layer is extremely sensitive to oxygen exposures as demonstrated in Fig. 11: only 0.08 monolayers of oxygen quench the surface plasmon almost completely (compare the two lowest curves in Fig. 11), presumably as a result of transfer of free-electron like charge to oxygen atoms [38].

The last example confirmed that reflection EELS is an excellent monitor of changes in the surface electronic structure, and so it can be used to plot the progress of reactions at solid surfaces. The interaction of hydrogen with a Ce(001) single crystal surface, as studied by Rosina et al. [39], is shown in Fig. 12. After exposing the clean Ce(001) surface to 300 L H_2 (1L = 10^{-6} torr. sec) a loss at 3.1 eV, due to a H induced interband transition, is observed. For exposures of $\gtrsim$ 600 L H_2 a very sharp peak at 1.6 eV loss energy shows up, which grows further in intensity at higher exposures. This unusually sharp feature is associated with a highly damped low-energy plasmon excitation characteristic of a surface dihydride phase [39]. The fact that this plasmon is still dominant in the loss spectrum recorded with E_p = 1000 eV (top curve of Fig. 12), where the inelastic mean free path of electrons is 3-4 nm, indicates that the dihydride phase extends beyond

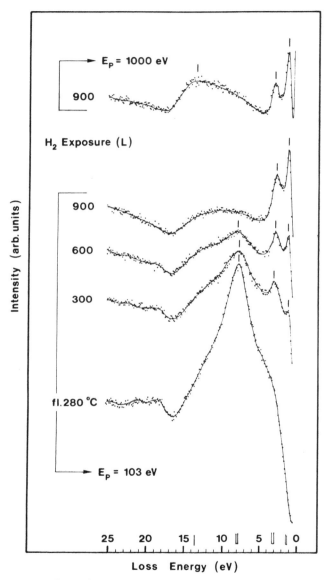

Fig. 12. EELS of Ce(001) as a function of hydrogen dosage.

3-4 nm into the bulk. So, EELS of valence excitations is very useful to yield detailed information on the initial stages of hydride formation.

5. Core Excitations

As a result of their strong binding to the nucleus collective effects are unimportant in inner-shell excitations and a single atom description is often appropriate. The differential scattering cross section is then proportional to an atomic transition matrix element M(ΔE), which is weighted by the density of final states in the case of transitions to band like states [40]

$$\frac{d\sigma}{d(\Delta E)} \propto |M(\Delta E)|^2 N(E) \qquad (23)$$

Within the validity of dipole selection rules $\Delta l = \pm 1$, with the $\Delta l = + 1$ transitions predominating [41], N(E) has to be interpreted as a symmetry projected density of states. In a one-electron approximation where all other electron wave functions but that of the excited electron remain unchanged the structure at the ionisation edge is represented by a simple step function. In practice, the conduction electrons in the solid screen the Coulomb field of the excited core and the screening response introduces a many-body problem, the famous X ray threshold problem [2]. The experimental core excitation spectra exhibit therefore more complicated near-edge structure, but in reflection electron loss the structures at ionisation edges are often not spectacular if dipole selection rules confine final states to delocalised band like states.

The situation is different, however, if localised final states are involved. If initial state and final state wave functions have similar spatial extent maximum wave function overlap may result causing large transition matrix elements and high oscillator strength near the ionisation threshold. The arena to investigate these localised resonance phenomena in the present article will be the rare earth metals with their partly filled 4f states, which provide excellent candidates for localised final states in core excitations. The empty 4f states in the rare earths are compact, atomic like and of similar spatial extent as the core states. This is illustrated in Table 1 where the mean radii of several orbitals in the ground state of free Ce are compared. Strong wave function overlap of core states with the 4f states is therefore expected, and indeed intense resonances are seen in X ray absorption and EELS near the 4d and 3d excitation thresholds of the rare earths. In the following 4d and 3d excitations of the rare earths will be discussed in some detail, and the similarities and differences between X ray absorption (XAS) and electron loss will be emphasised. I will also show that 4p excitations of the rare earths yield localised resonance effects, and that these may be understood in terms of 4p → 4f quadrupole resonances. Finally, the 2p excitations in sodium oxide will be briefly mentioned.

5.1 4d and 3d Excitations in the rare earths

XAS experiments [43] at the end of the 1960's revealed a series of complicated peaks near the 4d ionisation threshold of the lanthanides and intense resonances which extend up to 20 eV into the continuum. These excitations were ascribed to

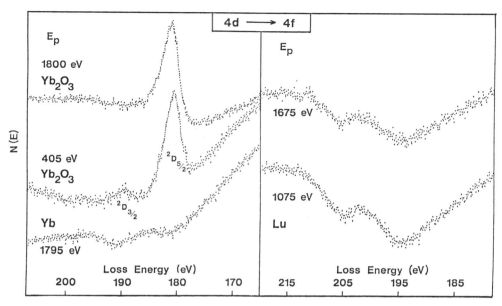

Fig. 13. EELS of Lu and Yb metal and Yb_2O_3 at a medium high and a low primary energy (Strasser et al. [45]).

atomic like $4d^{10}4f^n \rightarrow 4d^9 4f^{n+1}$ transitions [44]. The lanthanides show this complicated structure because the unfilled 4f shell is collapsed, and therefore not subject to large interaction with the environment, and because of the large electrostatic interaction between the 4d and the 4f shell. The resulting large exchange interaction splits the $4d^9 4f^{n+1}$ final state configuration and raises some multiplets up to 20 eV above the configuration average, thus driving them far into the continuum; autoionisation to $4d^9 4f^n \varepsilon f$ configurations then broadens the high-energy levels. It is useful to begin the discussion at the end of the lanthanide series with Lu and Yb, where the spectra may be interpreted in one electron terms, to move on to La and Tm, where two particle 4d — 4f states control the spectral structure, and finally to consider more complex configurations.

In Lu and Yb metal the 4f shell is filled and no 4d → 4f resonance is possible; EELS in Fig. 13 therefore shows only weak $4d_{5/2}$ and $4d_{3/2}$ ionisation edges [45]. Yb_2O_3, however, has a $4d^{10}4f^{13}$ ground state and can support transitions of the type $4d^{10}4f^{13} \rightarrow 4d^9 4f^{14}$. At medium-high energy (E_p = 1800 eV in Fig. 13) only $^2F_{7/2} \rightarrow {}^2D_{5/2}$ transitions at 181 eV are dipole allowed and the sharp resonance agrees well with photon absorption [45] and shows a Fano-type profile resulting from the interference between the discrete transition and various direct recombination channels [47]. At lower primary energy (E_p = 405 eV) the <u>dipole forbidden</u> transition $^2F_{7/2} \rightarrow {}^2D_{3/2}$ ($\Delta J = 2$) gains significant intensity. The spectral profile of this system is particularly simple because the 4f shell is filled in

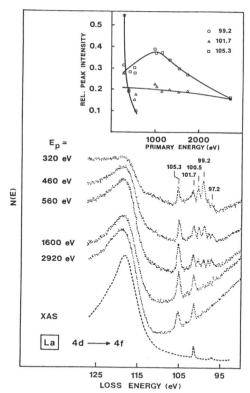

Fig. 14. 4d → 4f excitations of clean La in EELS for various E_p's. An XAS spectrum is shown for comparison (bottom curve). The inset displays the intensities of sharp peaks, normalised to the maximum of the main resonance, as a function of E_p (Netzer et al.[48]).

the final state and multiplet splitting does not occur. The optically forbidden $^2D_{3/2}$ state is therefore unambiguously identified[45].

In La $4d^{10}4f^0(^1S_0) \rightarrow 4d^94f^1$ transitions are involved and in principle $^1P_1$, $^3P_{2,1,0}$, $^1D_2$, $^3D_{3,2,1}$, $^1F_3$, $^3F_{4,3,2}$, $^1G_4$, $^3G_{5,4,3}$, $^1H_5$ and $^3H_{6,5,4}$ final states are possible. Within dipole selection rules all the oscillator strength would be concentrated in the $^1P_1$ state in the LS coupling limit, but weak spin-orbit interaction transfers a small amount of dipole strength to the $^3P_1$ and $^3D_1$ states. XAS of La is readily understood in these terms, but the EELS at medium-high primary energy ($E_p \approx 3$ keV) is markedly different from XAS [48] (Fig. 14). The peak in the position of the $^3D_1$ at 101.7 eV is enhanced and a new sharp feature at 105.3 eV appears which is not present in XAS (compare the two lowest curves in Fig. 14). Multiplet analysis reveals that these states correspond to ^1H (approximately degenerated with ^3D) and ^1F. An understanding of why these two particular excitations break the dipole embargo at relatively high incident energies follows from a group theoretical examination of the scattering problem [49]. Accordingly, for $4d^{10}4f^n \rightarrow 4d^9 4f^{n+1}$ transitions in predominantly forward scattering and with no spin change involved, ΔL can be 1,3 or 5. Thus, the $\Delta L = \pm 1$ selection rule becomes relaxed at intermediate primary energies. As E_p is reduced the cross section for exchange scattering becomes appreciable, and the complex structure that appears in the spectra (Fig. 14) is consistent with a theory in which all final state multiplets are allowed[48]. La is an ideal system for emphasising such non-dipole excitations. The strong 4d-4f exchange interaction drives the main $^1P_1$ dipole allowed multiplet to <u>high</u> energy and the other multiplet states lie <u>below</u> the 4d ionisation edge; they are sharp, well localised and superimposed on a smooth background of secondary electrons.

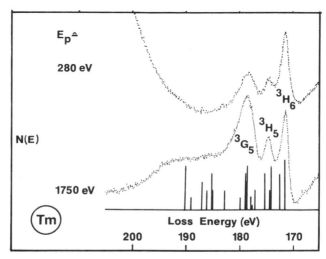

Fig. 15. 4d → 4f excitations of clean Tm in EELS (Strasser et al. [45]).

At the other end of the lanthanide series the dipole allowed transitions lie at energies <u>lower</u> than the principal non-dipole components. In Tm (Fig. 15) $4d^{10}4f^{12}$ ($^3H_6$) → $4d^9 4f^{13}$ ($^3G_5$, $^3H_5$ and $^3H_6$) transitions between 170 and 180 eV are allowed optically. Multiplet analysis [45] suggests that many additional multiplets should appear between 180 - 190 eV for low E_p. None are in fact observed (see Fig. 15 for E_p = 285 eV). This may be due to difficulties with the secondary electron background, but it should be noted that these non-dipole allowed final states lie well above the $4d_{3/2}$ ionisation threshold and may be broadened considerably.

We have so far discussed only the simplest 4d excitation systems, but the trend through the rare earth series can be readily understood within the range of ideas so far considered. This is so because the 4d-4f interaction is strong compared to the 4f-4f interaction. Following Sugar[50], going from La

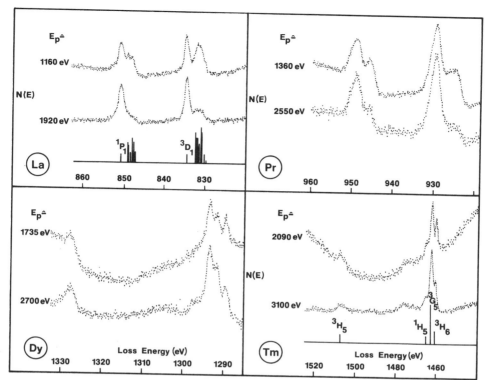

Fig. 16. 3d → 4f EELS of several rare earth elements. Spectra are shown for $E_p \gtrsim 2000$ eV and for E_p close to threshold (Strasser et al. [45]).

($4d^9 4f^1$) to Ce ($4d^9 4f^2$), Pr ($4d^9 4f^3$) etc. by successively adding electrons leads to multiplets having well defined $4d^9 4f^1$ ^1P parentage. At high primary energies, the EELS agree generally well with XAS, but at low E_p it is the <u>low energy</u> multiplets, which have stronger $4d^9 4f^1$ ^3H etc. parentage, that gain most in intensity as in La. At the other end of the lanthanide series it is possible to use the Tm $4d^9 4f^{13}$ multiplets as the parent structure, and then successively add 4f holes. The EELS of Er and Ho follow the trend observed in Tm, but in the middle of the series d-f parentage ideas tend to break down and oscillator strengths are more widely spread. Spectral changes for E_p close to threshold still occur, but the effects are generally less spectacular [45].

As for 4d excitations, 3d excitations may also be described as atomic like transitions of the type $3d^{10} 4f^n \rightarrow 3d^9 4f^{n+1}$. There are, however, important differences between 4d and 3d: the spin-orbit interaction is dominant in 3d excitations and the exchange interaction is reduced. As a result the $3d^9 4f^{n+1}$ multiplets are separated into two groups of structures, which may be loosely associated with $3d_{5/2}$ and $3d_{3/2}$ hole states; the multiplet splitting within each group of states is much smaller than in the 4d case.

The 3d EELS profiles at higher incident energies show in general remarkable agreement with the XAS spectra of Thole et al. [51], but some extra intensity is noted in EELS at the lower energy side of the main peaks, particularly in the first members of the rare earth series (Fig. 16) [45,52]. Reduction of E_p then yields intense satellites in that energy region, as seen in Fig. 16 for the case of La. The multiplet analysis of Esteva et al. [54] on La confirms that the main dipole allowed transitions in each group ($^3D_1$ for $3d_{5/2}$ and $^1P_1$ for $3d_{3/2}$) occur at high energy relative to the configuration average, while the multiplets of high J tend on average to be low in energy. This is shown in Fig. 16 for La, where the lines in the bar diagram are in the calculated positions and have length 2J+1. Even at E_p = 1.9 keV there is substantial non-dipole excitation, and when E_p = 1 keV there is a very large cross section in the high J region.

The 3d XAS of Tm has three dipole allowed $(3d_{5/2}^{-1})$ $4f^{13}$ final states ($^1H_5$, $^3G_5$ and $^3H_6$) and one $(3d_{3/2}^{-1})$ $4f^{13}$ state ($^3H_5$). As for 4d $^3H_6$ lies low in energy, and gains relatively in intensity in EELS as E_p is reduced (Fig. 16). It appears, therefore, that the final state with the highest multiplicity 2J+1 is being favoured at low E_p, and this trend is observed throughout the rare earth series [45].

Due to their atomic like character the spectral profile of 3d and 4d excitations of the rare earths are determined mainly by the multiplet structure of the $3(4)d^94f^{n+1}$ final states, and not by the chemical environment; d core excitation spectra may therefore be used to determine the occupation of the 4f shell. 4d excitations are particularly useful as fingerprints of the 4f occupation because of their characteristic multiplet structure. In the case of a valence change in rare earth atoms the occupation of the 4f shell changes and electrons are transferred from the localised, chemically inert 4f shell into delocalised valence states or vice versa. Core EELS, in particular of 4f states, can therefore be used to monitor valence changes at rare earth surfaces during surface reactions, taking advantage of the surface sensitivity of reflection EELS. Fig. 17 shows 4d excitation profiles of clean and oxidised Eu, compared to those of clean and oxidised Gd [55]. In Gd the 4d excitation curves remain virtually unchanged between metal and oxide, but in Eu pronounced changes are observed upon oxidation. The main resonance shifts by 2.5 eV to higher energy, and the preresonance fine structure is markedly different. The spectrum becomes similar to that of divalent Sm in the gas phase, and manifests the valence change from divalent metal to trivalent oxide.

5.2 4p Excitations in the Rare Earths

The case of 4p excitation is complicated by breakdown of the one-electron approximation for 4p core holes through $4p^{-1} \leftrightarrow 4d^{-2} \varepsilon f$ giant Coster-Kronig coupling [56]. This is strongest for $4p_{1/2}$ holes, which are characterised by broad ill defined peaks in photoemission, but $4p_{3/2}$ shows a conventional peak. In EELS the unstable $4p_{1/2}^{-1}$ excitations lead to ill-defined structure in the corresponding energy region, but sharp peaks are observed near the $4p_{3/2}$ ionisation threshold [57,58]. Two possibilities may be considered here: dipole allowed 4p → 5d transitions may occur or quadrupolar 4p → 4f transitions. It ought to be possible to distinguish these processes by E_p variation, but the changes that are observed are not conclusive. The

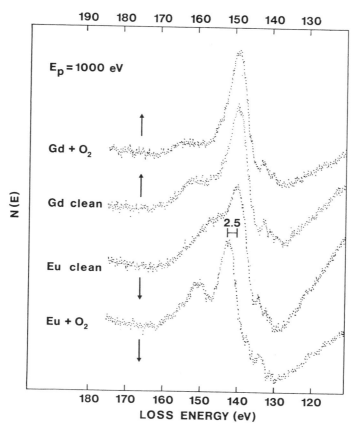

Fig. 17. 4d → 4f giant resonances of Gd and Eu metal and corresponding spectra of the oxidised surfaces (Strasser et al.[55]).

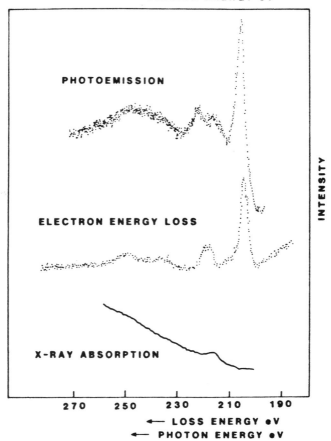

Fig. 18. Comparison of XAS, EELS and photoemission of Ce in the region of the 4p ionisation threshold (Matthew et al. [60]).

problem is that 4p and 4f wave functions may overlap very strongly, while 4p-5d overlap is likely to be much weaker (Table 1). This implies that quadrupole excitation may dominate dipolar excitation throughout the primary range available in reflection EELS [59].

Recently, Matthew et al. [60] have compared 4p XAS of Ce with EELS and photoemission. As shown in Fig. 18 EELS yields a large peak that shows primary energy dependence [57], and lies below the $4p_{3/2}$ threshold, and smaller features approximately co-inciding with satellite photoemission peaks that may form part of the $4p_{1/2}$ spectrum. In contrast XAS has no structure at the energy of the main EELS feature, but does show a weak peak in the region of 218 eV. This provides conclusive evidence that the main EELS peak is non-dipole in character - a giant 4p-4f quadrupole resonance. Such an interpretation is supported by Hartree-Fock calculations which confirm the strong 4p-4f exchange coupling and the dominance of the 4p-4f quadrupole matrix element over the 4p-5d dipole matrix element [60]. A systematic comparison of 4p EELS and XAS throughout the

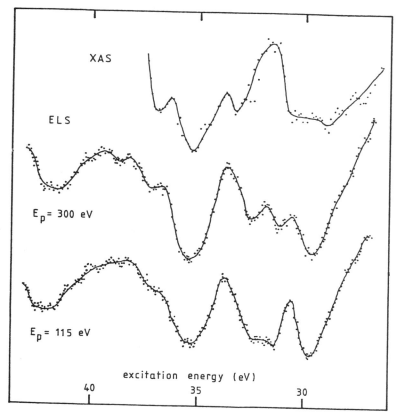

Fig. 19. 2p EELS (E_p = 115 and 300 eV) and XAS of NaO_2 (Bertel et al. [61]).

lanthanide series has yet to be made, and should be accompanied by calculations that account for at least some of the many body character of these systems.

5.3 2p Excitations in Sodium Oxide

The 2p excitations in thin films of NaO_2 have been investigated recently by EELS and photon absorption by Bertel et al. [61]. Whereas the alkali metals show only weak p threshold features in reflection EELS, the oxides are characterised by pronounced structures indicative of localised resonances. Fig. 19 compares XAS and EELS recorded at primary energies 300 eV and 115 eV. The main structures in XAS may be readily understood following the ideas of Åberg and Dehmer [62] in terms of atomic like dipole allowed 2p → 3s (31.7 eV) and 2p → 3d (34 eV and 36.5 eV) transitions, the latter ones split into t_{2g} and e_g type final states as a result of ligand field splitting. The EELS spectra between 30 eV and 38 eV reproduce the XAS structures reasonably well, but the pronounced EELS feature at threshold (30.5 eV) has no correspondence in the XAS spectrum; its non-dipole character is emphasised at the lower E_p. We conjecture that dipole forbidden excitations into the antibonding $O_2^- \tilde{\pi}g$ orbital are responsible for this transition [61].

6. Summary

The merits of electron energy loss spectroscopy to elucidate electronic structure of solids have been discussed, and the surface character of experiments performed in reflection geometry has been emphasised. Excitations of the valence electron manifold and of core electrons have been treated separately, and the theoretical description in terms of a collective response function and of atomic like transitions, respectively, has been outlined. The examples given to illustrate valence excitations range from simple metals to transition metals, and from clean surfaces to thin overlayer films and surface reactions. In the case of core excitations localised resonance phenomena have been exemplified with use of inner-shell transitions of the rare earth elements, where electron excitation has been discussed comparatively to photon excitation. It is shown that the well known dipolar resonances are accompanied by a range of non-dipole transitions, which can be disentangled by variation of the electron primary energy.

Table 1: Orbital size of the ground state of atomic Ce [42]

Orbital	$\langle r_{nl} \rangle$ au
4s	0.67
4p	0.69
4d	0.74
4f	0.95
5d	2.64

Acknowledgements

I wish to acknowledge stimulating discussions with J.A.D. Matthew, University of York, England and E. Bertel, now at the Max-Planck-Institut für Plasmaphysik, München, which have helped to shape my views of EELS. The scientific programme that has sustained my interest in the field has been supported by the Fonds zur Förderung der Wissenschaftlichen Forschung of Austria.

References

1. C.E. Brion, S. Daviel, R. Sodhi and A.P. Hitchcock, in "X ray and Atomic Inner-Shell Physics 1982", B. Crasemann, ed., AIP Conference Proceedings No. 94, p. 429
2. S.E. Schnatterly, Solid State Physics, Vol. 34, Academic Press (1979), p. 275
3. J.A.D. Matthew and S.M. Girvin, Phys. Rev. B 24: 2249 (1981)
4. R.J. Celotta and R.H. Huebner, in "Electron Spectroscopy: Theory, Techniques and Applications" Vol. 3, C.R. Brundle and A.D. Baker, eds., Academic Press (1979), p. 41
5. R.A. Bonham, ibid. p. 127
6. G.K. King and F.H. Read, in "Atomic Inner-Shell Physics", B. Crasemann, ed., Plenum Press (1985), p. 317
7. S. Trajmar, Acc. Chem. Res. 13: 14 (1980)
8. A. Kuppermann, W.M. Flicker and O.A. Mosher, Chem. Rev. 79: 77 (1979)
9. H. Raether, "Excitations of Plasmons and Interband Transitions by Electrons", Springer Tracts in Modern Physics Vol. 88, Springer Verlag (1980)
10. H. Froitzheim, in "Electron Spectroscopy for Surface Analysis", H. Ibach, ed., Topics in Current Physics Vol. 4, Springer Verlag (1977), p. 205
11. Ph. Avouris and J.E. Demuth, Ann. Rev. Phys. Chem. 35: 49 (1984)
12. F.P. Netzer and J.A.D. Matthew, in "Handbook on the Physics and Chemistry of Rare Earths" Vol. 10, K.A. Gschneidner et al., eds., North Holland (1987)
13. F.P. Netzer, J.A.D. Matthew and E. Bertel, in "Advances in Spectroscopy: Spectroscopy of Surfaces", R.J.H. Clark and R.E. Hester, eds., John Wiley and Sons (1988)
14. M.P. Seah and W.A. Dench, Surface Interface Anal.1: 2 (1979)
15. H. Ibach and D.L. Mills, "Electron Energy Loss Spectroscopy and Surface Vibrations", Academic Press (1982)
16. N.R. Avery, Surface Sci. 111: 358 (1981)
17. F.P. Netzer and M.M. El Gomati, Surface Sci. 124: 26 (1983)
18. G. Meister, P. Giesert, H. Hölzl and L. Fritsche, Surface Sci. 143: 547 (1984)
19. H. Bethe, Ann. Physik 5: 325 (1930); Z. Physik 76: 293 (1932); Handbuch der Physik, Vol. 24-1, H. Geiger and K. Scheel, eds., Springer Verlag (1932), p. 273
20. M. Inokuti, Rev. Mod. Phys. 43: 297 (1971)
21. M. Inokuti, Y. Itikawa and J.E. Turner, Rev. Mod. Phys. 50: 23 (1978)
22. R.D. Leapman, in "Electron Beam Interactions with Solids for Microscopy, Microanalysis and Microlithography", D.F. Keyser et al., eds., SEM Inc. AMF O'Hare, Chicago (1982), p. 217
23. M. Inokuti and S.T. Manson, ibid. p. 1
24. J.A.R. Samson, in "Electron Spectroscopy: Theory, Techniques and Applications" Vol. 4, C.R. Brundle and A.D. Baker, eds., Academic Press (1981), p. 361
25. R.F. Egerton, "Electron Energy Loss Spectroscopy in the Electron Microscope", Plenum Press (1986)
26. C.J. Powell, in "Electron Beam Interactions with Solids", SEM, Inc., AMF O'Hare, Chicago, p. 19
27. A. Chutjian and D.C. Cartwright, Phys. Rev. A23: 2178 (1981)
28. J.K. Rice, A. Kuppermann and S. Trajmar, J.Chem. Phys. 40: 945 (1968)
29. G.W. Rubloff, Solid State Commun. 26: 523 (1978)

30. F.P. Netzer and J.A.D. Matthew, Solid State Commun. 29: 209 (1979)
31. F.P. Netzer, E. Bertel and J.A.D. Matthew, Surface Sci. 92: 43 (1980)
32. J.C. Riviere, in "Practical Surface Analysis by Auger and X ray Photoelectron Spectroscopy", D. Briggs and M.P. Seah, eds., Wiley and Sons (1983), p. 17
33. P.M.Th. Van Attekum and J.M. Trooster, Phys. Rev. B 18: 3872 (1978)
34. C. Colliex, M. Gasgnier and P. Trebbia, J. Physique 37: 397 (1976)
35. J. Daniels, C.v. Festenberg, H. Raether and D. Zeppenfeld, Springer Tracts in Modern Physics, Vol. 54, Springer Verlag (1970)
36. J.A.D. Matthew, E. Bertel and F.P. Netzer, Surface Sci. 184: L389 (1987)
37. H. Froitzheim, H. Ibach and D.L. Mills, Phys. Rev. B11: 4980 (1975)
38. L. Surnev, G. Rangelov, E. Bertel and F.P. Netzer, Surface Sci. 184: 10 (1987)
39. G. Rosina, E. Bertel and F.P. Netzer, Phys. Rev. B34: 5746 (1986)
40. S.T. Manson, in "Topics in Applied Physics" Vol. 26, Springer Verlag (1978), p. 135
41. U. Fano and J.W. Cooper, Rev. Mod. Phys. 40: 441 (1968)
42. A.D. McLean and R.S. McLean, Atomic Data and Nuclear Data Tables 26: 197 (1981)
43. T.M. Zimkina, V.A. Fomichev, S.A. Gribovskii and I.I. Zhukova, Sov. Phys. Solid State 9: 1128, 1163 (1967); R. Haensel, P. Rabe and B. Sonntag, Sol. State Commun. 8: 1845 (1970)
44. J.L. Dehmer, A.F. Starace, U. Fano, J. Sugar and J.W. Cooper, Phys. Rev. Lett. 26: 1521 (1971); J.L. Dehmer and A.F. Starace, Phys. Rev. B5: 1792 (1972); J. Sugar, Phys. Rev. B5: 1785 (1972)
45. G. Strasser, G. Rosina, J.A.D. Matthew and F.P. Netzer, J. Phys. F: Met. Phys. 15: 739 (1985)
46. L.I. Johansson, J.W. Allen, I. Lindau, M.H. Hecht and S.B. Hagström, Phys. Rev. B21: 1408 (1980)
47. U. Fano, Phys. Rev. 124: 1866 (1961)
48. F.P. Netzer, G. Strasser and J.A.D. Matthew, Phys. Rev. Lett. 51: 211 (1983)
49. W.A. Goddard III., D.L. Huestis, D.C. Cartwright and S. Trajmar, Chem. Phys. Lett. 11: 329 (1971)
50. J. Sugar, Phys. Rev. B5: 1785 (1972)
51. B.T. Thole, G. Van der Laan, J.C. Fuggle, G.A. Sawatzky, R.C. Karnatak and J.-M. Esteva, Phys. Rev. B32: 5107 (1985)
52. J.A.D. Matthew, G. Strasser and F.P. Netzer, Phys. Rev. B27: 5839 (1983)
53. F.P. Netzer, G. Strasser and J.A.D. Matthew, in "Science with Soft X-Rays", F.J. Himpsel and R.W. Klaffky, eds. Proc. SPIE 447: 34 (1984)
54. J.-M. Esteva, R.C. Karnatak, J.C. Fuggle and G.A. Sawatzky, Phys. Rev. Lett. 50: 910 (1983)
55. G. Strasser, E. Bertel, J.A.D. Matthew and F.P. Netzer, in "Valence Instabilities", P. Wachter and H. Boppart, eds., North Holland (1982), p. 169
56. G. Wendin, "Photoelectron Spectra", Structure and Bonding 45, Springer Verlag (1981)
57. G. Strasser, F.P. Netzer and J.A.D. Matthew, Sol. State Commun. 49: 817 (1984)

58. F.P. Netzer, G. Strasser, G. Rosina and J.A.D. Matthew, J. Phys. F: Met. Phys. 15: 753 (1985)
59. J.A.D. Matthew, in "Giant Resonances in Atoms, Molecules and Solids", Nato ASI Les Houches 1986, J.P. Connerade et al., eds., Plenum Press (1987)
60. J.A.D. Matthew, F.P. Netzer, C.W. Clark and J.F. Morar, Europhysics Lett. 1987, in press
61. E. Bertel, G. Rosina, H. Saalfeld, G. Rangelov and F.P. Netzer, Phys. Rev. B, to be published
62. T. Åberg and J.L. Dehmer, J. Phys. C 6: 1450 (1973)

PROPERTIES OF SYNCHROTRON RADIATION

George S. Brown

Stanford Synchrotron Radiation Laboratory
Stanford University
Stanford, California 94305

INTRODUCTION

Physics students are taught at an early age that the electromagnetic field mediates the interaction of charged particles. Furthermore, the acceleration of charged particles generates an electromagnetic wave which propagates at a speed c which is independent of the frequency of the radiation. As a practical matter, electrons or positrons are the favored particles to accelerate for the generation of electromagnetic fields, because of their low mass. As a further practical matter, useful electromagnetic fields are generated by four techniques: direct manipulation, as in electron tubes, klystrons, and magnetrons (often called physical electronics); atomic transitions, as in incandescent sources, lasers, and x-ray generators; bremsstrahlung, as in continuum x-ray sources and high energy physics sources; and nuclear transitions, resulting in quanta from a few KeV to hundreds of MeV.

There is no small irony in the emergence, from high energy physics research, of a new type of physical electronics, called the electron storage ring, generating electromagnetic radiation, called synchrotron radiation. The irony emerges from the fact that this radiation will probably set an upper limit to the energies available in electron storage rings for high energy physics research, with the prevailing opinion being that the LEP storage ring at CERN will probably be the highest energy electron storage ring to ever be built. The irony also emerges from the fact that these storage rings are designed expressly to produce virtual photons of energies from about one GeV to about one hundred GeV (in the case of LEP), but that they are turning out to be superb sources of real photons in the energy region of a few electron volts to perhaps a million electron volts (also in the case of LEP). This stems from the fact that synchrotron radiation has a characteristic wavelength which is just the radius of curvature of the particle's orbit, ρ, divided by γ^3, where γ is the kinetic energy of the electron divided by its rest energy. In the following discussion, I would like to make some qualitative arguments about how this comes about, followed by some more quantitative discussion and then some practical information about how this extraordinary source might be utilized in, for example, x-ray spectroscopy.

QUALITATIVE ASPECTS OF SYNCHROTRON RADIATION

The basic spectral properties of synchrotron radiation can be deduced from elementary considerations of special relativity theory. Consider, for example, a particle moving with relativistic velocity along a path of uniform curvature. For convenience, we may analyze the problem in a reference frame in which the particle is instantaneously at rest. Since, by definition, the particle is stationary for some brief period, the motion of the particle is nonrelativistic, and we may use classical electromagnetic theory to analyze the emitted radiation. In particular, it is well known that the radiation emitted is well-described by a dipole antenna, resulting in a radially propagating radiation field with the qualitative features shown in figure 1. In particular, the radiation vanishes along the direction of the acceleration vector of the particle, and peaks along a propagation direction normal to the acceleration vector. This property is

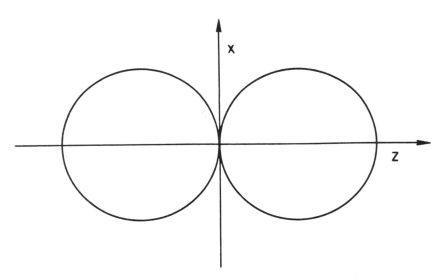

Fig. 1 The radiation pattern from an oscillating dipole with zero average velocity

quite general and independent of the details of the time dependence of the particle's motion. Of course, the actual functional form of the angular pattern will depend upon the details of the motion.

The angular properties of the radiation in the laboratory frame can be understood by invoking the Doppler transformation:

$$\tan\theta = \frac{\sin\theta}{(\cos\theta + \beta)}$$
$$\omega' = \gamma\omega(1 + \beta\cos\theta) \tag{1}$$

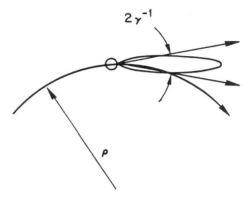

Fig. 2 The radiation from an oscillating dipole in instantaneous circular motion.

From these two equations, one can immediately conclude that a forward-emitted lob of dipole radiation will be collapsed into a cigar-shaped lobe with characteristic angle $1/\gamma$, and that the frequency in the co-moving frame is boosted by the factor γ. However, since the motion in the co-moving frame is not periodic, the observed spectrum will be rather broad. An estimate of this spectrum can be made by reference to figure 2. The lobe of radiation emitted by a particle moving in instantaneously circular motion will sweep by an observation point O in a time $t = \rho/c\gamma^3$. From elementary considerations of Fourier theory, one can conclude that the radiation pulse will have frequencies clustered in the neighborhood of $\gamma^3 c/\rho$, implying wavelengths in the neighborhood of ρ/γ^3, as mentioned in the introduction. Thus, an electron with an energy of a few GeV moving in a circular orbit of radius 10 meters or so will emit radiation that has a significant component in the x-ray spectrum.

QUANTITATIVE CONSIDERATIONS

The detailed spectral and angular characteristics of synchrotron radiation have been worked out by many authors (1-4), and so will not be treated in detail here. For reference purposes, however, we shall present the results of such calculations, and leave the details to the interested student. The basic result of the analysis is eq. (2):

$$\frac{d^3n}{d\Omega(d\omega/\omega)} = \frac{3\alpha\gamma^2(\omega/\omega_0)^2}{4\pi^2}(1 + \theta^2\gamma^2)^2 \cdot$$
$$(K^2_{2/3}(\eta) + \frac{\gamma^2\theta^2}{1 + \theta^2\gamma^2}K^2_{1/3}(\eta)) \qquad (2)$$
$$\eta = (\gamma/2)(\omega/\omega_c)(1 + \theta^2\gamma^2)$$
$$\omega_c = \gamma^3(3c/2\rho)$$

Fortunately, the interpretation of this formula is rather straightforward. The first term accounts for radiation emitted with polarization vector parallel to the plane of the particle's orbit, and the second term accounts for radiation emitted perpendicular to the orbital plane. By inspection, the perpendicular component vanishes on axis, resulting in 100% plane polarized radiation in the plane of the orbit. An angular and energy integration of the two terms in eq. (2) yields a ratio of power in the two polarizations to be about 7 to 1 favoring the parallel polarization; for this reason, most polarization-sensitive experiments utilize the parallel component.

A detailed evaluation of the profile of the parallel component of the radiation shows that the angular width of the radiation is weakly dependent upon the frequency. If we define a characteristic angle θ_c, then this angle has the following frequency dependence:

$$\theta_c \simeq (1/\gamma)(\omega_c/\omega)^{1/3} \qquad \omega \ll \omega_c$$
$$\theta_c \simeq (1/\gamma)(\omega_c/3\omega)^{1/2} \qquad \omega \ll \omega_c \qquad (3)$$

From these expressions we see that for frequencies of order of the critical frequency, the opening angle is of order $1/\gamma$, and that the characteristic angle slowly diminishes for increasing photon energy.

Many, if not most, experiments integrate over the full vertical profile of the synchrotron radiation. For these experiments, it is useful to calculate the angle-integrated spectrum, which is given by the following expression:

$$\frac{d^3n}{d\phi(d\omega/\omega)} = (3/4\pi^2)^{1/2}\alpha\gamma(I/e)(\omega/\omega_c) \quad dx K_{5/3}(x) \qquad (4)$$

This formula confirms the broadband nature of the photon spectrum alluded to in the introduction. For reference purposes, the limiting forms of formula (4) are given below:

$$\frac{d^2n}{d\phi(d\omega/\omega)} = 4^{1/3}(1/\Gamma(1/3))\alpha\gamma(I/e)(\omega/\omega_c)^{1/3} \qquad \omega<<\omega_c$$

$$\frac{d^2n}{d\phi(d\omega/\omega)} = (3/2\pi)^{1/2}\alpha\gamma(I/e)(\omega/\omega_c)e^{-\omega/\omega_c} \qquad \omega>>\omega_c$$

(5)

Thus, it can be seen that for $\omega<<\omega_c$, the spectrum is proportional to the 1/3 power of the frequency, while for high frequencies, the spectrum is roughly exponential in ω/ω_c. This latter form is useful for estimating intensities well above the critical frequency.

WIGGLER MAGNETS AND SPECTRA

The previous discussion has applied to particles moving in instantaneously circular orbits, as from the bending magnets in a synchrotron. However, it is obvious that this sort of motion can be generated in regions of the storage ring where the particles would ordinarily drift in straight lines, provided that the "wiggles" comprise an orbit that does not introduce any net displacement or angle to the beam. Then, a series of N quasi-periodic wiggles will introduce an N-fold increase in the intensity per unit horizontal angle. The spectrum will be identical to the spectrum of the single bend described above, provided that the angular deflection of the particle in its periodic orbit is significantly greater than the characteristic angle $1/\gamma$. Such a magnet is called a "wiggler" magnet.

UNDULATOR MAGNETS AND SPECTRA

Recall that the fundamental assumption leading to the derivation of the bend magnet or wiggler spectrum was the fact that the photon beam swept by the observation point for a brief interval of order $1/\omega_c$, to return again only after a much longer time. Now for the wiggler magnet described above, it is possible to imagine that the cone of synchrotron light continuously illuminates the observation point during the wiggles, because the angular amplitude of the motion is small compared to $1/\gamma$. In this case, a new time scale is introduced, related to the periodicity of the motion. Consider, for example, the sinusoidal orbit described in figure (3). An observer situated at a polar angle of observation θ with respect to the undeflected motion will observe a periodic wave train approaching him with a wavelength that is given by the formula

$$\lambda = (\lambda_u/2\gamma^2)(1 + K^2/2 + \theta^2\gamma^2) \tag{6}$$

where K is the so-called strength parameter given by

$$K^2 = <(dx/dz)^2>\gamma^2 \tag{7}$$

We see that K is a measure of the excess path length that the particle must travel over the rectilinear path, λ_u. For simplicity, consider the case where K is near zero, and where the radiation is viewed on-axis. The new characteristic wavelength is now just $1/2\gamma^2$. For radiation emitted off-axis, the radiation is shifted to a longer wavelength, dependent only upon the polar angle. This effect may be regarded simply as the Doppler effect, well known from classical electrodynamics.

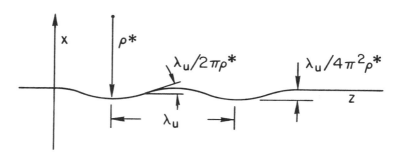

Fig. 3   The electron trajectory in the laboratory frame.

If the number of periods of the undulator is finite (as it must be!) then the strict correlation between angle and wavelength (or energy) is no longer valid. An easy way to understand this is to imagine the radiation pattern in a Lorentz reference frame moving with a velocity equal to the average velocity of the particle. In this frame, the particle emits quasi-monochromatic radiation with the characteristic dipole pattern, but the frequency spread is roughly 1/N, with the detailed spectrum depending upon the way the trajectory is cut off at the ends. Thus, when this radiation is viewed at some unique observation angle, the frequency spread is again 1/N.

The reader might now well inquire as to whether this radiation is intense enough to be useful. To answer this question, we first quote the power radiated by such an insertion device, a result that is easily derivable from semiclassical arguments:

$$P = (2/3)\alpha hc\gamma^4 <\rho^{-2}> L(I/e) \tag{8}$$

A quick inspection of formula (8) tells us that for typical currents, energies, and insertion device lengths, powers of order of kilowatts will be radiated! But what about the spectrum? For most experiments, the main interest will lie in the spectrum at the fundamental frequency ω_1 or at odd multiples of this frequency (more about these so-called harmonics later). The intensity per unit solid angle and the angle-integrated spectrum at ω_1 can be expressed in a simple form:

$$\frac{d^2 n}{d\Omega (d\omega/\omega)} = 4\alpha N^2 (I/e)(1 + K^2/2)^{-1} (J_1(\xi) - J_0(\xi))^2$$

$$\frac{dn}{(d\omega/\omega)} = 4\pi\alpha N(I/e)\xi (J_1(\xi) - J_0(\xi))^2 \tag{9}$$

where α is the fine structure constant (1/137.04), and $\xi = (K^2/4)(1+K^2/2)$.

If the motion is simple harmonic in the co-moving frame described above, why are there harmonics? Under circumstances where the amplitude is small (K<<1) then the motion is indeed simple harmonic. However, a little reflection will convince the reader that a relativistic particle executing sinusoidal oscillations in the lab frame (or any other periodic oscillation), does not execute simple transverse simple harmonic motion in the co-moving frame. The reason is that the particle is moving very close to the speed of light, with a constant speed (provided that the deflecting fields are magnetic), and thus longitudinal velocities in the center mass are necessary. These motions are in the form of a figure-eight, as shown in figure (4); they occur at even harmonics of the fundamental frequency; and they are by no means simple harmonic. A detailed analysis is beyond the scope of this article, and the interested reader is referred to the literature.

CONCLUSION

In this short space we have endeavored to show how relativistic particles, in particular electrons or positrons, under the deflection by quite modest magnetic fields, emit copious quantities of x-radiation,

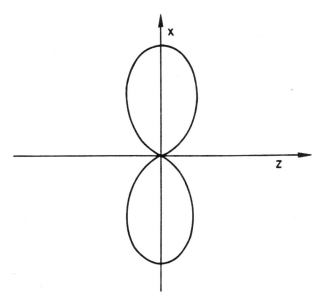

Fig. 4 The electron trajectory in the co-moving reference frame.

whose properties can be tailored to the experiment of interest. The intensities are many orders of magnitude greater than can be achieved with conventional x-ray generators, making possible experiments that have been hitherto impossible.

ACKNOWLEDGEMENT

The work reported herein was done at SSRL which is supported by the Department of Energy, Office of Basic Energy Sciences; and the National Institutes of Health, Biotechnology Resource Program, Division of Research Resources.

REFERENCES

1. John David Jackson, <u>Classical Electrodynamics</u>, John Wiley and Sons, Inc., New York.
2. D.F. Alferov, Yu. A. Bashmakov, and E.G. Bessonov, Zh. Tekh. Fig. <u>43</u>, 2126 (1973).
3. S. Krinsky, M.L. Perlman, and R.E. Watson, Characteristics of Synchrotron Radiation and of its Sources, in <u>Handbook on Synchrotron Radiation</u>, Vol. I, edited by E.E. Koch, North-Holland Publishing Co., Amsterdam.
4. S. Krinsky, IEEE Transactions on Nuclear Science, <u>NS-30</u>, 4(1983).

GASEOUS X-RAY DETECTORS

A.J.P.L. Policarpo

Physics Department
University of Coimbra
3000 Coimbra
Portugal

ABSTRACT

A brief selective review is made of gaseous X-ray detectors, mainly intrinsic parameters are referred (energy and position resolution, bandwidths, high rate capabilities, etc) and no particular applications in any field of research are considered. A rather unusual relevance is given to detectors based on the photon flux and to multi-step chambers, as it is felt that their potentialities are not yet fully realized. Particular examples reflecting the state-of-the-art are given both for charge and light devices.

INTRODUCTION

The main purpose of this work is to call attention to the potentialities of some principles of detection and to give an overall view of the field of gaseous X-ray detectors, although some important and recent developments are only very briefly referred. The general principles of the induced charge read-out method, examples of very good position resolution that reflect details of the X-ray interaction, the main characteristics and some recent information on the performance of multi-step chambers, gas scintillation proportional counters and the electron counting technique, as well as high counting rate devices are considered.

INDUCED SIGNALS

Although planar multiwire chambers have been developed more than three decades ago and the main features of the mechanisms of electron drift, avalanche and electric signals development were well stablished, MWPC associated to induced signals read out techniques are in current use only during the last decade. A contribution to this field is the work 1 where it was shown that the shape of the avalanche around thin wires memorizes to some extent the azimuth of the arrival of a punctual cloud of primary electrons and an upper limit $\sigma=6\mu m$ in position accuracy was obtained, related to electronic noise, accuracy of digitization and fluctuations in the charge distribution of the avalanche. Better intrinsic position resolutions have been obtained recently (see next section).

Fig. 1 illustrates the relevant principles of MWPC read out based on measurements of the centroids of the charges induced on cathode strips x_i and y_i [1]. x_i designates also the x coordinate of strip x_i. A, B, C, X_i and Y_i are the charges induced in the electrodes designated in the figure by the corresponding small letters. Let $E_x = \Sigma X_i$, $E_y = \Sigma Y_i$ and $x' = \Sigma X_i \, x_i / \Sigma X_i$.

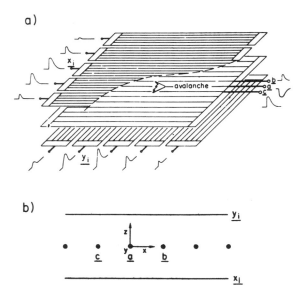

Fig. 1. a) Diagram of the MWPC showing the position of the strips x_i and y_i, main anode wire a, and neighbouring wires b and c, as well as the pulses on these electrodes; b) System of reference.

Using events obtained by bombarding uniformly with Al X-rays the cell that corresponds to wire a, the usual position resolution associated with avalanches around one wire is displayed in Fig. 2a. Using convenient coordinates the same events are displayed also in Fig. 2b, showing not only that much better position resolutions are inherent to the read out technique, but also that the avalanches do not develop more or less isotropically around the anode wire. Indeed each dot in Fig. 2b, corresponding to an individual avalanche, has x, y and z coordinates that are related to the point in space where the X-ray interacted in the medium [1].

This is clearly illustrated in Fig. 3 where avalanches on a single wire are used providing neverthless bidimensional information. In this figure the coordinates are B/C and $\Sigma Y_i \, y_i / \Sigma Y_i$. Position information is then obtained, of course in the wire direction, but also in the direction orthogonal to the anode wires [1].

Progress during the last years using information from induced charges have been made looking for optimum geometry for strip cathodes, defining resolution limits due to electronic noise, analysing the non-linearity problems associated to finite sampling size, etc [2]. Great progress was made in the hardware computation of centroids.

Although the MWPC geometry is by far the most used, the essential principles associated to induced charge imaging can be applied to a variety of geometries. A particularly simple device is shown in reference 3 where a point anode is surrounded by four planar pickup cathode electrodes.

This imaging is proceeded by a convertion drift space for X-ray positioning, and a spatial resolution better than 1mm FWHM over a field of 30mm diameter was obtained with an energy resolution of 20% for 5.9keV X-rays.

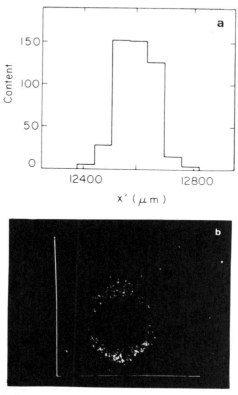

Fig. 2. a) x' position spectrum for Al X-rays in a xenon + isobutane + methylal mixture; b) E_y/E_x vs B/C for the same events.

Fig. 3. Bidimensional radiograph with Al-K X-rays. Letters are 1.5mm is height, 50µm in width, cut out in a copper mask.

POSITION DETECTION WITH MWPC

There are intrinsic factors that determine the attainable position resolution, namely the range of the energy deposition, diffusion effects and fluctuations of the avalanche along the anode wire. The electronic noise of the position scanning device is also inherent to the detector as a whole.

Range of Energy Deposition

Electrons range. Let us consider processes in which, following the absorption of an X-ray, fluorescence emission does not take place. Then the energy of the X-ray, after the interaction, emerges shared by the photoelectron and Auger electron. A general view of this energy sharing process is displayed in Fig. 4 [4].

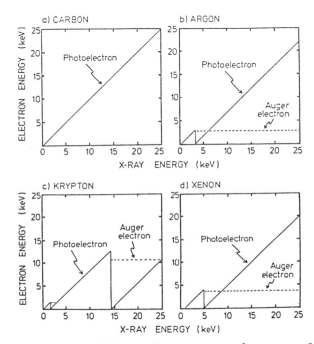

Fig. 4. Photoelectron and Auger electron energies, as a function of X-ray energy, after X-ray absorption by atom of a) carbon, b) argon, c) krypton and d) xenon. Electron shell brinding energies: C_K:0.28keV; Ar_K:3.2keV; Kr_K:14.3keV; $\overline{Kr_L}$:1.8keV; $\overline{Xe_L}$:5keV. Emitted electron energies are for X-ray absorption only in most likely shell of atom. Energy of only the most energetic Auger electron is shown. Fluorescence emission is not taken into account.

The position resolution of the detector is determined by the average range of the energy deposition in the medium. Following reference 4, assuming isotropy and uncorrelation for the emission of the photoelectron and the Auger electron, and on average uniformity of ionization density along the practical ranges, R_p, of the electrons, the following can be shown. If most of the X-ray energy is transfered to only one electron the centroids of ionization arising from many interactions, projected in one dimension, give rise to a rectangular distribution of width R_p. Smaller full widths half-maximum arise for this distribution, that changes shape, if the X-ray energy is shared more evenly by the photoelectron and the Auger electron, reaching a minimum for equal energy sharing.

Fig. 5 [4] corresponds to measurements done at atmospheric pressure, selecting only photopeak events and using a detector such that effects that can deteriorate the position resolution, diffusion, avalanche fluctuations and electronic noise are negligible compared to those arising from electron ranges. These data are is agreement with the previous considerations.

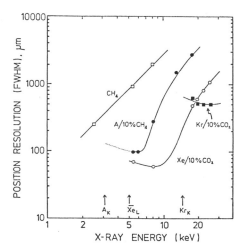

Fig. 5. Position resolution, limited only by photoelectron and Auger electron ranges, as a function of X-ray energy, in CH_4 (open squares), Ar/10% CH_4 (full circles), Kr/10% CO_2 (full squares) and Xe/10% CO_2 (open circles), all at atm pressure.

For CH_4 the FWHM position resolution varies as $Ex^{1.78}$, a similar power law relating Rp and Ex [5], in agreement with Fig. 4a). The same trend of the variation of the FWHM position resolution with Ex can be seen both for Ar/10% CH_4 and Xe/10% CO_2, when the photoelectron energy is much larger then the energy of the Auger electron, and the minima of the position resolutions arising from equal energy sharing for these electrons appear at the proper energies (see Fig. 4b) and d) respectively). Data for Kr/10% CO_2 are also in agreement with Fig. 4c).

Fluorescence radiation. If one is able to keep the electron range effects dominating the position resolution, a variation of this quantity with 1/p is expected. This is seen comparing the data of Fig. 5 with the data of Fig. 6 [4], obtained at 10 atmosphere, for the curves Xe/10% CO_2 and Kr/10% CO_2 (photopeak), that display the same general behaviour arising from energy sharing between the photoelectron and Auger electron, the FWHM position resolution being a factor ~10 smaller.

In the energy range displayed in Fig. 6, and unlike what happens with xenon and argon, most interaction with krypton are followed by fluorescence emission. The X-ray interaction in krypton, mainly in the K shell, gives rise to a photoelectron of Ex-14.3keV and in 64% of the cases, the excited atom deexcites by emission of a fluorescent photon of 12.8keV that has,

in general, a large probability of escaping out of the detector, these
events originating the escape peak. The smaller range of the low energy
photoelectron, Ex-14.3keV, is responsible for the very good position
resolution displayed in Fig. 6. Around 25keV the position resolution of the
photopeak is smaller than that of the escape peak, due to the fact that the
photoelectron and the Auger electron share equally the energy available
(see Fig. 4c)) in the case of the photopeak event.

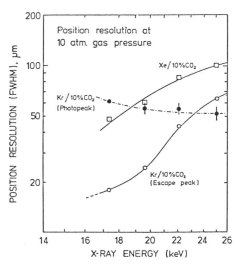

Fig. 6. Optimum position resolution vs X-ray energy for 17-25keV X-rays
in Kr/10% CO_2 (photopeak events (full circles); escape peak
events (open circles)) and Xe/10% CO_2 (open squares).

MULTI-STEP AND PARALLEL-PLATE CHAMBERS

In a conventional MWPC, X-ray conversion, primary electron drift and
avalanche development take place in the same structure defined by the three
planes of electrodes. In the multi-step avalanche chamber MSC [6], the X-ray
conversion takes place in a separated region and the charge amplification
process is of a multi-step nature and occurs in physically distinct regions
(see Fig. 7 [7]). Primary electrons produced in the conversion region drift
to the high field preamplification gap, a fraction (~20%) is transfered
to the transfer region, and a further amplification takes places on the
last amplification gap. This last structure can be a parallel plate structure
(and then fragile thin wires are avoided) or a MWPC, and is provided with
positioning capabilities. Stable gains, well in excess of 10^5 are obtained
with MSC.

The transversal spread of the avalanches tends to give a non modulated
position response associated to the anode wires if the last amplification
stage is a MWPC, and lower local gains are involved as the total gain is
divided in two steps. Avalanche instabilities are then reduced and lower
space charge densities are involved, both effects favouring energy and
position resolution. Apart from these characteristics MSC feature possibilities of correction of parallax errors[7], can be operated at low pressures
with large conversion gaps [8], proved useful as fast ultraviolet photon
detectors (single electron detection with sub-milimetric x,y,z position

resolution [9]), the signals from the several electrodes available can be used due to the progress in low-cost, low-noise, high-gain amplifiers, and are finding applications in several fields of physics. They have the disadvantage of the mechanical accuracy effects associated to parallel-plate amplification.

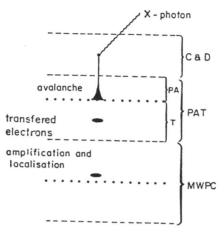

Fig. 7. A conversion and drift space (C and D) is followed by a preamplification (PA) and transfer (T) elements, a further amplification taking place in the last region, a MWPC in this case.

The association of multi-step avalanche chambers with Penning gas mixtures provides good overall performances for X-ray detection [10]. The use of Penning mixtures improves the energy resolution (11% at 6keV) of conventional proportional counters but this was achieved with gas gains less than 50 [11]; see also [12]. Using a MSC with an active diameter of 180mm (energy range 0.1-8keV), and a mixture of Ar + ~1% C_2H_2, preliminary results yield an energy resolution of ~25% FWHM and the position resolution over the full field of view is 200μm FWHM, for 1.5keV X-rays [10]. These values are a good compromise of the characteristics of both MWPC and GSPC concerning energy and position resolution: typically 36% FWHM energy resolution for MWPC and >1mm position resolution for GSPC.

This is a promising field. Using argon and xenon quenched with 2% isobutane an energy resolution of 12.8% for 5.9keV was obtained from the preamplification gap at a gas gain of 10^4 in this gap, and very good position resolutions have been obtained using information from the second gap, this one providing the amplification needed for positioning purposes [13].

Parallel-plate counters with large areas and position sensitive feature simplicity of construction compared with MWPC and MSC and are currently used for the detection of heavily ionizing particles, with very good localization and timing properties. But they provide only small gas gains before breakdown, ~10^3-10^4, being thus of difficult use for soft X-rays.

Higher gains, similar to those obtained with MSC in conditions of good energy and position resolutions are still available in a parallel-plate configuration if special care is taken to avoid edge breakdown, by increasing the distance between the electrodes near the frame's edge [14].

The thickness of the amplification gap together with the high uniform electric field involved, provide reduced avalanche spread, fast signals (~10ns full width at the base for the electronic signal) and fast collection of the positive ions. The last electrode is made of a printed circuit board, with internal (anode) and external orthogonal strips for localization. For a 20x20cm^2 with a mechanical accuracy of ± 30μm, a maximum ± 15% gain variation across the whole sensitive area was obtained. Using signals ~100ns local energy resolution measurements gave 20% for 6keV and the position resolution was 280μm FWHM for 8keV, including of course electron diffusion (6mm drift space) and photoelectron ranges [14].

OPTICAL DETECTORS

Main Principles

Most gas scintillation proportional counters, GSPC, providing information both in energy and position incorporate two planar regions defined internally using grids or planes of wires: the absorption or drift region where low E/p values are used and the scintillation region (~1.5cm), where E/p is set just below the threshold for charge multiplication.

The characteristic molecular UV emission (at ~1700Å for Xe, by far, up to now, the most common filling) is viewed through an appropriate window (currently CaF$_2$ or spectrosil) either by an array of photomultipliers (the Anger Camera GSPC), by microchannel plates (the Microchannel Plate GSPC) or by multiwire proportional chambers, MWPC, or imaging proportional counters with fillings that have additives with low photoionization potential like TEA and TMAE (the Photoionization GSPC).

The main factors that determine the favourable energy resolution of GSPC are the very high efficiency with which the electric field energy of the applied field is converted into a photon flux through a electroluminescence process and the small statistical spread in the number of photons associated to the drift of one electron between the cathode and the anode of the detector. Both these characteristics are displayed in Fig. 8 [15] for xenon at atmospheric pressure.

Let H be the number of electronic excitations associated to the drift of one electron, g the fraction of the energy supplied by the field that is used in excitations and let J be defined by the relationship $(\sigma_{H/\bar{H}})^2 = J/\bar{H}$. In the expression for the energy resolution, fluctuations in H appear associated to the Fano factor through the term $F + J/\bar{H}$. Noting that $F \sim 0.12$ and taking into account the information from Fig. 8, for favourable values of E/p (just below the threshold for charge multiplication), $J/\bar{H}$ can be orders of magnitude smaller than the Fano Factor, and this intrinsic parameter essentially limits the energy resolution of GSPC.

Apart intrinsic parameters, in general terms, the measured energy resolutions, approximately a factor of two better than with proportional counters, are determined by UV photon detection efficiency and the position resolution, worse than for proportional detectors, by the degree of sampling of the UV light distribution from the scintillation region, the uniformity of these characteristics over large areas being very important.

A more recent development to achieve good energy and position resolution, although at present not yet achieved simultaneously, the electron counting technique (ECT) [16], also known as the optical avalanche detector [17], relies in individually counting and/or positioning the primary electrons by detection of fast light pulses (~1ns) emitted by avalanches, each initiated by a single electron. This method tends to reach then the best possible information.

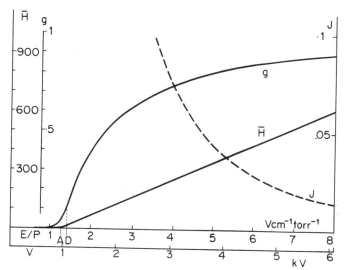

Fig. 8. For xenon, g and J are represented as a function of the reduced electric field E/p. Assuming a scintillation gap of 1cm and 760 torr the mean number of photons associated with the migration of one electron, $\bar{H}$, is also represented as a function of the voltage V across the gap.

The intrinsic characteristics of the GSPC concerning energy, time and position information arise naturally from the facts that essentially only the lower excited states of the noble gas medium are involved and then faster atomic transitions are absent, the evolution of the excited species up to deexcitation is slow, and low drift velocities and large diffusion coefficients, specially the transverse are involved.

In the ECT method high charge gains are used such that large photon yields are available per primary electron and it is then possible, by using additives that lead to fast collisional processes, to supress all the slow emission mechanisms and still have, useful, fast, photon information. A wide choice of suitable additives leads also to a simple control of the drift velocity and diffusion effects, providing higher drift velocities and lower diffusion than with noble gases and allowing the use of medium or even poor conditions of purity. This is an advantage relatively to the stringent requirements of cleanliness and purity of GSPC.

The basic geometry is of an imaging proportional counter, IPC, coupled to a P.M. tube or an image intensifier, (see Fig. 9 [18]). Using high fields on the thin avalanche region the development time of the avalanche arising from a single primary electron is ~1ns in a basic mixture Ar-CH_4 with addition of CO_2 and N_2 [19]. These additives reduce diffusion, increase the light output and through the collisional quenching of the Ar and N_2 excited states, give rise to fast photon emission (~1ns), the main emission arising from atomic states of Ar at ~7000 Å and from de second positive band of N_2, ~3500 Å. Then the duration of the light flash associated to the avalanche initiated by one primary electron is of the order of a few ns.

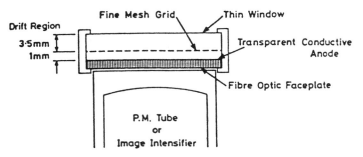

Fig. 9. Schematic of the imaging proportional counter from [18]. Copyright c 1981 IEEE.

Individual avalanches corresponding to the primary electron associated to an X-ray interaction can be counted electronically, using fast photomultipliers and electronics, by using low fields in the drift region (see Fig. 9): the size of the cloud of primary electrons is made to be such that on average these electrons arrive separated by a few tens ns at the avalanche region. Results obtained using the ECT, for energy resolution, are shown in Fig. 10 [19] the performance of the photomultiplier and electronic counting system limiting the ECT to energies below ~2keV.

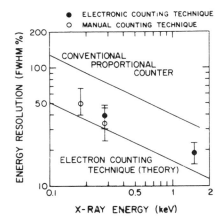

Fig. 10. Energy resolution versus energy for conventional proportional counters and using the electron counting technique from [19]. Copyright c 1983 IEEE.

The large diffusion needed to allow for good energy resolution is, in principle, incompatible with good position resolution as this is essentially limited by transverse diffusion effects (photoelectrons ranges are negligible compared to diffusion). By minimizing diffusion (fields on the drift region ~300V/cm; $Ar-CH_4$ (5%) - CO_2 (10%) mixture), resolutions of 350μm FWHM were obtained for C-K and Al-K X-rays using an image intensifier without centroiding [19], similar to the resolutions obtained with current centroiding parallel plate detectors.

Energy and Position Resolution and Bandwidth

Fig. 11, a reproduction of a table from reference 17, summarises the basic characteristics, predicted performances, potentials problems, etc, of the three types of GSPC considered, as well as the optical avalanche detector (or electrons counting technique). Common to all GSPC there are the problems of UV window technology and the gas purity requirements (for low energies, high vacuum technology does not solve problems associated with thin windows), both not present for the optical avalanche detector. This has the handicap of small energy bandwith, probably limited to a few keV, in contrast with the very large bandwidth of the GSPC. At least the Anger Camera and the Photoionization GSPC have the potentiality to grow to large areas without loss of simultaneously providing good energy and position information.

Detector type	Anger camera GSPC [4,8]	Microchannel plate GSPC [10,12]	Photoionisation GSPC [16]	Optical avalanche detector [24]
Gas	xenon	xenon	xenon (+P20+TMAE)	$Ar+CH_4+CO_2+N_2$
Gas purity requirement	very good [a] impurities ≤ 10 ppm?	very good	very good	medium to poor?
Achieved spatial and energy resolution	1.7 mm 6 keV 9% 6 keV	5 mm 6 keV 24% 6 keV	0.9 mm 6 keV 8% 6 keV	~0.35 mm 1.5 keV ~19% 1.5 keV
Demonstrated bandwidth	0.2–8 keV [4] 1.1–>40 keV [8]	(6 keV only)	0.1–6 keV	0.1–1.5
Demonstrated open aperture	6.5–10 cm	(5 cm) (MCP)	~5 cm	3 cm
Predicted spatial and energy resolution	~1 mm 6 keV ~9% 6 keV	$\leq 500\,\mu m$ 6 keV ~8% 6 keV	~250 μm at 6 keV ~8% at 6 keV	~0.1–0.2 mm 1.5 keV ~19% 1.5 keV
Future areas of study	larger sizes? non-linearities? electron gain?	prove predicted resolutions	larger size improved performance	methods of achieving simultaneous position and energy readout
Potential "problems"	PM tube: gain vs temperature	construction lifetime of MCP	lifetime? temperature control?	simultaneous readout? broad band width?
Common technological problems		reliable gas flow and circulation system UV window technology		

[a] Exact purity requirements are not known.

Fig. 11. Some achieved and predicted characteristics of optical detectors.

Examples of several GSPC using photomultipliers and photoionization detectors can be seen in reference 20. See reference 21 for information on microchannel plates GSPC. It should be referred that energy and position resolutions of 1.9 keV FWHM and 3 mm for 60 keV (3 atm) have been obtained [22] with Anger Camera GSPC and that systems are considered for hard X-ray and gamma ray detection with read out based on wave shifter fibers, well suited for high pressures (~40 atm) and large areas (~1 m^2) [23].

Timing and Driftless GSPC

The scintillation mechanisms and its time evolution are reasonably well known as well as the drift velocity and diffusion coefficients. The FWHM timing resolution using the first photoelectron detected by photomultipliers and arising from the secondary scintillation, determined by the rate of excitation and the efficiency of UV photon detection, can then be calculated [24]. Following this work, the FWHM timing resolution was computed as a function of energy for 1.9% UV detection efficiency, a current value, for three values of E/p in the scintillation region [20]. The GSPC is not a very fast detector for the low energies considered ~100ns for 0.1keV and ~30ns for 6keV (for higher energies the primary scintillation can be used) but these time resolutions clearly show that driftless detectors, in which detection is made on the scintillation region, are quite possible as in practice the secondary scintillation takes several µs, providing then the typical good energy and position information with reasonable timing characteristics [20,24].

A detailed study of several problems affecting both the energy and position resolution of GSPC, like loss of electrons to the window, the gas and the grid between the drift and scintillation region, the effect of electrons photoelectrically liberated from the detector window, etc, may imply some improvement of its instrumental characteristics if driftless detectors are used [21,25].

HIGH COUNTING RATES

Generalities

If the pulse corresponding to the detection of an X-ray is essentially related to a photon flux its intrinsic duration is dependent on electron drift velocities, on the over-all mechanisms associated to deexcitation of the medium (collision rates, life-times) on the size of the primary electron cloud (photoelectron range for example), and diffusion effects. And of course some of these parameters are also of relevance if a charge pulse is used. Medium highly electroluminescent feature low drift velocities and long deexcitation times, and then the need may arise of using fast mixtures from both point of view. In these mixtures the low electroluminescent yields imply the need of electron avalanche development.

If the output of the detector is a charge pulse the need of narrow pulses implies the use of only a fraction of the total charge pulse and electronic noise forces large gas gains, and then space charge effects arise. Typically in a MWPC configuration ~10% of the total charge pulse, corresponding to the ion motion during the first few ns in the high field region, is used, although a few tens µs are needed for the ions to reach the cathodes. The corresponding space charge creates a dead region ~200µm during ~20µs (dependent of the anode-cathode spacing), and loss of local amplification arises at rates of ~10^4 per mm of wire [26]. If the energy resolution is not the important parameter such that only a reasonable discrimination threshold is involved higher rates can be handled.

Short pulse duration is of course needed to handle high counting rates. If large detection areas are involved also some localized read-out method is essential, otherwise pulse pile-up effects and dead time losses can make the system unoperational. For example in MWPC one is then forced to read individual anode wires or individual cathode strips, rather then using the simpler delay line readout technique.

Photon Information

Although up to now light pulses have not yet been used to provide simultaneously positioning and high rates handling for large area detectors, a few examples are given of principles and devices that may contribute towards this aim.

Using the spherical anode counter GSPC with pure xenon and no charge gain, obtaining then energy resolutions about half the typical proportional counter, the effect of the counting rate on the peak position and energy resolution was studied, see Fig. 12 [27]. Although xenon is a slow detection medium, up to 10^5 c/s no detectable intrinsic (after correcting for photomultiplier effects) peak shift within 0.4% was detected and the energy resolution deterioration was less than 14%. In that geometry as in any other in which the thickness of the scintillating gap is essentially controlled by the applied voltage, a compromise between energy resolution and pulse duration can be achieved, and the timing associated to the scintillating mechanism in xenon would eventually limit the counting rates.

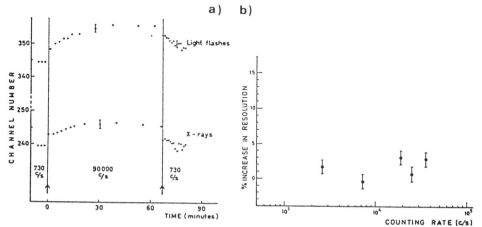

Fig. 12. a) Experimentally determined peak position as a function of counting rate. The light flashes from a gallium phosphide diode were used to measure the photomultiplier contribution; b) Increase of energy resolution (percent) as a function of counting rate.

Pure noble gases and mixtures between them or with nitrogen were used looking for high counting rates [28]. Fig. 13 displays single rate plateaux for high fluxes at a fixed threshold of detection, the shift in the plateaux position with rate indicating a space charge effect. At the highest rate available, 2×10^6 c/s from well collimated 1mm² beam from an X-ray generator (~8keV) no shift was observed. Charge gains were estimated up to a few hundred [28].

High counting rates capabilities can be associated with positioning, and MWPC configurations are then a natural approach. In this configuration, using anode wires 400μm diameter and a mixture of 10% N_2 in argon, the energy resolution and the mean pulse height from the photomultiplier were studied up to 10^7 c/s using a X-ray generator which produced a narrow 6.9keV X-ray beam [29]. All events were detected on one wire. The main results are summarized in Figs. 14 and 15 were space charge effects are clearly observed. It is seen that rates up to 10^7 c/s can be accepted, although the mixture used is not particularly fast (a few tens ns luminescence decay times) and cathode to anode distances, that determine space charge effects, are relatively large (~6mm).

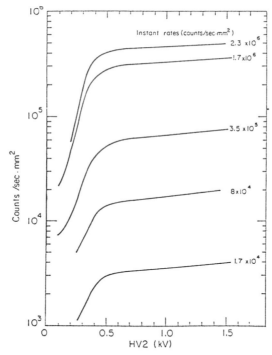

Fig. 13. Singles plateaux for an X-ray generator beam of 8keV collimated to 1mm^2, with a 20% duty cycle, for increasing intensities. No effect is observed on the plateau positions at instant rates > 2×10^6/s mm^2.

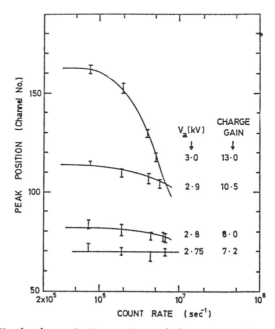

Fig. 14. Variation of the peak position, as a channel number, with count rate for different sense wire voltages.

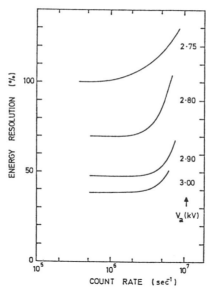

Fig. 15. The effect of count rate on energy resolution at different anode voltages.

Much faster electroluminescent mixtures as those associated to the electron counting technique have been used in work that may be relevant to high rate and positioning devices [30]. Using a MWPC configuration (10μm wires at 4mm spacing, 1cm distance between anode and cathode planes), using a mixture of Ar (78%), N_2 (10%) and CO_2 (12%) pulses from 1.5keV and 5.9keV X-rays seen by a photomultiplier were about 5 and 12ns duration respectively, with moderate charge amplification. Moreover the use of thin and narrow light converters to transport the light from large surface detectors to batteries of photomultipliers was shown to be feasible, without change of the time characteristics of the signals.

Charge Devices

Following [31] three types of one dimensional position-sensitive detectors may be considered: the delay line type, operating up to about 5×10^5 c/s over the whole detection field; the MWPC type, able to handle rates up to about 2×10^5 c/s wire (each wire having a preamplifier, discriminator and scaler); and the charge integrating type featuring higher maximum attainable counting rates than the previous ones. Only pulse counting techniques will be considered as they feature larger dynamic ranges in counting rates.

Most fast charge devices give pulses <10ns full duration, minimizing the input capacitance per channel and using fast delay line clipping techniques, associated with pole/zero networks in the shaping amplifier, and using interelectrode distances as small as possible, typically around 1mm (to reduce local space charge saturation). Relatively short drift spaces (~4mm) are used together with mixtures featuring high electron drift velocities (to take full profit of the rise time of the shaping amplifier).

In the previous conditions with mixtures of Xe/5% CO_2/10% CF_4 (CF_4 increases the drift velocity and CO_2 avoids polymerization products arising from organic quencher), and with noise levels of ~1500 electrons rms for each channel, rates in excess of 10^6/s wire were obtained [32], corresponding to rates in excess of 10^8/s over the whole active area of 12.7x2 cm^2.

Although for purposes other than X-ray detection, attention should be given to [33], where using gas mixtures of CF_4 and hydrocarbons at atmospheric pressure and low noise fast signal processing, rates of about 10^8/s cm^2 with position resolution below 1mm (FWHM) were obtained; see also reference 34. In reference 35 a curved multiwire proportional chamber, parallax-free one dimensional device is considered.

As was referred in a previous section of this work MSC and parallel-plate chambers can be well suited for high rate detection. Using fast current amplifiers they feature pulses of 10ns [4,36], and the mechanism of the avalanche originates small widths for the induced charge profile. The much simpler amplifier-discriminator circuit, rather than the expensive charge-recording electronics, could be used on each strip, with submillimeter localization accuracy.

CONCLUSIONS

Several relevant detection techniques have not been considered in this work or were only briefly referred. This is the case of topics such as Penning mixtures, the method of soft X-ray spectrum reconstruction, the problem of long living fillings, special geometries adequate to particular applications, the self-quenching streamer mode, etc.

Neverthless the overall view given in this work is enough to show that gaseous detectors will go on playing an important role in the field of X-ray detection, specially if good profit is taken of the progresses in the physical structure of chambers, in electronics and computers, and enough work is invested in the search of better fillings and in a deeper understanding of the fundamental physical processes involved.

ACKNOWLEDGEMENTS

Thanks are due to M.A.F. Alves. Financial support from Instituto Nacional de Investigação Científica - Centro de Fisica da Radiação e dos Materiais da Universidade de Coimbra, is acknowledged.

REFERENCES

1. G. Charpak, G. Peterson, A. Policarpo and F. Sauli, Progress in High-Accuracy Proportional Chambers, Nucl. Instr. and Meth. 148:471(1978).
2. E. Gatti, A. Longoni, H. Okuno and P. Semenza, Optimum Geometry for Strip Cathodes or Grids in MWPC for Avalanche Localization along the Anode Wires, Nucl. Instr. and Meth. 163:83(1979).
3. J. E. Bateman, The Imaging Pin Detector - a Simple and Effective New Imaging Device for Soft X-Rays and Soft Beta Emissions, Nucl. Instr. and Meth. A240:177(1985).
4. J. Fischer, V. Radeka and G. C. Smith, X-Ray Position Detection in the Region of 6μm RMS with Wire Proportional Chambers, Nucl. Instr. and Meth. A252:239(1986).
5. L. Katz and A. S. Penfold, Range-Energy Relations for Electrons and the Determination of Beta-Ray End-Point Energies by Absorption, Rev. Mod. Phys. 24:28(1952).
6. G. Charpak and F. Sauli, The Multistep Avalanche Chamber: a New High-Rate, High-Accuracy Gaseous Detector, Phys. Lett. 78B:523(1978).
7. G. Charpak, Parallax-Free, High Accuracy Gaseous Detectors for X-Ray and VUV Localization, Nucl. Instr. and Meth. 201:181(1982).

8. G. Charpak and F. Sauli, Use of TMAE in a Multistep Proportional Chamber for Cherenkov Ring Imaging and other Applications, Nucl. Instr. and Meth. 225:627(1984).
9. F. Sauli and G. Charpak, Ultraviolet Photon Detection in TMAE using a Multistep Proportional Chamber, IEEE Trans. Nucl. Sci. NS-32:663(1985).
10. H. E. Schwarz and I. M. Mason, Development of the Penning Gas Imager, IEEE Trans. Nucl. Sci. NS-32:516(1985).
11. H. Sipilä, Energy Resolution of the Proportional Counter, Nucl. Instr. and Meth. 133:251(1976); The Statistics of Gas Gain in Penning Mixtures, IEEE Trans. Nucl. Sci. NS-26:181(1979).
12. J. P. Sephton, M. J. L. Turner and J. W. Leake, The Quenching of Penning Mixtures in a Cylindrical Proportional Counter, Nucl. Instr. and Meth. A256:561(1987).
13. B. D. Ramsey and M. C. Weisskopf, The Performance of a Multistep Proportional Counter for Use in X-Ray Astronomy, IEEE Trans. Nucl. Sci. NS-34:672(1987).
14. A. Peisert and F. Sauli, A Two-Dimensional Parallel-Plate Chamber for High-Rate Soft X-Ray Detection, Nucl. Instr. and Meth. A247:453(1986).
15. A. J. P. L. Policarpo, Ionization Scintillation Detectors, Nucl. Instr. and Meth. 196:53(1982).
16. O. H. W. Siegmund, J. L. Culhane, I. M. Mason and P. W. Sanford, Individual Electrons Detected after the Interaction of Ionizing Radiation with Gases, Nature 295:678(1982).
17. M. R. Sims, A. Peacock and B. G. Taylor, The Gas Scintillation Proportional Counter, Nucl. Instr. and Meth. 221:168(1984).
18. O. Siegmund, P. Sanford, I. Mason, L. Culhane, S. Kellock and R. Cockshott, A Parallel Plate Imaging Proportional Counter with High Background Rejection Capability, IEEE Trans. Nucl. Sci. NS-28:478(1981).
19. O. H. W. Siegmund, S. Clothier, J. L. Culhane and I. M. Mason, Improved Energy Resolution Capability of an Imaging Proportional Counter using Electron Counting Techniques, IEEE Trans. Nucl. Sci. NS-30:350(1983).
20. A. J. P. L. Policarpo, Charge and Scintillation Gaseous Detectors for Low Energy X-Rays, in: "Springer Series in Optical Sciences", vol. 43 (X-Ray Microscopy), G. Schmahl and D. Rudolph, eds., Springer-Verlag, Berlin, Heidelberg (1984) p. 172.
21. D. G. Simons, P. A. J. de Korte, A. Peacock, A. Smith and J. A. M. Bleeker, Performance of an Imaging Gas Scintillation Proportional Counter with Microchannelplate Read-out, IEEE Trans. Nucl. Sci. NS-32:345(1985).
22. H. Nguyen Ngoc, J. Jeanjean, H. Itoh and G. Charpak, A Xenon High-Pressure Proportional Scintillation-Camera for X and Gamma-Ray Imaging, Nucl. Instr. and Meth. 172:603(1980).
23. B. Sadoulet, R. P. Lin and S. C. Weiss, Gas Scintillation Drift Chambers with Waveshifter Read-out for Hard X-Ray Astronomy, IEEE Trans. Nucl. Sci. NS-34:52(1987).
24. H. P. von Arb, J. Böcklin, F. Dittus, R. Ferreira Marques, H. Hofer, F. Kottmann, R. Schaeren, D. Taqqu and M. Wälchli, A Large Area Xenon Gas Scintillation Proportional Counter (GSPC) with Timing Information for the Detection of Low X-Rays, Nucl. Instr. and Meth. 207:429(1983).
25. A. Smith, A. Peacock and T. Z. Kowalski, A Gas Scintillation Proportional Counter for the X-Ray Astronomy Satellite SAX, IEEE Trans. Nucl. Sci. NS-24:57(1987).
26. G. Charpak, F. Sauli and R. Kahn, On some Factors Controlling High-Accuracy Measurement of X-Ray Quanta Positions with Multiwire Proportional Chambers, Nucl. Instr. and Meth. 152:185(1978).
27. A. J. P. L. Policarpo, M. A. F. Alves, M. Salete S. C. P. Leite and M. C. M. dos Santos, Detection of Soft X-Rays with a Xenon Proportional Scintillation Counter, Nucl. Instr. and Meth. 118:221(1974).

28. G. Charpak, S. Majewski and F. Sauli, The Scintillating Drift Chamber: A New Tool for High Accuracy, Very High Rate Particle Localization, Nucl. Instr. and Meth. 126:381(1975).
29. S. S. Al-Dargazelli, T. R. Ariyaratne, J. M. Breare and B. C. Nandi, The Performance of a Gas Scintillation Counter at High Count Rates, Nucl. Instr. and Meth. 200:341(1982).
30. D. Anderson and G. Charpak, Some Advances in the Use of the Light Produced by Electron Avalanches in Gaseous Detectors, Nucl. Instr. and Meth. 201:527(1982).
31. Koh-Ichi Mochiki and Ken-Ichi Hasegawa, Charge Integrating Type Position-Sensitive Proportional Chamber for Time-Resolved Measurements Using Intense X-Ray Sources, Nucl. Instr. and Meth. A234:593(1985).
32. J. Fischer, V. Radeka and G. C. Smith, Developments in Gas Detectors for Synchrotron X-Ray Radiation, Nucl. Instr. and Meth. A246:511(1986).
33. J. Fischer, A. Hrisoho, V. Radeka and P. Rehak, Proportional Chambers for Very High Counting Rates Based on Gas Mixtures of CF_4 with Hydrocarbons, Nucl. Instr. and Meth. A238:249(1985).
34. R. Henderson, W. Frazer, R. Openshaw, G. Sheffer, M. Salomon, S. Dew, J. Marans and P. Wilson, A High Rate Proportional Chamber, IEEE Trans. Nucl. Sci. NS-34:528(1987).
35. P. Pernot, R. Kahn, R. Fourme, P. Leboucher, G. Million, J. C. Santiard and G. Charpak, A High Count Rate One-Dimensional Position Sensitive Detector and a Data Acquisition System for Time Resolved X-Ray Scattering Studies, Nucl. Instr. and Meth. 201:145(1982).
36. A. Peisert, The Parallel Plate Avalanche Chamber as an Endcap Detector for Time Projection Chambers, Nucl. Instr. and Meth. 217:229(1983).

VACANCY DISTRIBUTION STUDIES AT MULTIPLE IONIZATION IN HIGH-ENERGY ION-ATOM COLLISIONS

I. Kádár, S. Ricz, D. Varga,
B. Sulik, J. Végh and D. Berényi

Institute of Nuclear Research of the Hungarian
Academy of Sciences, H-4001 Debrecen, Pf 51

Preliminary results of the detailed analysis of the high resolution (cca 1 eV) neon K - Auger spectra from the 5.5 MeV/u H^+, N^{2+}, Ne^{3+}, Ne^{10+}, Ar^{6+} and Ar^{16+}- neon collisions are presented. The spectra have been measured at the beam of the heavy - ion cyclotron U-300 in Dubna, except for the H^+- neon one measured at the Debrecen cyclotron. The evaluation and identification is based on the measured series of spectra rather than on the evaluation of a representative one.

The task to determine the position and intensity of all the components of these complex spectra is not well defined in the case of a single spectrum. To overcome this difficulty we try to use the advantage given by the fact that we have measured a long series of spectra, i. e. from the simplest Ne K - Auger spectrum from a H^+ - Ne collision up to the most complicated one obtained from the Ne^{10+} - Ne collision. These spectra contain the transitions from different vacancy configurations in different proportion, so one may hope to extract useful information from them concerning the production cross section of different configurations and possibly also of some multiplet terms.

The detailed evaluation procedure is in progress. The results until now: identification of most of the Li-like, Be-like and KL-LLL ionization satellites, as well as the diagram transitions (example: Table 1 - Li-like transitions). All the intensive lines of the spectra are assigned to a configuration, so we could establish the L-vacancy distribution in the case of two projectiles. From the average number of L-vacancies vacancy production probability values have been deduced for one L-shell electron in the case of these two collisions, namely 0.19 in the case of the 5.5 MeV/u Ne^{3+}-Ne and 0.33 in the case of Ne^{10+}-Ne collision in agreement with the identical data determined from the diagram/total ratio [1].

Table 1 The group of Ne K-Auger transitions from Li-like initial states

Transition*	Energy		Relative intensity
	Present	Theory[2]	Total int.=1
120 ^{2}S => 200 ^{1}S	652.23(12)	656.4	0.046 (2)
111 ^{4}P => 200 ^{1}S	656.49 (7)	656.2	0.346(41)
Li-like core +	660.10 (9)		0.007 (2)
+ spectator	662.06 (6)		0.008 (3)
	665.71(21)		0.026 (8)
111 ^{2}P+ => 200 ^{1}S	668.44(19)	668.7	0.074(18)
111 ^{2}P- => 200 ^{1}S	673.45(11)	675.8	0.222(46)
102 ^{4}P => 200 ^{1}S	674.11(12)	673.9	0.137(40)
102 ^{2}D => 200 ^{1}S	680.61 (8)	682.3]	0.135(18)

* An abbreviated notation is used in the tables ($1s^1 2s^2 2p^5$ = 125)

References
[1] Kádár I, Ricz S, Shchegolev V A, Sulik B, Varga D, Végh J, Berényi D and Hock G, 1985 J. Phys. B 18 275
[2] Maurer R J and Watson R L, 1986 At. Nucl. Data Tables 34 185

M-SUBSHELL X-RAY PRODUCTION CROSS SECTIONS OF URANIUM BY 0.2 - 0.6 MeV PROTON IMPACT

T. Papp

Institute of Nuclear Research of the Hungarian Academy of Sciences (ATOMKI), Debrecen, H-4001 Pf.51, Hungary

The goal of the present experiment was to study the projectile energy dependence of the M-subshell X-ray production cross sections.

For this purpose M-subshell X-ray production cross sections in thin (4 µg/cm^2) uranium target were measured for 0.2 - 0.6 MeV protons. The targets were obtained by vacuum evaporation onto carbon backings. The purity of the targets was tested by PIXE method. The X-ray spectra were detected by a Si(Li) detector and were smoothed by fast Fourier transform smoothing algorithm and than fitted by approporiate number of gaussian peaks and a quadratic background. The M-subshell X-ray production cross sections were determined from the $M_1O_{2,3}$, M_2O_4, M_3O_5, M_4N_6 and M_5N_7 line intensities using the radiative transition rates of Chen and Crasemann [1].

The obtained absolute X-ray production cross sections and cross section ratios were compared with the predictions of RPWBA-DHS and RPWBA-BC theories [2]. The measured data for M_5-subshell agree with the RPWBA-DHS predictions, and for M_4-subshell are between the RPWBA-DHS and RPWBA-BC values. In the cases of M_1- and M_2-subshell the measured points agree with the RPWBA-BC values. The M_1 and M_2 cross section ratio can be seen in fig. 1 as a function of the bombarding energy. The calculations describe well the energy dependence of the measured cross section ratios.

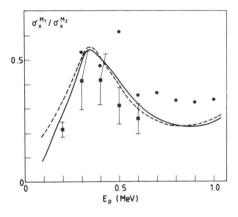

Fig.1 Ratios of M_1-to M_2-subshell X-ray production cross sections of U by protons, as function of the projectile energy. Experimental points: ■ present work, ● data from ref. [3]; theoretical curves [2]: --- RPWBA-DHS, —— RPWBA-BC.

REFERENCES

[1] Chen M.H. and Crasemann B., 1984, Phys. Rev. A30, 170.
[2] Chen M.H., Crasemann B. and Mark H., 1983, Phys. Rev. A27, 2358.
[3] de Castro Faria N.V., Freire F.L. Jr., de Pinho A.G. and da Silveira 1983, Phys. Rev. A28, 2770.

AN X-RAY CRYSTAL SPECTROMETER FOR ION-INDUCED MULTIPLE IONIZATION STUDIES, PARTICULARLY FOR M SATELLITE STRUCTURE

I. Török and B. Tóth

ATOMKI – Institute of Nuclear Research of the Hungarian Academy of Sciences, Debrecen, Pf. 51, H-4001, Hungary

ABSTRACT

A Soller-type flat-crystal spectrometer was developed with rhombic goniometer. The angular divergence of the Soller-slits is $0.3°$. The detection of the X-rays is done with a flow proportional counter using P10 gas. We have 8 different analyzer crystals ranging in 2d from 2.8 to 26 Å. The excitation of samples can be performed by X-ray or ion-beam. A single-card microcomputer is controlling the measurements and a special 4XIK channel analyzer, built in our institute. It can work either in PHA or MSC mode. The goniometer is driven by a $1.8\ °$ step stepping motor. The instrument was installed for ion-excited X-ray measurements on the 5 MV Van de Graaff generator of the ATOMKI. Hundreds of test spectra were taken using different targets and projectiles. Three interesting preliminary results are shown in the followings.

A) Relative intensities of aluminium K_α and K_β satellites were determined from the measured spectra bombarding an aluminium thick target by 3.2 MeV He^+ ions. Our results for $K_\alpha L^i$ fit well between the results of former investigators.

B) A byproduct of Ca $K_\alpha L^1$ investigation (bombarding natural chalk by 2 MeV protons) was a spectrum containing in close vicinity the first order reflection of MgK_α and the third order reflection of CaK_α. It offers the possibility to determine small Mg contents in the presence of much Ca, e.g. to measure the dolomitization grade of chalks, or the Ca/Mg ratio in concretes – which is important for their corrosion resistance.

C) Significant satellite intensities were observed in Ta M_α and M_β lines. The contribution of the satellite lines to the total M_α intensity is about 50 per cent of the intensity of the diagram line in the case of 2 MeV proton bombardment of thick tantalum metal target, and about 100 per cent for 1.6 and 3.2 MeV He^+ bombardment. In the later case probably not only the known M_α and M_β satellite lines are produced but other satellites of higher energies. The resolution of the measuring system at this energy range was demonstrated by taking a $SiK_\alpha L^i$ spectrum at the same conditions as the Ta M spectra, obtained at 3.2 MeV He^+ bombardment.

THE M SPECTRA OF La (57) AND Ce (58)

A.F. Burr, M.B. Chamberlain and R.J. Liefeld

New Mexico State University
Las Cruces, NM USA

ABSTRACT

This paper presents the M X-Ray spectra of lanthum and cerium. Features discussed include the energy of the lines, their excitation curves, and their relative intensities. Special features of the spectra are also mentioned (1,2).

The results reported here were recorded with a vacuum, two crystal monochromator. Potassium acid phthalate crystals were used as diffracting elements and the soft X-Ray detector was a flowing gas (P-10) proportional counter. Spectra was obtained by observing detected intensity as a function of monochromator setting. Elements of the monochromator were angularly stepped in increments ranging from 0.01 degrees to 0.06 degrees (Bragg angle) to generate the spectra, with fixed counting time at each position and constant voltage and current in the X-Ray tube source. For all spectra, the anode voltage was maintained within 50 mV of a constant DC value. The operating pressure in the X-Ray tube was in the 10 torr region. Currents up to 100 mA were supplied to the sample over an area of 2 square centimeters.

The M spectral region shown contained six lanthanum and seven cerium lines. Two for each metal of which have not been studied before. The following table summarizes the results.

Table 1
M Series Characteristic Emission Lines of La and Ce

LEVELS			EXPECTED ENERGY[a]	OBSERVED ENERGY	PREVIOUS VALUE
M2 N4			Ce 1162.8 ± 0.5	1162.2 ± 1.0	NEW
M3 N4,5	γ		La 1024.6 ± 0.6	1024.5 ± 0.8	1027 ± 4
			Ce 1075.4 ± 0.5	1075.4 ± 0.6	1074.9 ± 1.0
M4 N3			La 657.1 ± 0.7	654.1 ± 1.5	NEW
			Ce 694.1 ± 0.6	695.9 ± 1.0	NEW
M4 N6	β		La 848.4 ± 0.6	851.1 ± 1.5	854 ± 3
			Ce 901.2 ± 0.6	900.7 ± 1.0	902 ± 2
M5 N3	ζ		La 640.3 ± 0.7	637.5 ± 1.5	638 ± 2
			Ce 676.1 ± 0.6	676.0 ± 1.0	676 ± 2
M5 N6,7	α		La 831.7 ± 0.3	834.6 ± 1.5	833 ± 3
			Ce 883.2 ± 0.3	882.2 ± 1.0	883 ± 1
M5 O3			La 817.3 ± 1.0	812.0 ± 1.5	NEW
			Ce 863.5 ± 1.0	859.5 ± 1.0	861.6 ± 3

(a) J.A. Bearden and A.F. Burr, Rev. Mod. Phys. 39, 125 (1967)

REFERENCES

(1) R.J. Liefeld, A.F. Burr and M.B. Chamberlain, Phys. Rev. A 9, 316 (1974).

(2) M.B. Chamberlain, A.F. Burr and R.J. Liefeld, Phys. Rev. A 9, 663 (1974).

HIGH RESOLUTION X-RAY SPECTROSCOPY

G.L. Borchert, D. Gotta, T. Rose, R. Salziger
and O.U.B. Schult

IKP, KFA
Jülich, Germany

The use of high resolution photon diffraction techniques in X-Ray spectroscopy yields additional information about atomic and nuclear properties, that cannot be achieved by other methods. As an illustration four different series of experiments are discussed, where optimised crystal diffraction devices are preferably used:

1) Performing high resolution studies of antiprotonic X-Rays information about fundamental properties as mass and magnetic moment of the antiproton and the range of the strong interaction can be achieved. Discussing the specific conditions of such an experiment at the antiproton stage in LEAR an optimized Johann type douple focussing spectrometer using a position sensitive detector: MWPC or microchip detector, is shown in detail.

2) Using a Dumond type crystal spectrometer on-line at the cyclotron of the IKP, first results of a study of multihole creation due to light ion bombardment of metallic targets is reported.

3) To push th precision of the focussing transmission diffractometer further a new source technique has been developed, which is shown in detail. A relative accuracy of $\Delta E/E = 10$ has been achieved that allowed to reveal three effects that shift the energy of X-Ray transitions significantly, when comparing electron capture sources to photofluorescence sources: the hyperfine shift, the dynamic shift and the atomic structure shift, that are discussed with some results from the rare earth elements.

4) The same setup as described in 3 has been used to study the intensity of K X-Ray transitions as function of the chemical bond. Significant intensity shifts are reposted for several K-O transitions.

THEORETICAL AND EXPERIMENTAL ASPECTS OF X-RAY ABSORPTION SPECTROSCOPY

Guang Zhang

National Biostructures Participating Research Team
Brookhaven Synchrotron Light Source, Beam Line X-9
and Institute for Structural and Functional Studies
University City Science Center
Philadelphia, Pennsylvania 19104 U.S.A.

The Biostructures PRT is funded as a NIH Research Resource, and administered by the Institute for Structural and Functional Studies of the University City Science Center. This PRT provides the biochemical and biophysical community with facilities and training for X-ray biological structure studies using synchrotron radiation. The principal site of operation is Beam Line X-9 at the National Synchrotron Light Source at the Brookhaven National Laboratory. Major research activities include synchrotron radiation studies of biomolecules and related materials, instrumental developments for time resolved X-ray absorption experiments, as well as theoretical approaches to EXAFS and XANES.

Multiple scattering XANES is of both theoretical and practical interest for understanding XANES and EXAFS. The K-edge XANES of the tetrahedral molecular gases $GeCl_4$, GeH_3Cl, and GeH_4 have been experimentally studied and the results indicate the ratio of the multiple scattering amplitude to the single scattering amplitude drops to less than 10% beyond 40 eV above the midpoint of the edge. Theoretical simulations significantly overestimate the size and range of multiple scattering contributions to XANES.

Several apparatuses have been developed for the time resolved X-ray absorption experiments. The flow flash apparatus allows the X-ray absorption studies of Hemoglobin-CO photolysis to be studied at room temperature. With time sharing optical monitoring X-ray absorption studies of the photolysis products of biomolecules can be conducted at cryogenic temperatures. Intermediates from biological reactions can be frozen for X-ray absorption structural studies by means of the rapid mixing and freeze quenching apparatus.

The Biostructures PRT welcomes scientists from all over the world to join the team.

THE ELECTRONIC STRUCTURE OF WC(0001)

STUDIED BY ARUPS

D.C. Jain, J. Kanski and P.O. Nilsson

Physics Department
University of Technology
Göteborg, Sweden

ABSTRACT

Energy distribution curves (EDC˜s) have been recorded at various emission angles along Γ- M azimuth with the light incidence angle of 45 deg. To examine the effect of polarization some measurements were performed at 15 deg. incidence angle. Present photoelectron spectra at normal emission are in some agreement with Stefan et. al. at 16.8 eV photon energy, but at 21.2 eV photon energy our results are quite different. The most notable difference is the prominent peak at 2 eV below the Fermi level, completely absent in ref.1. Structure plots of our data show only little dispersion suggesting an interpretation in terms of projected density of states as concluded by Stefan et al. Present EDC˜s show strong intensity variations with polar angles indiceting that direct transitions are operative. The strong evidence for the importance of direct transitions is found in the 21.2 eV spectrum about 2 eV below the Fermi level. Our results support the calculated photoelectron spectra by Larsson and Pendry (ref.2) using an adjusted non-self consistent Potential.

REFERENCES

1. P.M. Stefan, M.L. Shek, W.E. Spicer, L.I. Johansson, F. Herman, R.V. Kasowski and G. Brogren, Phys. Rev. B29, 5423 (1984).

2. C.G. Larsson and J.B. Pendry, Surf. Sci. 162, 19 (1985).

DOUBLE K-SHELL PHOTOIONIZATION IN COPPER SULPHATE

D. Glavic, A. Kodre, and M. Hribar

J. Stefan Institute, Jamova 39
61111 Ljubljana, Jugoslavija

The double K photoionization in copper has recently been studied by Salem and Kumar [1]. Following a comment on apparent suppression of the multiple photoionization cross section in solid target [2] we looked for the double K photoeffect on copper atoms in copper sulphate. The absorption of a thin crystal platelet across the double K-edge of copper was measured to determine the cross section. The simultaneous scans of the counts of incident and transmitted x-rays in steps of ~15 eV across the double K-edge region are recorded and repeatedly superposed to until 10^6 counts per channel are accumulated. The thickness of the copper culphate absorber was determined in a separate absorption experiment. By interpolation the value of $\mu d = 1.40$ at the double K-edge is extracted.

The relative jump $\Delta I/I$ of the absorption scan at the double K-edge is, in the linear approximation, proportional to the relative increase $\Delta\mu/\mu$ of the total absorption cross section due to the new reaction channel. For the KK process the relative jump is obtained as $\Delta I/I = 0.026 + 0.013$ and the middle point of the edge $18.24 + 0.03$ keV. The energy of the double K-edge is in agreement with published data [1] within the resolution of the monochromator. Using the tables [3] the ration σ^{KK}/σ^{K} is obtained as $(4.1 \pm 2.0)10^{-3}$. This value is close to the $(3.20 \pm 0.45)10^{-3}$ measured in metallic copper [1].

Further measurements are required.

REFERENCES

[1] S.I. Salem, A. Kumar, J. Phys. B: At. Mol. Phys. 19, 73 (1986).
[2] M. Deutsch, M. Hart, Phys. Rev. A 29, 2946 (1984).
[3] E. Storm and H.I. Israel, Nuclear Data Tables A7, 565-681, (1970).

TRACES ELEMENTS ANALYSIS BY X-RAY FLUORESCENCE

USING SYNCHROTRON RADIATION OF L.U.R.E ORSAY

P. Chevallier, J.X. Wong and L. Scotee

L.P.A.N., Université Paris VI
L.U.R.E. 91405 Orsay

Fluorescence analysis is based on the photoionization of atoms and the subsequent emission of a characteristic X-Ray of the element studied.

X-Ray fluorescence analysis using synchrotron radiation, monochromatized by reflection from a 0,4° curved crystal, is performed with the sample placed at 45° to the incident beam. Only a standard experimental set-up is used. For an energy dispersive analysis, the emitted X-radiation is detected by a Si(Li) diode. In the wavelength dispersive mode, the emitted radiation is detected using an analysing crystal and a position sensitive detector.

Compared to atomic excitation using charged particles analysis with synchrotron radiation can attain lower concentrations: $10^{-1} - 10^{-2}$ ppm. This improved detection limit is due partly to the large photoionization cross section. In practice, in the usually available conditions used in an X-Ray analysis, that is, photons of 5-50 KeV, electrons of 10-50 KeV, or protons 1-3 MeV, the photoionisation cross section of photons is often several order of magnitude superior to those to the charged particles.

Another important advantage of using photons, and particularly monochromatic synchrotron radiation, is the small energy deposition in the sample. Thus the method is truly non-destructive, and can be recommended for analysing fragile, precious or volatile samples.

Another aspect of this method is the simplicity of the sample preparation. For example, a sample can often be ground and pressed into a pellet of several millimiter in diameter and analysed in this form.

Synchrotron radiation has been applied to numerous problems. To cite only two, in biology, serums have been analysed for Se.

In archeology, multielement analyses followed by statistical data treatment have allowed differentation of groups of objects. Here Rb and Sr are excellent traces. In a study of Gallo Roman ceramics from the region of Narbonne, ceramics of three different origins, have been identified on the basis of their Rb and Sr analyser.

ABSORPTION CORRECTION IN EDXRF USING FILTERED RADIATION FOR SAMPLES OF INTERMEDIATE THICKNESS VIA THE COHERENT/INCOHERENT SCATTERED X-RAYS

M. Fátima Araújo, P. Van Espen and R. Van Grieken

Department of Chemistry, University of Antwerp (UIA)
B-2610 Antwerp - Wilrijk, Belgium

A method has been developed to calculate sample absorption corrections for the energy-dispersive X-Ray fluorescence analysis of samples with an unknown intermediate thickness by using the backscattered X-Rays. The incident beam is produced by a Rh anode and filtered through a Rh thin filter (0.05 mm Rh) on a Tracor X-Ray Spactrace 5000.

A filter with the same composition as the tube anode allows the passage of the characteristic lines, but it absorbs the Bremsstrahlung above these lines, reducing to a large extent the continuum radiation in the excitation spectrum. Correlations between atomic number of elements and the scatter factors have been established and these allow the determination of the effective sample thickness. Samples are considered to be composed of a light matrix and a heavy fraction. The mass of the high-Z elements can be obtained from their lines, and their contribution to the backscattered radiation can be subtracted from the total.

Mass absorption coefficients of elements, mixtures and compounds of low-Z elements and the experimental ratio of the coherent/incoherent scatter intensities are highly correlated. This relation does not depend on the thickness of th sample and can be used as a basis of an absorption correction method. Absorption corrections are calculated from the characteristic and scatter peaks by an iterative process using the previously determined total mass and mass absorption coefficients.

The procedure has been tested with mixtures of different thickness of Cu and H_3BO_3. Percentages of Cu and low-Z elements (H_3BO_3) have been determined using this method with an error of less than 10%, including heterogeneity effects.

Optimization of the analytical conditions to qualify low-Z elements ($11 < Z < 20$) is being studied.

PHOTON INDUCED X-RAY EMISSION AND MULTIVARIATE TECHNIQUES

Carlos M. Rojas- Physics Department
Universidad de Santiago de Chile

Paulo Cartaxo N.- Physics Department
Universidad de São Paulo, Brazil

Réné Van Grieken- Chemistry Department
University of Antwerp, Belgium

ABSTRACT

During the summer of 1987, 51 pairs of aerosol samples were collected at the Universidad de Santiago de Chile Planetarium using a virtual dichotomous sampler . The sampling inlet cut-off was 15 μm; the fractioning aerodynamic diameter was 2,5 μm. The samples were analysed by X-Ray fluorescence spectrometry for 11 elements (Al, Si, P, K, V, Mn, Fe, Cu, Zn, Br, Pb) in the fine fraction, i.e., d<2,5 μm, and for 15 elements (Mg, Al, Si, P, S, K, Ca, Ti, V, Mn, Fe, Cu, Zn, Br, Pb) in the coarse fraction (>2,5μm). The data set consisting of concentrations was subjected to diverse receptor models such as: Chemical Mass Balance and Multivariate Techniques in order to determine the major sources of the atmospheric trace elements, to extract the source profiles, and to calculate the contributions of the various sources to the observed elemental concentrations.

SOFT X-RAY SPECTROSCOPY OF NICKEL-ALUMINA CO-PRECIPITATED CATALYSTS

P. Mc Cluskey and D.S. Urch

ABSTRACT

Soft X-Rays ($10\text{Å} < \lambda < 200\text{Å}$) have escape depths of a micron or less from a solid. A spectroscopy based on such photons is therefore ideally suited for the chemical analysis of surfaces, and should bridge the gap between X-Ray photo-electron spectroscopy which analyses only the outermost few atomic layers of a sample and X-Ray fluorescence which is ideal for bulk analysis.

The potential areas of application of surfaces techniques are manifold.

Catalysis- Distinction between surface and bulk techniques, poisoning, sintering and carbon deposition.

Corrosion- Thin film analysis of corrosion layers and protective films.

Adhesion- Nature of bonding in coatings.

SOFT X-RAY FLUORESCENCE

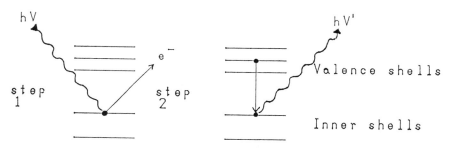

Two step process

Excitation - X-Rays

Emission - X-Rays

Escape depth $\leqslant$ 1um

Vacuum required $\leqslant$ 0,1 torr

Elemental range $>$ 100 eV

Sampled area $\simeq$ 1 cm^2

Information - Elemental composition chemical enviroment (shift).

THE NON-CORRELATION OF XPS AND XES

P. Mc Cluskey and D.S. Urch

ABSTRACT

The main features that are observed in both X-Ray Photoelectron Spectroscopy (XPS) and X-Ray Emission Spectroscopy (XES) can be rationalised using one electron models. For XPS the model is due to the photoemission of an electron from a specific orbital while for XES it is for an allowable electron transition which is governed by very simple selection rules. Since XPS can be used to measure the initial and final states in the X-Ray emission process the two types of spectra are complementary and the reasoning of satellite peaks observed in one will have implications for the other. For example if a shake-up satellite is observed in XPS then relaxation of that excited state should be observed in XES or if exchange splitting is postulated as is the case for Mn3s (Mn2+) then corresponding splitting might be observed in the XES spectrum.

This paper briefly surveys a few situations where such corresponding features might be expected to show up in XPS and XES and concludes surprisingly in fact that they are often not observed at all.

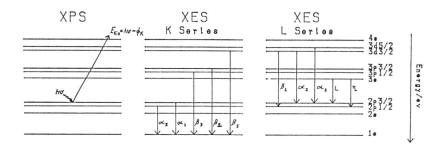

TECHNIQUES IN SOFT X-RAY SPECTROSCOPY

S.R. Luck and D.S. Urch

Chemistry Department, Queen Mary College
University of London

Using multilayers or organic crystals with large 2d spacings it is possible to carry out soft X-Ray spectroscopy using commercially available equipment. An open window X-Ray tube is used for excitation. NKα, BeKα and Si L2,3M X-Ray spectra have been observed from a wide range of samples and, even for the multilayers, peak shifts were observed. The bombardment of a sample of anodised aluminium with a beam of electrons whose energy could be varied, permitted the depth profile chemical analysis of the thin oxide layer.

SILCON SURFACES FOR EPITAXIAL GROWTH

John Thornton

ABSTRACT

In order to understand and improve the preparation of Si for epitaxial growth, HF acid etching has been carried out on Si(100) wafers which had been oxidised in several different ways (e.g. Native, RCA, Thermal and UV--Ozone).

HF etching forms part of the RCA etch, which leaves a thin (15Å) oxide on Si, removable by heating in ultra high vacuum. It is also hoped to use HF etching in conjunction with UV-Ozone cleaning in order to obtain ultimate carbon contamination.

XPS spectra are shown for an etched native oxide, which shows on oxide, with carbon and oxygen physisorbed contamination at less than 1/10 monolayer. A (1x1) LEED pattern was obtained from the same unheated sample. Further experiments on the surface (e.g. a LEELS investigation of the assumed hydrided surface) are being carried out.

CALCULATION OF THE ELECTRON MOMENTUM DISTRIBUTION
IN Zr AND ZrH$_2$ BY THE APW METHOD

N.I. Papanicolaou- Department of Physics, University of Ioannina
45332 Ioannina, Greece

N.C. Bacalis- Research Center of Crete, P.O. Box 1527
71110 Heraklion, Crete, Greece

D.A. Papaconstantopoulos- Naval Research Laboratory
Washington DC 20375, USA

ABSTRACT

Recently a growing number of measurements have been performed to study the electronic properties of solids by means of Compton scattering technique. The investigation of the momentum density in transition metals and metal-hydrogen systems is of particular importance, because of their interesting electronic structure and technological applications.

We have calculated the electron momentum distribution in Zr (fcc) and ZrH$_2$ using the self-consistent augmented-plane-wave (APW) method within the local density approximation of Hedin-Lundqvist. The calculation was scalar-relativistic, i.e. neglecting spin-orbit coupling.

The behaviour of the electron momentum distribution in Zr and ZrH$_2$ along [100], [110] and [111] directions, as well as the remarkable differences which appear by the introduction of hydrogen can be understood in terms of the respective energy bands, Fermi-surface topology and character of the wave functions.

SOFT X-RAY SPECTROSCOPY WITH AN ANALYTICAL WAVELENGTH DISPERSIVE SPECTROMETER

M.O. Figueiredo- Centro Crist. Min., IICT, Al. Afonso Henriques 41-4º Esq., 1000 Lisboa, Portugal

Z. Melo- LNETI, Az. Lameiros
1699 Lisboa Codex, Portugal

M.J. Basto- INIC & LAMPIST, Av. Rovisco Pais
1096 Lisboa Codex, Portugal

ABSTRACT

Emission and absorption X-Ray spectroscopy are powerful means of investigating the electronic structure of solids as both shape and characteristic energy on X-Ray lines corresponding to transitions from valence to core levels are affected by the bonding state of the emitting element (1,2). However, the required equipment is usually beyond the intrumental facilities of most chemical laboratories, and it appeared useful to test the capability of an analytical wavelength-dispersive X-Ray fluorescence spectrometer for the assessment of the bonding state of analysed chemical elements.

Light elements - namely, aluminium, silicon, phosphorus and sulphur - were studied in a first approach. For purely analytical purposes, the more intense $K\alpha$ line should be preferably used to improve detection limits and accuracy, but the $K\beta$ band is by far more sensitive to chemical bond effects.

The results attained with an automated WDS-XRF Philips 1400 spectrometer are critically evaluated. Instead of the usual output of accumulated intensity counts at a fixed wavelength, slow-speed scans of $K\beta$ spectra were run under optimized instrumental conditions. To improve wavelength resolution, a germanium dispersing crystal was used for S and P, and a penta-eryrhritol crystal for Si and Al. A "splitting index" is defined to characterize the shape of $K\beta$ doublet; the obtained experimental values range from unit in tottaly unresolved peaks to about three in quite individualized lines. The so-called "chemical-shift" of $K\beta$ line -peak centroid drift (3)- as well as the intensity and the positioning of its $K\beta'$ longer wavelength satellite, are compared for various bonding, valence and coordination states of the studied light elements.

An aluminium foil was the probe for metallic state; formal valence 3+ was studied in α-Al_2O_3 and Kaolinite, $Al_2Si_2O_5(OH)_4$ (octahedral coordination), also in mullite, Al_2SiO_5 (tetrahedral plus octahedral coordination) and feldspar, $KAlSi_3O_8$ (tetrahedral coordination alone). Feldspar also served to scan the $K\beta$ band for silicon in tetrahedral sites and formal valence 4+, along with quartz, trigonal SiO_2. Crystalline silicon was the probe for valence state zero. A thin film of nickel-phosphorous deposited over aluminium (P^0), various phosphates (tetrahedral coordination of P^{5+}) and phosphites (pyramidal coordination of P^{3+}) were used for phosphorus. Yellow sulphur (S^0), zinc blende, Zn S (S^- single anion), pyrite, FeS_2 (S_2^- polyanion), some sulphates (S^{6+}) and sulphites (S^{4+}) were scanned for sulphur in various bonding states. A fairly good agreement was found between actual results and the published experimental values obtained for P under similar instrumental conditions (4), for Al and Si under primary excitation by an electron beam in a microprobe (5), and for S with a specially designed X-Ray fluorescence spectrometer (6).

The potentialites of the analytical WDS-XRF spectrometer towards the assessement of the chemical bonding of light elements (7) in natural compounds and industrial materials therefore seems promising. Scans of the $K\alpha$ main peaks and non-diagram lines are presently under study, and further extension to elements of the first transition metal series is also in progress.

REFERENCES

(1) M. BROGLIE (1920) Compt. Rend. Acad. Sci., Paris, 170, 1245, 1344
(2) A. FAESSLER (1931) Zeit. f. Physik, 72, 734
(3) K.I. NARBUTT (1980) Phys. Chem. Minerals, 5, 285
(4) T.C. YAO & J.J. HOLST (1967) Spectrochim. Acta, 23B, 19
(5) C.G. DODD & G.L. GLEN (1969) Amer. Miner. 54, 1299
(6) L.S. BIRDS & J.V. GILFRICH (1978) Spectrochim. Acta, 33B, 305
(7) D.S. URCH (1973) X-Ray Spectrometry, 2, 3

THE DOUBLE IONIZATION OF THE K SHELL BY PROTONS

V. Cindro

J. Stefan Institute, Jamova 39
61111 Ljubljana, Yugoslavia

ABSTRACT

Characteristic X-Rays from thick Ca and Ti targets induced by 1,3-1,7MeV protons were analysed with a Bragg spectrometer combined with position sensitive detector. The energy resolution was 10 eV at titanium $K\alpha$ line. The hypersatellite line due to the double K-shell ionization was observed between $K\alpha$ and $K\beta$ line. The cross-section for double K-shell ionization was determined from the measurements at different proton energies.

The cross section was also calculated using impact parameter formalism (ref.1). The probability P(b) for removing an electron from the K-shell with the projectile having an impact parameter b was calculated in SCA (ref.1) and BEA (ref.2) approximation. The comparison of the experimental and theoretical results for 1,6 MeV protons is shown on table I. The results of SCA calculations (corrected for binding) are in much better agreement with measured results than simple BEA calculations.

Table I: The comparison of theoretically and experimentally determined ratio $\sigma_{K\alpha H}/\sigma_{K\alpha}$

target	SCA	BEA	Experiment
Ca	$3 \cdot 10^{-4}$	$5.1 \cdot 10^{-4}$	$2.3 \cdot 10^{-4}$
Ti	$2 \cdot 10^{-4}$	$3.5 \cdot 10^{-4}$	$2.2 \cdot 10^{-4}$

REFERENCES

1. J.M. Hansteen et al.: ADNDT 15, 305-317 (1975)
2. M. Gryzinski: Phys. Rev. 138 A (1985) 336

APPROXIMATE CALCULATIONS IN THICK TARGET PIXE ANALYSIS

Z. Smit

J. Stefan Institute
E. Kardelj University of Ljubljana, Yugoslavia

X-ray yields induced by protons in infinitely thick targets depend significantly upon stopping of the protons along their trajectories and upon absorption of the X-rays along their escape path lengths. For numerical evaluation of the thick target X-ray yields an integration procedure is required. Calculations are simplified if a power approximation for the energy dependence of the X-ray production cross section and of the proton stopping power is asumed. Thick target X-ray yields are then proportional to a two parameter thick target correction function which provides a straightforward procedure for the evaluation of elemental concentrations from the measured data[1]. Secondary fluorescence X-ray yields are calculated accordingly from a power series expansion[2]. The thick target X-ray yields as obtained exactly and by the present approximation were compared for the elements 11 Z 51 and for proton energies between 1 and 4 MeV. The mean difference between both types of data was found to be 0.6%, the maximum difference not exceeding 2%[3]. Recently, variation of the mean X-ray path length was considered for a rough surface target. The correction proposed may be used for the surface roughness up to one half of the proton range[4].

REFERENCES
(1) Z. Smit, M. Budnar, V. Cindro, V. Ramsak and M. Ravnikar, Nucl. Instr. and Meth. B4 (1984) 114.
(2) same, Nucl. Instr. and Meth. 228 (1985) 482.
(3) Z. Smit, Nucl. Instr. and Meth. B17 (1986) 156.
(4) same, to be published in Nucl. Instr. and Meth. B.

ON THIN TARGET PIXE EFFICIENCY

M. Budnar

J. Stefan Institute
University E. Kardelj, Ljubljana, Yugoslavia

Thin target PIXE efficiency can be measured if thin, homogeneous targets are irradiated by protons in measuring chamber (1). It can be also defined through fundamental parameters as:

$$F_i(E_p, Z_i) = \frac{\delta_i^x \cdot \varepsilon \cdot T \cdot \Delta\Omega/4\pi}{\cos\alpha \times M/N_L}$$

Here, δ_i^x is X-ray production cross section for particular chemical element i at proton energy E_p; ε is X-ray detector efficiency; $\Delta\Omega/4\pi$ is detector solid angle; T is transmission of emitted X-rays through the absorbers on the way to the detector; angle α measures inclination of target regarding to proton beam and M/N_L is mass of element measured.

Experimental and calculated PIXE efficiencies were compared and influence of precision and accuracy in the fundamental parameters was determined (2). Better than 10% overall agreement was achieved for $K\alpha$ lines of elements with 16 Z 41 and also $L\alpha$ lines of elements 50 Z 92. Observed differences were attributed mainly to ± 5% precision in target thicknesses.

The agreement obtained proves so far the choice of calculated efficiency for PIXE system calibration when thin targets are concerned. For light elements (Z 15, E_x 2.5 KeV) the above approach is partly limited in our case as intrinsic Ge detector was used.

REFERENCES

(1) M. Budnar, M. Kregar, U. Miklavzic, V. Ramsak, M. Ravnikar, Z. Rupnik and V. Valković, Nucl. Instr. Meth., 179 (1981) 249-258.
(2) M. Budnar, to be published.

CHARGE STATE DISTRIBUTIONS PRODUCED IN ATOMIC AND MOLECULAR TARGETS BY 40MeV Ar^{13+} ION IMPACT

R.J. Maurer, B.B. Bandong, C. Can and R.L. Watson
Texas A&M University
Cyclotron Institute and Department of Chemistry
College Station, Texas 77843, USA

ABSTRACT

The ionization and fragmentation patterns for Ne and O_2 produced in collisions with 40MeV Ar^{13+} projectiles undergoing no charge exchange, one-electron capture and two-electron capture have been identified by time-of-flight mass spectroscopy. The highest charge state of Ne produced by direct ionization is 6+, whereas for O_2, the highest fragment ion charge state is about half that of Ne, thus indicating the importance of L-shell electron rearrangement in molecular ions prior to dissociation. The Ne recoil ion charge state distributions resulting from collisions with 40MeV Ar^{13+} ions having one- and two-electron capture are bell-shaped distributions with average recoil ion charge states of +4,1 and +5,6, respectively. These values are considerably less than those estimated from K X-Ray and Auger measurements which give an average charge state of +7. The discrepancy is apparently due to the collision conditions imposed by K X-Ray and Auger measurements which select only K-shell capture events requiring small impact parameter collisions, whereas in the present measurements, there is no such restriction on the impact parameter and capture can occur from both the K- and L-shells of Ne.

For O_2, however, velocity matching conditions favor electron capture predominantely from the oxygen K-shell and not the L-shell as was observed for Ne. The average fragment ion charge state for O_2 following collisions with 40MeV Ar^{13+} ions having one- and two-electron capture are +2,7 and +3,9, respectively. The oxygen fragment ion distribution produced following electron capture is discussed in terms of a mechanism involving electron rearrangement in the L-shell, but not in the K-shell. Also, from the peak splitting in the q≥1 charge states of oxygen atomic fragments, average kinetic energies of the ions gained in a Coulomb explosion were deduced. Oxygen fragments with q=5 charge state were detected having a kinetic energy of 154 eV, presumably originating from the dissociation of O_2^{10+} molecular ions.

METAL-COATED CAPILLARY USED AS X-RAY WAVEGUIDE

P. Engström, H. Riedl and A. Rindby

Department of Physics
Chalmers University of Technology
Göteborg, Sweden

ABSTRACT

It is well known that X-rays coming from a "less" dense media into a "dense" media can undergo total reflexion if the incident angle is small enough and the surface of interface is sufficiently flat. Thus capillaries of glass can be used as X-ray waveguides in the same way as fibres are used in the optical range. This was demonstrated in a previous paper.

In order to increase the intensity of X-rays propagating inside the capillary, we have tried to increase the reflection coefficient of the capillary wall. This has been done by coating the wall with layers of heavy metal which has high reflectivity due to the high density of electrons at the metal surface.

Although the heavy metal surface will increase the reflectivity of the capillary, the increased absorption of X-rays in the metal (compared to the absorption in the capillary glass wall) will somewhat counterbalance this effect.

A complete calculation of the intensity spectrum out from such a capillary have been performed taken the absorption into account.

The experimental results gave a higher intensity in certain regions as compared with the calculated spectrum. However, taken the grain structure of the wall-material into account a reasonable agreement between theory and experiment has been achieved. The results indicates that the capillary technique would be useful to construct effective X-ray band-pass filters and X-ray micro beams.

PARTICIPANTS

BELGIUM
   Araújo, M.F.   Department of Chemistry, Antwerp
   Palma, C.R.   U.I.A., Antwerp

FRANCE
   Simionovici, A.   L.P.A.N., Paris
   Said, E.   L.P.A.N., Paris
   Tolentino, H.   Lure, Orsay
   Chermette, H.   I.P.N., Lyon
   Wang, J.X.   L.P.A.N., Paris
   Scotee, L.   L.P.A.N., Paris
   Cherifa, M.   I.P.N., Lyon

GREECE
   Anagnostopoulos, F.D.   Ioannina
   Papanikolaou   Ioannina

PORTUGAL
   Costa, A.M.   C.F.A., Lisboa
   Cardoso, C.P.   C.F.A., Lisboa
   Guimarães, D.   Univ. Aveiro
   Parente, F.   C.F.A., Lisboa
   Ribeiro, J.P.   C.F.N., Lisboa
   Serrão, J.M.   F.C.U.L., Lisboa
   Salgueiro, L.   C.F.A., Lisboa
   Alves, L.C.   L.N.E.T.I., Sacavém
   Jesus, M.A.   C.F.N., Lisboa
   Martins, M.C.   C.F.A., Lisboa
   Marques, M.I.   C.F.A., Lisboa
   Monteiro, M.L.   C.F.M., Lisboa
   Carvalho, M.L.   C.F.A., Lisboa
   Figueiredo, M.O.   I.I.C.T., Lisboa
   Lima, M.T.   C.F.A., Lisboa
   Pinheiro, M.T.   L.N.E.T.I., Sacavém
   Ramos, M.T.   C.F.A., Lisboa
   Conde, O.   C.F.M.C., Lisboa
   Melo, Z.   L.N.E.T.I., Lisboa

SWEDEN
- Jain, D. — C.U.T., Gothenburg
- Antonsson, I. — I.P., Uppsala
- Oblad, M. — C.U.T., Gothenburg
- Keane, M. — I.P., Uppsala
- Bjorneholm, O. — I.P., Uppsala
- Engstrom, P.S. — C.U.T., Gothenburg

TURKEY
- Kendi, E. — H.U., Ankara
- Ozbey, S. — H.U., Ankara

U.S.A.
- Burr, A.F. — N.M.S.U., Las Cruces
- Zhang, G. — I.S.F., Philadelphia
- Murphy, R. — P.D., Pennsylvania
- Maurer, R. — T.A. & M.U., Texas

U.K.
- Sharland, A.P. — U.C.C., Cardiff
- Thornton, J. — U.C.C., Cardiff
- Mecluskey, P.J. — Q.M.C., London
- Horn, R. — Q.M.C., London
- Luck, S.R. — Q.M.C., London

WEST-GERMANY
- Borchert, G. — I.J.K., Julich
- Jaegermann, W. — H.M.I., Berlin

YUGOSLAVIA
- Glavic, D. — J.S.I., Ljubljana
- Budnar, M. — J.S.I., Ljubljana
- Cindro, V. — J.S.I., Ljubljana
- Smit, Z. — J.S.I., Ljubljana

INDEX

Absorption spectroscopy, 107, 110, 401
Adsorbates, 345
Aerosol
    atmosferic, 323
    sampling, 324
AES, 218
Alloys, 255
Analysis
    multielemental, 301
    non destructive, 32
    X ray emission, 319
APW method, 411
Auger
    electrons, 1, 15, 27, 58, 155, 378
    electron
        ion coincidence, 23
        process, 25
        spectroscopy, 15
    process, 1, 15, 157, 229
    resonance, 2, 10
    spectra, 10, 163, 177, 222
    spectroscopy, 1, 29, 51, 159
    transition, 1, 25, 47, 227
ASF, 15
Born-Beth formalism, 339
Capture, double electron, 54
Chemical effects, 155, 163, 165, 177
Chemisorbed molecules, 215
Compton spectrum, 279
Core
    excitation, 228, 353

Core (continued)
    hole excited states, 215
    ionisation energies, 165, 177, 196, 218
Coster Kronig transitions, 26, 170
CSF, 15
Detectors, 237, 251
    bandwidths, 375
    optical, 352
    position sensitive, 5, 251, 400
    resolution
        energy, 251, 375
        position, 375
    X ray, 307, 375
    gaseous, 375
DES, 218
Dielectric, formalism, 341
Dipolar resonances, 358
Double K shell ionisation, 414
EDXRF, 405
EELS, 335
Electron
    correlation, 19
    energy loss spectrum, 279, 335
    spectrometer, 3
Electronic
    decay, 215, 229
    relaxation effects, 201, 208
    structure, 177, 279, 402
Epitaxial growth, 410
EXAFS, 85, 107, 253, 401

421

FMS, 92
Fluorescence spectroscopy, 237, 412
Fragmentation of molecules, 227
Free molecules, 215
GSPC, 352
IMS, 92
Ion-atom collision, 393, 417
Ionisation cross section, 301
Instrumental analytics, 301
Interatomic, 201
Intra-atomic, 201
LCAO, 178
LEE IXS, 196
MCDF, 15
MDL, 238
Microprobe
   photon, 250
   proton, 314
Molecular orbital theory, 174, 177
Molecular X ray process, 201
Monochromators, 244
MOs, 178
Multicenter approximation, 201, 204
Multichannel, 69
Multistep chamber, 375
Multiple
   ionisation, 28, 40, 301, 393, 397
   scattering, 69, 112, 115
MWPC, 375, 400
Near edge core hole spectroscopy, 111
NEXAFS, 111
One centre approximation, 204
Orbital model
   frozen, 203
   relaxed, 208
PAX, 155, 157, 177, 181
PES, 218
Plasmons, 279
Photoabsorption, 67

Photo electron, 155, 378
   emission, 155, 177
   spectroscopy, 155
Photoemission X ray, 68, 157
Photoionization, 255, 403
PIXE, 302, 406, 415, 416
   interdisciplinary applications, 301
   method, 301
PS, 155
Rydberg
   excitation, 214
   orbitals, 217
   states, 228
Satellite structure, 34, 166, 397
Satellites, 157, 161, 189, 193
SEXAFS, 107, 120
Scattering, inelastic, 279
Shake off (SO), 20, 164
Shake up (SU), 20, 164
Soft X ray, 227, 255, 407, 409, 412
SR, 108
SRXFA, 237
SS, 92
Synchrotron radiation, 108, 237, 367, 404
SXS, 255
Trace elements, 253, 301, 404
   in archeology, 328
   in biomedical studies, 327
   in geology, 328
   in material science, 328
Transition probabilities, 21, 24, 201
TOF, 228
Undulators, 371
UP, 157
Valence
   band, 165, 177, 181
   excitation, 347
XANES, 97, 253, 401
XAS, 107, 353
XES, 157, 201, 408

XP, 157
XPS, 155, 159, 177, 181, 255, 408
XRF, 248, 412
X-ray
   cross sections, 395
   emission, 155, 177, 301, 398
   fluorescence, 237, 404
   Raman, 279
   resonant Raman, 279
   waveguide, 418
Wiggler, 371